Progress in Mathematics

Volume 241

José Seade

On the Topology of Isolated Singularities in Analytic Spaces

Birkhäuser Verlag
Basel · Boston · Berlin

Author:

José Seade
Instituto de Matemáticas
Unidad Cuernavaca
Universidad Nacional Autónoma de
México (UNAM)
Av. Universidad s/n, col. Lomas de
Chamilpa, Ciudad Universitaria
C.P. 62210 Cuernavaca, Morelos
México
e-mail: jseade@matcuer.unam.mx

2000 Mathematics Subject Classification Primary 14Bxx, 32Sxx, 53Cxx, 55A25, 57C45; Secondary 11Fxx, 32L30, 34Mxx, 37F75

A CIP catalogue record for this book is available from the Library of Congress, Washington D.C., USA

Bibliographic information published by Die Deutsche Bibliothek
Die Deutsche Bibliothek lists this publication in the Deutsche Nationalbibliografie; detailed bibliographic data is available in the Internet at <http://dnb.ddb.de>.

ISBN 3-7643-7322-9 Birkhäuser Verlag, Basel – Boston – Berlin

© 2006 Birkhäuser Verlag, P.O. Box 133, CH-4010 Basel, Switzerland
Part of Springer Science+Business Media
Printed on acid-free paper produced of chlorine-free pulp. TCF ∞
Printed in Germany
ISBN-10: 3-7643-7322-9   e-ISBN: 3-7643-7395-4
ISBN-13: 978-3-7643-7322-1

9 8 7 6 5 4 3 2 1   www.birkhauser.ch

Ferran Sunyer i Balaguer (1912–1967) was a self-taught Catalan mathematician who, in spite of a serious physical disability, was very active in research in classical mathematical analysis, an area in which he acquired international recognition. His heirs created the Fundació Ferran Sunyer i Balaguer inside the Institut d'Estudis Catalans to honor the memory of Ferran Sunyer i Balaguer and to promote mathematical research.

Each year, the Fundació Ferran Sunyer i Balaguer and the Institut d'Estudis Catalans award an international research prize for a mathematical monograph of expository nature. The prize-winning monographs are published in this series. Details about the prize and the Fundació Ferran Sunyer i Balaguer can be found at

http://www.crm.es/FSBPrize/ffsb.htm

**This book has been awarded the
Ferran Sunyer i Balaguer 2005 prize.**

The members of the scientific commitee of the 2005 prize were:

Hyman Bass
  University of Michigan

Antonio Córdoba
  Universidad Autónoma de Madrid

Paul Malliavin
  Université de Paris VI

Joseph Oesterlé
  Université de Paris VI

Oriol Serra
  Universitat Politècnica de Catalunya, Barcelona

To Swami Chidvilasananda and the Siddha lineage,
whose light turned my life into a joyful path;
and to Teresa, whose love is my everything.

Contents

Preface

Singularity theory stands at a cross-road of mathematics, a meeting point where many areas of mathematics come together, such as geometry, topology and algebra, analysis, differential equations and dynamical systems, combinatorics and number theory, to mention some of them. Thus, one who would write a book about this fascinating topic necessarily faces the challenge of having to choose what to include and, most difficult, what not to include. A comprehensive treatment of singularities would have to consist of a collection of books, which would be beyond our present scope. Hence this work does not pretend to be comprehensive of the subject, neither is it a text book with a systematic approach to singularity theory as a core idea. This is rather a collection of essays on selected topics about the topology and geometry of real and complex analytic spaces around their isolated singularities.

I have worked in the area of singularities since the late 1970s, and during this time have had the good fortune of encountering many gems of mathematics concerning the topology of singularities and related topics, masterpieces created by great mathematicians like Riemann, Klein and Poincaré, then Milnor, Hirzebruch, Thom, Mumford, Brieskorn, Atiyah, Arnold, Wall, Lê Dũng Tráng, Neumann, Looijenga, Teissier, and many more whose names I cannot include since the list would be too long and, even that, I would leave aside important names. My own research has always stood on the shoulders of all of them. In taking this broad approach I realize how difficult it is to present an overall picture of the myriad of outstanding contributions in this area of mathematics during the last century, since they are scattered in very many books and research articles. This work is a step in that direction and our first main purpose is to put together some of these gems of current mathematics in an accessible way, indicating in all cases present lines of research and appropriate references to the literature. This step occupies Chapters I to V of this monograph. I have made some contributions to the subject, either by myself or in collaboration with other colleagues, and I think they are interesting enough to be included here.

On the other hand, the theory of real singularities is much less developed than that of complex singularities, for various reasons. There are however interesting families of real analytic singularities having remarkable geometric and topological properties similar to those of complex singularities. These are related to complex differential equations and defined via complex geometry, and the second main purpose of this monograph is to explain some aspects of their geometry and similarities with complex singularities. I believe the study of these singularities in Chapters VI–VIII and the picture of complex singularities presented in Chapters I–V, together provide interesting new insights into the geometry and topology of singularities. This study also gives new methods for constructing smooth differentiable manifolds equipped with a rich geometry and defined by algebraic equations, which can be of interest on its own. For instance, at the end of Chapter VI we obtain explicit algebraic equations that provide analytic embeddings of oriented surfaces of all genera in the sphere $\mathbb{S}^3$; this can be useful for problems in differential geometry and physics.

The material that I present here is mostly contained in the literature but, as said before, it is spread over many different research papers and books, making it difficult to put together a complete picture. In some cases I have given simple proofs of well-known theorems; in other cases I have included the complete proofs in detail; yet in others I have preferred to content myself by giving only some of the key ideas to understand some specific theory. In all cases I have included appropriate references and indications for further reading. I have also added a number of open problems as extension of the text.

This monograph is constructed like a fan, where each individual chapter is to some degree independent of the others and can be read on its own, and yet as a unity, spreads out to give the reader a taste of the richness of this fascinating topic. I am sure that everyone who reads it, regardless of whether he or she is a graduate student, an expert in singularity theory, or a researcher in any other field of geometry and topology, will find interesting things in it.

Introduction

Chapter I sets up the foundations for the rest of the book. Here we review some of the classical results of Milnor and others about the topology of real and complex analytic spaces around an isolated singular point. Two of these results are key for us. One is the theorem about the local conical structure of analytic sets. This has a long history, starting with Brauner (1928) and others in their local study of complex plane curves, noticing that every complex curve in $\mathbb{C}^2$ with an isolated singularity at the origin meets transversally every sufficiently small sphere around 0; the intersection is a classical link in the sphere (i.e., a collection of pairwise disjoint embedded circles), and therefore knot theory can be used to study complex plane curves.

In 1961 Mumford proved a very surprising result in this direction: he showed that one cannot find a counter-example to the classical Poincaré conjecture using complex singularities; more precisely he proved that if the link (i.e., the boundary of a small neighborhood) of a point in a complex surface is simply connected, then the point is actually smooth and the link is the usual 3-sphere. May I recall at this point that a few years earlier (1956), Milnor had found the first examples of exotic differentiable structures on spheres. Hence these two theorems (of Milnor and Mumford) drove the attention of several great mathematicians towards studying the topology of links of isolated singularities in analytic spaces. In his classical book [168] Milnor proved that if V is a (real or complex) analytic space in $\mathbb{A}^n$, where $\mathbb{A}$ is either $\mathbb{R}^N$ or $\mathbb{C}^N$, and if V has an isolated singular point at the origin, then V intersects transversally every sufficiently small sphere $\mathbb{S}_\varepsilon(0)$ around 0; the intersection $M = V \cap \mathbb{S}_\varepsilon(0)$, the link, is a smooth manifold of real dimension $m - 1$, where m is the real dimension of V, and the topology of V and of its embedding in $\mathbb{A}^n$ are completely determined by the embedding $M \hookrightarrow \mathbb{S}_\varepsilon(0)$; more precisely, the pair $(\mathbb{B}_\varepsilon(0), V \cap \mathbb{B}_\varepsilon(0))$ is homeomorphic to the cone over the pair $(\mathbb{S}_\varepsilon(0), V \cap \mathbb{S}_\varepsilon(0))$, where $\mathbb{B}_\varepsilon(0)$ is the closed ball bounded by $\mathbb{S}_\varepsilon(0)$. The manifold M is *the link* of 0 in V.

The second result in Chapter I which is key for this work is the fibration theorem of Milnor. The first part of it holds for real singularities in general and is just an easy extension of Ehresmann's fibration theorem. This says that if a real analytic map $f : (\mathbb{R}^{n+k}, 0) \to (\mathbb{R}^k, 0)$ is a submersion at every point in a punctured neighborhood of the origin, then for every sufficiently small sphere $\mathbb{S}_\varepsilon(0)$ around 0, and for every $\delta > 0$ sufficiently small with respect to ε, the restriction of f to the tube $f^{-1}(\mathbb{S}_\delta^{k-1}) \cap \mathbb{B}_\varepsilon(0)$, where $\mathbb{S}_\delta^{k-1}$ is a sphere in $\mathbb{R}^k$ around 0, is the projection of a (locally trivial) smooth fibre bundle; and moreover, the tube $f^{-1}(\mathbb{S}_\delta^{k-1}) \cap \mathbb{B}_\varepsilon(0)$ can be "inflated" to become (diffeomorphically) the complement of the link M in $\mathbb{S}_\varepsilon(0)$, and therefore one has a fibre bundle projection $\mathbb{S}_\varepsilon(0) - M \longrightarrow \mathbb{S}_\delta^{k-1}$. This is, so to say, the "weak" form of Milnor's fibration theorem (see [167]). The strongest form of this theorem is for complex singularities: if f is now a complex analytic map $(\mathbb{C}^{n+1}, 0) \to (\mathbb{C}, 0)$ with a (possibly non-isolated) critical point at 0, then the above projection map $\mathbb{S}_\varepsilon(0) - M \longrightarrow \mathbb{S}_\delta^1$ is given by $\phi = f/|f|$. If $0 \in \mathbb{C}^{n+1}$ is an isolated critical point of f then one gets an open book decomposition of the sphere (see the text for the definition), where the binding is the link M and the pages are the fibres of ϕ; moreover, each fibre $F_\theta = \phi^{-1}(e^{i\theta})$ has the homotopy type of a bouquet of spheres of middle dimension. A corollary of Milnor's work is that for these singularities the link is a highly connected manifold which bounds a parallelizable manifold (the Milnor fibre). We include here a result of [201, 269] extending Milnor's fibration theorem to the case of meromorphic functions f/g. This is related to the results in Chapter VIII.

The prototype of the singularities to which Milnor's theorem applies are the famous singularities:

$$f(z_0, \ldots, z_n) = z_0^{a_0} + \cdots + z_n^{a_n} \; ; \; a_i \geq 2 \, , \, i = 0, \ldots, n \, ,$$

studied by Pham and Brieskorn. Inspired by his knowledge of physics, where it was customary to think of the tangent bundle of the n-sphere as given by the equation $z_0^2 + \cdots + z_n^2 = 1$, Pham proved in 1965 that given any polynomial f as above, its non-singular level $z_0^{a_0} + \cdots + z_n^{a_n} = 1$ contains a canonical polyhedron of real dimension n, now called the *join of Pham*, which is a deformation retract of it. This implies, in particular, that this hypersurface has the homotopy type of a bouquet of spheres of middle dimension: a strong indication pointing towards Milnor's theorem about the topology of the fibres, proved a couple of years later. On the other hand Brieskorn proved in 1966 that in many cases the links of these singularities were exotic spheres, i.e., manifolds homeomorphic but not diffeomorphic to the usual $2n - 1$-sphere, $n > 3$. There were remarkable generalizations of this result by Brieskorn himself and by Hirzebruch, proving (among other results) that every homotopy sphere that bounds a parallelizable manifold is the link of an isolated complex hypersurface singularity. In particular all the 28 different differentiable structures on $\mathbb{S}^7$ can be obtained in this way.

Chapters II and III are a joint piece. These are about the beautiful relation between 3-dimensional Lie groups and 2-dimensional complex singularities. This began with Klein's theorem in 1884 giving a relation between the finite subgroups of $SU(2)$ and certain surface singularities in $\mathbb{C}^3$. In the simplest case, Klein's theorem establishes an isomorphism between $\mathbb{Z}_p \backslash \mathbb{C}^2$ and the singularity $z_1^2 + z_2^2 + z_3^p = 0$, where $\mathbb{Z}_p$ denotes the cyclic group of order p; hence the link of the latter is $\mathbb{Z}_p \backslash \mathbb{S}^3$; in particular this proved that the group $SO(3) \cong (\pm 1 \backslash \mathbb{S}^3)$ of motions of the Euclidean plane is the link of the quadric $z_1^2 + z_2^2 + z_3^2 = 0$, so it is the unit tangent bundle of the 2-sphere; for the binary icosahedral group $\Gamma = \langle 2, 3, 5 \rangle$ one gets that Poincaré's homology sphere $\Gamma \backslash \mathbb{S}^3$ is the link of $z_1^2 + z_2^3 + z_3^5 = 0$, and so on. We recall that Klein's program was to study geometry by looking at the corresponding groups of isometries of spaces.

This theorem of Klein was later completed by Milnor, Hirzebruch, Neumann and Dolgachev, giving the complete classification of the 2-dimensional, isolated complex surface singularities whose *link* (i.e., the smooth boundary of a small neighborhood of the singular point) is of the form $\Gamma \backslash G$, where G is a 3-dimensional Lie group and Γ is a discrete subgroup of G with compact quotient (these are called *uniform subgroups*). Actually their results, together with the classification of 3-manifolds of the form $M_\Gamma = \Gamma \backslash G$, where G and Γ are as above, i.e., a 3-dimensional Lie group G with a uniform subgroup Γ, give a very nice, unified view of these manifolds. There are, up to isomorphism, six 3-dimensional, simply connected Lie groups with uniform subgroups $\Gamma \subset G$. Given any such group G and a uniform subgroup Γ, there is associated to G a canonical complex 2-dimensional manifold X, equipped with a canonical holomorphic 2-form and a foliation $\mathcal{F}$ of X defined, in all cases but one, by an action of G. The manifold X is in all cases given by an automorphy factor of some line bundle. The quotient $\widetilde{V}_\Gamma^* = \Gamma \backslash X$ is a complex manifold, foliated by copies of $\Gamma \backslash G$, with a canonical never-vanishing holomorphic 2-form. This manifold $\widetilde{V}_\Gamma^*$ is actually an open cylinder $\Gamma \backslash G \times (0, 1)$, and one of its ends can be compactified by attaching to it a smooth divisor S_Γ, so that we get a complex analytic surface $\widetilde{V}_\Gamma = \widetilde{V}_\Gamma^* \cup S_\Gamma$ which may have isolated, normal singularities at S_Γ. In four of the six cases in question, this divisor can be blown down complex analytically and the result is a complex analytic surface V_Γ with a normal singularity P, whose link is the 3-manifold M_Γ; these are the cases envisaged by Klein, Milnor, Hirzebruch, Dolgachev and Neumann. In the remaining two cases the divisor S_Γ can only be blown down real analytically, so the quotient V_Γ is (homeomorphic to) a 4-dimensional real analytic space with an isolated singularity P, whose link is M_Γ and which has a complex structure away from P. In all cases one has on $V_\Gamma^* = V_\Gamma - P$ a canonical never-vanishing holomorphic 2-form that defines a trivialization of the tangent bundle of the link, and in all cases but one, we know that this trivialization lifts to a basis of the Lie algebra of the corresponding group G. I think this must also hold in the remaining case, but I do not know how to prove it.

The goal of Chapter IV is to show some of the ways that the general index theorem of Atiyah-Singer has had impact in singularities theory. Mostly I restrict the discussion to two "particular" cases of the index theorem: the Riemann-Roch formula and the Hirzebruch signature theorem. We consider also Rochlin's signature theorem.

The general philosophy is the following. The index theorem of Atiyah-Singer may be thought of as a beautiful and far-reaching generalization of the Hirzebruch-Riemann-Roch theorem, both in statement and in the spirit of the original proof. Given a closed, oriented manifold M, vector bundles E and F over M and an elliptic operator D from the sections of E to those of F, one has that both the kernel and the cokernel of D are finite-dimensional, and the difference of these dimensions is by definition the analytic index of D. The index theorem gives a description of this integer in terms of topological data implicit in the elliptic operator, the so-called topological index. This establishes a very deep connection between analysis/geometry and topology. Special cases are the signature theorem of Hirzebruch, the Hirzebruch-Riemann-Roch theorem, the Lefschetz fixed point formula, the relation of the Dirac operator with the $\widehat{A}$-genus for spin manifolds and several other fundamental theorems in mathematics. Rochlin's signature theorem can also be seen through index theory (see Sections 3 and 4 in Chapter IV).

If $(V, 0)$ is an isolated (real or complex) singularity germ in some affine space, then the diffeomorphism type of its link $M = V \cap S_\varepsilon$ depends only on the analytic structure of V, and not on the choices of defining equations for V or the radius of the sphere. Thus any invariant of closed manifolds gives automatically an invariant of singularities. This has led to a myriad of interesting results for complex singularities, using invariants coming from the Hirzebruch-Riemann-Roch theorem, or one of the signature theorems mentioned above (Hirzebruch or Rochlin). For instance the formula of Laufer (and its generalization to higher dimensions by Looijenga) for the Milnor number, and that of Durfee for the signature, are both obtained in this way, and we review these in the text.

We also give in this chapter a result of [76], which is a variant of Rochlin's signature theorem for the case when the spin^c manifold is actually a complex manifold. This theorem has the advantage of fitting naturally in the setting of algebraic geometry. Then, following [76], we give an interpretation of the geometric genus of normal, Gorenstein surface singularities in terms of the dimension of the space $H^0(-K, \mathcal{O})$, where K is the canonical divisor of the minimal resolution. This is interesting since the geometric genus is one of the key invariants for complex singularities, and it is related with the Seiberg-Witten invariants of the link (by [181, 182]).

Chapter V somehow provides a higher-dimensional analogue of Klein's theorem (in Chapter II) for the particular case of the quadric; this is based on [135]. We also prove an equivariant version of the Arnold-Kuiper-Massey theorem, saying that $\mathbb{CP}^2$ modulo conjugation is the 4-sphere; our proof is entirely analogous to that of Atiyah-Berndt in [17], though we only envisage the case of the complex projective plane and not the more general one that they consider. This is actually a byproduct for us, since our goal is to understand the geometry and topology of the pair $(\mathbb{C}^{n+1}, \widetilde{Q})$, where $\widetilde{Q}$ is the quadric $z_0^2 + z_1^2 + \cdots + z_n^2 = 0$. In fact Klein's theorem gives very precise information about the quadric $z_1^2 + z_2^2 + z_3^2 = 0$ using the group $SO(3)$. Here we look at the canonical action of the special orthogonal group $SO(n+1, \mathbb{R})$ on $\mathbb{C}^{n+1}$ and on $\mathbb{CP}^n$, the complex projective space, in order to study the pair $(\mathbb{C}^{n+1}, \widetilde{Q})$. This is of course related to the classical problem studied by Zariski [267] and others, of describing the topology of the complement of an affine algebraic hypersurface $V \subset \mathbb{C}^{n+1}$. We actually look at the projectivized situation. We begin by showing that the complement of a non-singular hyperquadric Q in $\mathbb{CP}^n$ is diffeomorphic to the total space of the tangent bundle of the real projective

n-space $\mathbb{R}P^n$,

$$\mathbb{C}P^n - Q \cong T(\mathbb{R}P^n).$$

Then we use the above observation on the topology of $\mathbb{C}P^n - Q$ to describe $\mathbb{C}P^n$ as the double mapping cylinder of the double fibration:

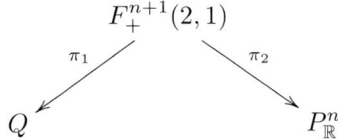

where $F_+^{n+1}(2,1) \cong SO(n+1,\mathbb{R})/(SO(n-1,\mathbb{R}) \times (\mathbb{Z}/2\mathbb{Z}))$, is the partial flag manifold of *oriented* 2-planes in $\mathbb{R}^{n+1}$ and *non-oriented* lines in these planes. The manifold $F_+^{n+1}(2,1)$ is diffeomorphic to the unit sphere normal bundle of Q in $\mathbb{C}P^n$, and it is also diffeomorphic to the unit sphere tangent bundle of $P_{\mathbb{R}}^n$. This decomposition is related to previous work by V. Vassiliev [254], J. Tits, C.T.C. Wall [262] and others, and we refer to [135] for details. Then we look more carefully at the decomposition of $\mathbb{C}P^n$ arising from the above double fibration. This describes $\mathbb{C}P^n$ as a 1-parameter family of codimension 1 submanifolds $F_+^{n+1}(2,1) \times \{t\}$, for $t \in (0,1)$, together with two "special" fibres: Q and a copy of the real projective space. We prove that these are the orbits of the natural action of $SO(n+1,\mathbb{R})$ on $\mathbb{C}P^n$, regarded as a subgroup of the complex orthogonal group $SO(n+1,\mathbb{C})$. This is an isometric action with respect to the Fubini-Study metric on $\mathbb{C}P^n$, and the principal orbits are the flag manifolds $F_+^{n+1}(2,1)$, which have codimension 1. The space of orbits is the interval $[0, \frac{\pi}{2}]$. The endpoints of this interval correspond to the two exceptional orbits, which are the quadric Q and the real projective space Π which is the fixed point set of the complex conjugation in $\mathbb{C}P^n$. Finally, in the last section of Chapter V we restrict the discussion to the case $n = 2$ and prove the equivariant Arnold-Kuiper-Massey theorem by constructing, via linear algebra, an explicit algebraic map $\mathbb{C}P^2 \to \mathbb{S}^4$ which is $SO(3)$-equivariant and commutes with complex conjugation in $\mathbb{C}P^2$.

The topology of isolated complex singularities has been long studied by many authors, as we know already, and it has a beautiful and well-developed theory. Chapters I–IV above are a sample of this, while in Chapter V we use real analytic group actions and foliations in order to study complex geometry. Chapter VI is the turning point, as thereafter we are concerned with the real counterpart of this theory, largely inspired by [168, 251]. The theme in Chapters VI–VIII is the interplay between complex geometry and real analytic singularities. Let us explain this with more detail.

In the last chapter of his book on singularities Milnor studied (germs of) real analytic mappings $(\mathbb{R}^{n+k}, 0) \xrightarrow{f} (\mathbb{R}^k, 0)$ which are submersions on a punctured neighborhood of the origin. For short we say that a map like this *satisfies the Milnor condition at* 0. He showed that such maps also define fibre bundle projections

$$\phi : (\mathbb{S}^{n+k-1} - M) \longrightarrow \mathbb{S}^{k-1},$$

where $\mathbb{S}^{n+k-1}$ is a small sphere around $0 \in \mathbb{R}^{n+k}$, $M = f^{-1}(0) \cap \mathbb{S}^{n+k-1}$ is the link and $\mathbb{S}^{k-1}$ is a small sphere around $0 \in \mathbb{R}^k$. The "weakness" of this theorem, as Milnor points out himself in his book, is that it is very hard to find examples of singularities satisfying these conditions, since the generic situation for $k > 1$ is to have a discriminant locus of dimension > 0 of points in the target over which the rank of the derivative drops. In fact Milnor asked whether there exist non-trivial examples of such maps (even for $k = 2$), other than the examples he gave in his book. This question was partially answered by Looijenga in 1971, and then by Church and Lamotke in 1975 used Looijenga's technique, to give the complete classification of the pairs (n, k) for which there exist maps satisfying Milnor's condition; in particular there are such maps for all n when $k = 2$. Those articles proved the existence of such maps but none of them exhibited explicit examples. An explicit example was constructed in 1973 by A'Campo for all n even and $k = 2$; this is given by the map from $\mathbb{R}^{2m+4} \cong \mathbb{C}^{m+2}$ into $\mathbb{R}^2 \cong \mathbb{C}$, $m \geq 0$, defined by

$$(u, v, z_1, \ldots, z_m) \longmapsto u\, v\, (\bar{u} + \bar{v}) + z_1^2 + \cdots + z_m^2 \,.$$

Thus an interesting problem is to find explicit families of real singularities satisfying the Milnor condition, and see which geometric properties they share with complex singularities. In fact Milnor's fibration theorem for real singularities has another important difference with the complex case. In the theorem for complex singularities the projection map ϕ of the fibre bundle $(\mathbb{S}^{2n-1} - M) \longrightarrow \mathbb{S}^1$ is given by the obvious map $\phi = f/|f|$; for real singularities this is false in general, as Milnor pointed out in his book. To fix the ideas, we say that the function f satisfies *the strong Milnor condition* at 0 (a notation taken from [214]) if it further satisfies that the projection map of the bundle $(\mathbb{S}^{n+k-1} - M) \longrightarrow \mathbb{S}^{k-1}$ is $\phi = f/|f|$. Similarly to complex singularities, a real singularity that satisfies the strong Milnor condition defines an open book decomposition of the sphere $\mathbb{S}^{n+k-1}$ and its link is a *fibred link*. In [111] Jacquemard considered real analytic maps $\mathbb{R}^{n+2} \to \mathbb{R}^2$ satisfying the Milnor condition, and he gave necessary conditions for such maps to satisfy the strong Milnor condition. It was shown recently in [218] that this is also related with the C-regularity studied by Bekka and others.

Chapter VI lays down the main ideas that we use in Chapter VII to construct and study infinite families of singularities satisfying the strong Milnor condition, and which have a rich geometry. Chapter VII is based in [88, 226, 227, 214], but it is largely inspired by the pioneering work of López de Medrano, Verjovsky and Meersseman (LVM for short) about new constructions of complex manifolds via holomorphic dynamics, and by the previous work of Camacho-Kuiper-Palis about the topology of linear vector fields in the complex domain.

As the previous comments suggest, the motivation for the constructions in these chapters comes from an entirely different setting to the one envisaged so far in this monograph. In fact, in [251] René Thom gave interesting ideas for the use of Morse theory to study foliations on smooth manifolds. In particular he noticed that given a foliation $\mathcal{F}$ on a smooth manifold M, and a Morse function $f : M \to \mathbb{R}^+ \cup \{0\}$, the restriction of f to the leaves of $\mathcal{F}$ has as critical set the critical points of f and the points where the level surfaces of f are tangent to the leaves of $\mathcal{F}$. He showed that the latter is "generically" a manifold, except that it may possibly have certain singularities, and called it the *variety of contacts* (*or polar variety*) of the two foliations ($\mathcal{F}$ and the one given by the level surfaces of f). Thom indicated how this polar variety could be used to study the foliation $\mathcal{F}$.

In the same vein, in [50] the authors look at linear vector fields

$$F(z) = (\lambda_1 z_1, \ldots, \lambda_n z_n),$$

in $\mathbb{C}^n$, where the λ_i are non-zero complex numbers, and look at the manifold V^* of points where the 1-dimensional holomorphic foliation $\mathcal{F}$ defined by F on $\mathbb{C}^n - 0$ is tangent to the foliation $\mathcal{S}$ given by the spheres around the origin (i.e., by the level sets of the Morse function "distance to 0", squared). It is an exercise to see that if $0 \in \mathbb{C}$ is not contained in the convex hull $\mathcal{H}(\lambda_1, \ldots, \lambda_n)$ of the λ_i, then V^* is empty. In this case the vector field is said to be in the *Poincaré domain*; in such a case all the leaves of $\mathcal{F}$ accumulate at the origin, being transversal to all spheres around 0. This situation is very important from the viewpoint of dynamical systems, but for this work the relevant case is when 0 is contained in $\mathcal{H}(\lambda_1, \ldots, \lambda_n)$, i.e., when F is a vector field *in the Siegel domain*. When this happens and the eigenvalues λ_i satisfy a certain genericity (the "weak hyperbolicity condition"), the polar variety V^* is actually a smooth manifold, defined by the complete intersection singularity,

$$V_\Lambda = \{\lambda_1 z_1 \bar{z}_1 + \cdots + \lambda_n z_n \bar{z}_n = 0\},$$

which is singular only at 0, of real codimension 2, and $V^* = V_\Lambda - 0$. The same constructions work if we consider several commuting linear vector fields (satisfying certain conditions). One obtains in this way real analytic complete intersection singularities of higher codimension, which have very rich and fascinating geometry and topology (see [149, 150, 151, 160, 268, 143, 161, 162, 31, 32], and also Chapter VI below). Alas, these singularities do not satisfy the Milnor condition at 0. However, one may replace the linear vector field $(\lambda_1 z_1, \ldots, \lambda_n z_n)$ by any non-linear vector field of the form:

$$F(z) = (\lambda_1 z_{\sigma_1}^{a_1}, \ldots, \lambda_n z_{\sigma_n}^{a_n}),$$

where $\{\sigma_1, \ldots, \sigma_n\}$ is any permutation of the set $\{1, \ldots, n\}$, $a_i \geq 2$ for all $i = 1, \ldots, n$ and the λ_i are arbitrary non-zero complex numbers; the corresponding polar variety V_F^* is again a smooth manifold, defined by the complete intersection

$$V_F = \{\lambda_1 z_{\sigma_1}^{a_1} \bar{z}_1 + \cdots + \lambda_n z_{\sigma_n}^{a_n} \bar{z}_n = 0\},$$

$V_F^* = V_F - 0$, which is singular only at 0, quasi-homogeneous, of real codimension 2. The remarkable fact is that these singularities do satisfy the strong Milnor condition at 0. Thus one has open book decompositions on the spheres analogous, but different, to those given by complex singularities. Notice that these singularities are reminiscent of the classical Pham-Brieskorn singularities

$$\lambda_1 z_1^{a_1} + \cdots + \lambda_n z_n^{a_n} = 0,$$

and if the permutation σ is the identity, then they turn out to be topologically (not analytically) equivalent to Pham-Brieskorn singularities, so we call them *twisted Pham-Brieskorn singularities* (see Chapter VII). As an example, our results in Chapter VII show that the famous Poincaré's homology 3-sphere Σ can be regarded as the set of points in the unit sphere $\mathbb{S}^5 \subset \mathbb{C}^3$ where $\mathbb{S}^5$ is tangent to the holomorphic foliation spanned by the vector field $F = (z_1^3, z_2^4, z_3^6)$; the points where F is tangent to $\mathbb{S}^5$ as a real vector field form the double of the E_8-manifold, containing Σ as an equator.

Along the way, in Chapter VI we look at the real hypersurfaces that arise when we consider a $\mathbb{C}$-valued holomorphic function f on an open set $\mathcal{U}$ in $\mathbb{C}^n$ and we compose it with the projection onto a real line through the origin in $\mathbb{C}$. We prove in particular that if f has an isolated critical point at 0, then the double of its Milnor fibre is diffeomorphic to the link of its real part. As a byproduct we get explicit real analytic embeddings in $\mathbb{S}^3$ of closed oriented 2-manifolds of all genera.

Chapter VIII is in part a return to the most basic situation: knots and links defined algebraically in the 3-sphere. The goal here is to present the results of [202] about the singularities

$$z_1^p \bar{z}_2 + z_2^q \bar{z}_1 = 0, \qquad (*)$$

which are of the class considered in the previous chapter, so we know they define open book decompositions of the 3-sphere "a la Milnor". This situation is very much reminiscent of the classical problem, dating back to Newton and others, of studying the topology of plane curves, i.e., sets in $\mathbb{C}^2$ defined by $f(z_1, z_2) = 0$, where f is a holomorphic function. One of the most classical ways for approaching this problem is via resolutions of the singularities. In fact, Max Noether (1883) proved that the singularities of every plane curve could be resolved by blowing ups; this means that given the germ at 0 of a curve $C \subset \mathbb{C}^2$ with a singular point at 0, one can find a smooth complex surface $\tilde{X}$ with a projection $\pi : \tilde{X} \to \mathbb{C}^2$ which is composition of a finite number of blow-ups, such that the divisor $E = \pi^{-1}(0)$ is connected, consists of a finite number $E_1, \ldots, E_r$ of copies of $\mathbb{CP}^1$ whose intersections are either empty or at ordinary double points (i.e., defined locally by $xy = 0$), π is biholomorphic away from E, and the closure $\tilde{C}$ in $\tilde{X}$ of $\pi^{-1}(C - 0)$ (which is called the strict transform of C) consists of a finite number of pairwise disjoint complex lines, as many as branches of C, which intersect E transversally at smooth points of E. It is customary to assign to such a resolution a *decorated plumbing graph*, also called *the resolution graph*, which describes its topology. Since the topology of C (as an embedded sub-variety) is determined by the topology of $C \cap \mathbb{S}_\varepsilon^3$, where $\mathbb{S}_\varepsilon^3$ is a small sphere around $0 \in \mathbb{C}^2$, and π is a diffeomorphism away from E, it follows that if we know the resolution graph of the singularity, then we know the topology of the plane curve.

This is what we aimed to do with the singularities $(*)$; the additional problem we face is that these are only real analytic and the usual process of resolutions via blow-ups gets stuck after the first step (i.e., the singularities are no longer simplified by the blow-ups). The trick here is to modify some of its branches by a homeomorphism in the second step of the resolution process, in order to make them become complex analytic singularities, which one can resolve in the usual way. In this way we get a "topological" resolution $\pi : X \to \mathbb{C}^2$ of the singularity $\{z_1^p \bar{z}_2 + z_2^q \bar{z}_1 = 0\}$, which determines the resolution graph and therefore the topology of these singularities.

However, just as for complex curves, this is not enough to determine the topology of the corresponding Milnor fibrations. One has to do more: the links that we get in this way are all Seifert links in the sense of [75], and they are *fibred*, defining open book decompositions of the 3-sphere which are *horizontal*, i.e., the pages are transversal to the Seifert fibres. Pichon in [200] studied these type of links; more generally she studied Waldhausen links which are horizontally fibred, and found a way for describing fully the topology of these fibrations, using the resolution graph and the orientations on the components of the link induced by the open book decomposition. In the particular case of Seifert links, which is simpler, one has that the monodromy of the corresponding fibration is cyclic and one can associate to it a *Nielsen graph*, which encodes all the information of the resolution graph, and which is also determined by the Seifert (or the resolution) graph. One has that the Nielsen graph of the monodromy determines the period of it and the topology of the fibres (using the Hurwitz formula). Thus we have the complete topological description of these fibrations.

The links of the singularities that we envisage in this chapter turn out to be isotopic to the links defined by the equations:

$$\bar{z}_1 \bar{z}_2 \left(z_1^{p+1} + z_2^{q+1} \right) = 0 \,,$$

which are a special type of singularities of the form $f\bar{g}$ with f, g holomorphic functions $\mathbb{C}^2 \to \mathbb{C}$. This type of singularities are already present in the work of N. A'Campo [2] and L. Rudolph [212]. In [199] Pichon proves that such a function $f\bar{g} : \mathbb{R}^4 \to \mathbb{R}^2$ has an isolated critical point (say at 0) iff the link $L_f \cup L_g$ is fibred (as an oriented link), where L_f, L_g are the links of f and g. She further proved that in this situation the projection map:

$$\phi : S_\varepsilon - (L_f \cup L_g) \to S^1$$

can be taken to be $\frac{f\bar{g}}{|f\bar{g}|}$ in a tubular neighborhood of the link. These results have been recently improved in [201], relaxing the condition of $f\bar{g}$ having an isolated critical point and showing that the projection ϕ can be taken as $\frac{f\bar{g}}{|f\bar{g}|}$ everywhere. We briefly explain this in the last section of Chapter VIII.

Acknowledgments

This monograph is the outcome of many years of working in the area of singularities, during which I have profited from the work of many great mathematicians, to whom I am deeply indebted. Their works have been to me the lights in my path as a mathematician, and have shown me the beauty of mathematics. My thanks to all of them. I would like to mention here two people who have had the strongest impact and influence on me during all this time, and whose work permeates all areas of mathematics that I have explored: Sir Michael F. Atiyah and Professor John W. Milnor.

During this time I have also profited from conversations with many people and I have had the support of several institutions which have either hosted me or given me financial support, or both. I am glad to have the opportunity of expressing here my gratitude to all of them. Although I am sure to leave out some important names, I want to mention those that I am able to remember.

First of all, I am most grateful to Nigel Hitchin, Lê Dũng Tráng and Alberto Verjovsky, for explaining to me many of the ideas and concepts included in this text, and for teaching me about the beauty of mathematics; talking to them has always been a source of inspiration for me. I have also profited a lot from conversations with a number of friends and colleagues, who have taught me a lot, especially: Jean Paul Brasselet, Hélène Esnault, Francisco González Acuña, Santiago López de Medrano, David Massey, A.J. Parameswaran, Anne Pichon, Elmer Rees, Maria Aparecida Ruas, V. Srinivas, Brian F. Steer, Tatsuo Suwa, Eckart Viehweg, and surely others whose name slips away from my memory. My heart-felt thanks to all of them.

I am very grateful to Juan Pablo Romero for helping me with the many intricacies of Latex and, mostly, for making the figures in this text. And to Elsa Puente for proof-reading the text and pointing out several misprints and corrections to be made.

I am indebted to my wife Teresa for her patience and for being the driving force that moves me forward.

Among the institutions that have hosted me and/or supported my work on the material of this text, I am most grateful to the Abdus Salam International Centre for Theoretical Physics at Trieste, Italy, for providing me the perfect atmosphere for concentrating on this work; I wrote the bulk of it while a Staff Associate of the ICTP. I am also very much indebted to the Institut de Mathématiques de Luminy (France), to the Mathematical Institute of the University of Oxford (U.K.), to Instituto de Matemática Pura e Aplicada de Rio de Janeiro (Brazil), University of Hokkaido (Japan) and Tata Institute of Fundamental Research (India).

During this time I have had several grants from CONACYT (México) and from the DGAPA of the National University of Mexico (UNAM), as well as from CNRS (France), SRC (U.K.), CNPq (Brazil) and from various other sources, and I am most grateful to all of them for this important support.

And, of course, I am most grateful and indebted to the Instituto de Matemáticas of UNAM and its Cuernavaca Unit, where I work, for its great support and stimulating atmosphere.

Chapter I

A Fast Trip Through the Classical Theory

In this chapter we review briefly some of the classical results of Milnor and others about the topology of real and complex analytic spaces around an isolated singular point. We review first the theorem of Milnor about the (embedded) local conical structure of analytic spaces; then we recall Ehresman's fibration theorem, the definition of open book decompositions; the fibration theorem of Milnor for real singularities in general and his improved version for complex singularities. Then we speak about the topology of the Milnor fibre, beginning with the case of the Pham-Brieskorn polynomials where the topology is given by the so-called "join of Pham". Finally we speak briefly about the work of Brieskorn, Hirzebruch and Milnor on exotic spheres and singularities.

I.1 An example: the Pham-Brieskorn polynomials

Let us begin by considering, as an example, the *Pham-Brieskorn* polynomials in $\mathbb{C}^{n+1}$:

$$f(z) = z_o^{a_0} + \cdots + z_n^{a_n},$$

where $z = (z_o, \ldots, z_n)$ and the a_i are integers ≥ 2, c.f. [40, 39, 198]. The derivative of f at a point z is:

$$Df(z) = (a_0 z_o^{a_0-1}, \ldots, a_n z_n^{a_n-1}).$$

Hence the origin 0 is the only critical point of f and therefore the variety

$$V = f^{-1}(0) = \{z_o^{a_0} + \cdots + z_n^{a_n} = 0\}$$

is a hypersurface (complex codimension 1) with an isolated singularity at $0 \in \mathbb{C}^{n+1}$.

In order to study the topology of V as an embedded subvariety of $\mathbb{C}^{n+1}$, let d be the least common multiple of the a_i, $i = 0, 1, \ldots, n$, set $q_i = d/a_i$ and consider the action of the non-zero complex numbers $\mathbb{C}^*$ on $\mathbb{C}^{n+1}$ given by:

$$t \cdot (z_0, \ldots, z_n) = (t^{q_0} z_0, \ldots, t^{q_n} z_n).$$

It is clear that if we restrict this action to $\mathbb{R}^+$ we obtain a real analytic flow in $\mathbb{C}^{n+1}$ satisfying:

(i) its flow lines (the orbits) are transversal to all the spheres around $0 \in \mathbb{C}^{n+1}$;

(ii) all the orbits accumulate at the origin as t tends to 0; and

(iii) the variety V is an invariant set of the flow, i.e., V is a union of orbits.

Notice this flow defines a 1-parameter group of diffeomorphisms $\{\phi_t\}$ of $\mathbb{C}^{n+1}$ that has 0 as a fixed point, V as an invariant set and it is strictly contracting for $t < 1$, while it is expanding for $t > 1$. This implies that:

(i) V intersects transversally every $(2n+1)$-sphere $\mathbb{S}_r(0)$ around the origin; hence the intersection $K_r = V \cap \mathbb{S}_r(0)$ is a smooth manifold of real dimension $2n - 1$ embedded as a codimension 2 submanifold of the sphere $\mathbb{S}_r(0)$;

(ii) the diffeomorphism type of K_r is independent of the choice of the sphere $\mathbb{S}_r(0)$; and

(iii) the embedded topological type of V in $\mathbb{C}^{n+1}$ is determined by the pair $(\mathbb{S}_r(0), K_r)$, i.e., the pair $(\mathbb{C}^{n+1}, V)$ is homeomorphic to the (global) cone over the pair $(\mathbb{S}_r(0), K_r)$.

The manifold $K = K_r$, for some $r > 0$, is called *the link* of the singularity, and it is known as *the Brieskorn manifold* $M_{(a_0, \ldots, a_n)}$.

Notice that we also have an $\mathbb{S}^1$-action on $\mathbb{C}^{n+1}$ obtained by restricting the above $\mathbb{C}^*$-action to the unit complex numbers. This $\mathbb{S}^1$-action on $\mathbb{C}^{n+1}$ is by isometries and therefore induces a locally free action on the links K_r. If $n = 1$ this means $K = K_r$, for some $r > 0$, is the union of a finite number of $\mathbb{S}^1$-orbits (as many as the irreducible components, or branches, of V). If $n = 2$ then K is a 3-manifold in $\mathbb{S}^5$ with an $\mathbb{S}^1$-action, so K is a Seifert manifold, whose Seifert invariants can be easily determined from the "weights" of the action, see for instance [194]. In this case the corresponding Brieskorn manifolds have very interesting topology and this will be explained later in the text. For instance the link of the polynomial $z_1^p + z_2^q + z_3^r$ is the lens space $\mathbb{S}^3/\mathbb{Z}_r$ when $p = q = 2$, and it is Poincaré's homology sphere for $(p, q, r) = (2, 3, 5)$. The topology of these manifolds for $n > 2$ has been studied by Brieskorn [40, 39], Milnor [171] and others, obtaining remarkable results (see Section 8 below).

As we mentioned above, for $n = 1$ the link is the union of a finite number of copies of the circle $\mathbb{S}^1$, so its topology is rather simple. Still, we can look at the embedded topological type of K in $\mathbb{S}^3$, which is a *knot* if it is connected, or a *link* in general (i.e., a disjoint union of knots). What type of knot or link is it? To answer this, assume for simplicity that $p = 2$ and $q = 3$ (essentially the same arguments hold whenever p, q are relatively prime). Notice that, up to isotopy, K

is defined by $\{z_1^2 = z_2^3\}$ and $\{|z_1|^2 + |z_2|^2 = 2\}$. Hence K is contained in the torus $\mathbb{T} = \{|z_1| = |z_2| = 1\}$ as the set of points given by:

$$2 \arg z_1 = 3 \arg z_2 .$$

Thus we can parametrize K by taking a point $t \in \mathbb{R}$ into the point $(e^{\frac{it}{2}}, e^{\frac{it}{3}}) \in K$, with $t \in [0, 12\pi]$. As we do so, the first coordinate wraps around the unit circle 3 times, while the second coordinate wraps around the circle 2 times. This means that the torus knot K goes around a parallel of $\mathbb{T}$ twice and it goes around a meridian of $\mathbb{T}$ thrice.

The same arguments show that if p, q are relatively prime, then K is always a *torus knot of type* (p, q), i.e., it is contained in a torus in $\mathbb{S}^3$ in such a way that it goes around a meridian q times and around a parallel p times. For this we may use the parameterization of K given by $\gamma(t) = (e^{it/p}, e^{it/q})$, which is periodic of period pq. More generally, if $p = sp', q = sq'$ with p', q' relatively prime, then it is an exercise to show that K consists of s disjoint copies of a torus knot of type (p', q'), so it is called a *torus link* of type (p, q). For instance, if the polynomial f is homogeneous of degree p, i.e., $p = q > 1$, then K consists of p fibres of the Hopf fibration $\mathbb{S}^3 \mapsto \mathbb{S}^2$, which are circles embedded in $\mathbb{T}$ of type $(1, 1)$, since they are given by the orbits of the standard $\mathbb{S}^1$-action on $\mathbb{C}^2$ defined by $e^{it} \cdot (z_1, z_2) = (e^{it} z_1, e^{it} z_2)$.

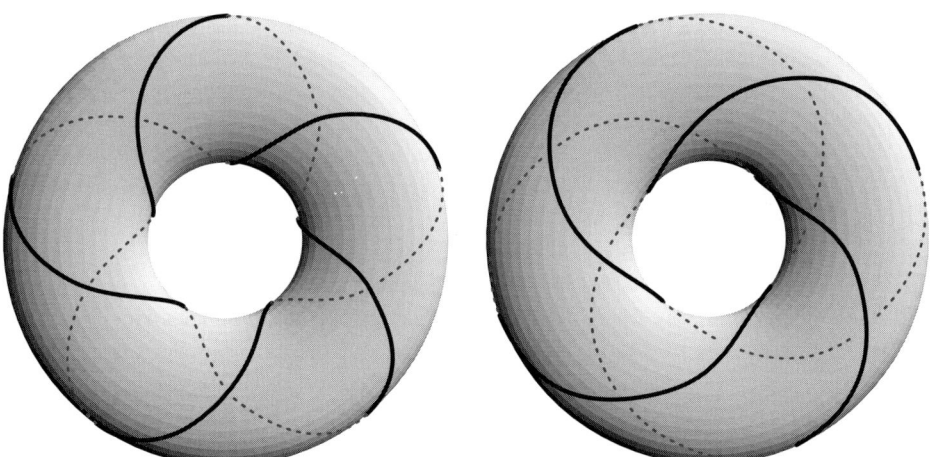

Figure 1: Toral knots of types $(2, 5)$ and $(3, 4)$.

1.1 Remarks on the topology of plane curves and curves in surfaces. The literature about this subject is vast, going back to work by Newton, Max Noether and others; Brauner [38] was the first one to study the topology of plane curves $\mathcal{C}$ in $\mathbb{C}^2$ (i.e., the zero set of a complex analytic map $\mathbb{C}^2 \to \mathbb{C}$ with an isolated critical point at $0 \in \mathbb{C}^2$) by looking at the intersection of $\mathcal{C}$ with a small sphere $\mathbb{S}_\varepsilon$ around 0; this intersection defines a knot (or link) K in the 3-sphere, and the pair $(\mathbb{S}_\varepsilon, K)$

determines the local topology of $\mathcal{C}$ near 0. The book of Milnor gives an account of this construction and more (see Chapter 10 in [168]), and this should become clear later in the text. The book [44] of Brieskorn and Knörrer contains everything we may pretend to say here about knots defined by complex plane curves, so we refer to that excellent text for those wanting to dive into this fascinating subject. This is closely related with the content of Chapter VIII below, so we shall return to plane curves in that chapter. We also refer to [249], where Teissier gives a very clear introduction to the subject. For a more advanced reading see the book [75] of Eisenbud and Neumann which describes important constructions of "complicated" links from simple pieces, and how to use graphs to study and describe this process. See also [128] for a short and clear presentation of the classic results on the topology of complex analytic plane curves from the differentiable viewpoint. The more recent article [134] studies the three-dimensional manifold M given by the complement in $\mathbb{S}^3_\varepsilon$ of an algebraic link K, i.e., a link in the 3-sphere defined by a complex plane curve. See also [200], which is concerned with knots in 3-dimensional manifolds which appear as links of complex normal surface singularities (see §2 below).

I.2   The local conical structure

We study here a construction that goes back to Brauner in [38] and was sketched in Section 1 above. Consider a (reduced, equidimensional) real analytic space V of dimension n, defined by a finite number of real analytic equations in an open ball $\mathbb{B}_r(0) \subset \mathbb{R}^N$ around the origin, and assume further that V contains the origin 0 and $V^* := V - \{0\}$ is non-singular, i.e., it is a real analytic manifold of dimension $n > 0$. We shall see that much of what we said in the previous section still holds in this more general setting provided we are sufficiently near the origin.

Consider the function $d : \mathbb{R}^N \to \mathbb{R}$ given by $d(x_1, \ldots, x_N) = x_1^2 + \cdots + x_N^2$, so that d is the square of the function "distance to 0". Its gradient vector field ∇d is given by $(2x_1, \ldots, 2x_N)$, so its solutions are the straight rays that emanate from the origin. Let us modify this vector field a little so that it "takes V into account".

In order to make the above statement precise, take the restriction d_V of d to V. At each point $x \in V^*$ the gradient $\nabla d_V(x)$ of d_V is obtained by projecting $\nabla d(x)$ to $T_x V^*$, the tangent space of V^* at x. Since $\nabla d(x)$ is never zero on V^* and it is a radial vector transversal to the sphere $\mathbb{S}_x$ passing through x and centred at 0, it follows that $\nabla d_V(x)$ vanishes if and only if $T_x V^* \subset T\mathbb{S}_x$. This means that the critical points of d_V, other than 0 itself, are the points in V^* where this manifold fails to be transversal to the sphere passing through the given point. In other words, a point $x \in V^*$ is a critical point of d_V iff V^* is tangent at x to the sphere passing through x and centred at 0. Just as in [168, Corollary 2.8], one has that d_V has at most a finite number of critical values corresponding to points in V^*, since it is the restriction of an analytic function on $\mathbb{B}_r(0)$. Hence for

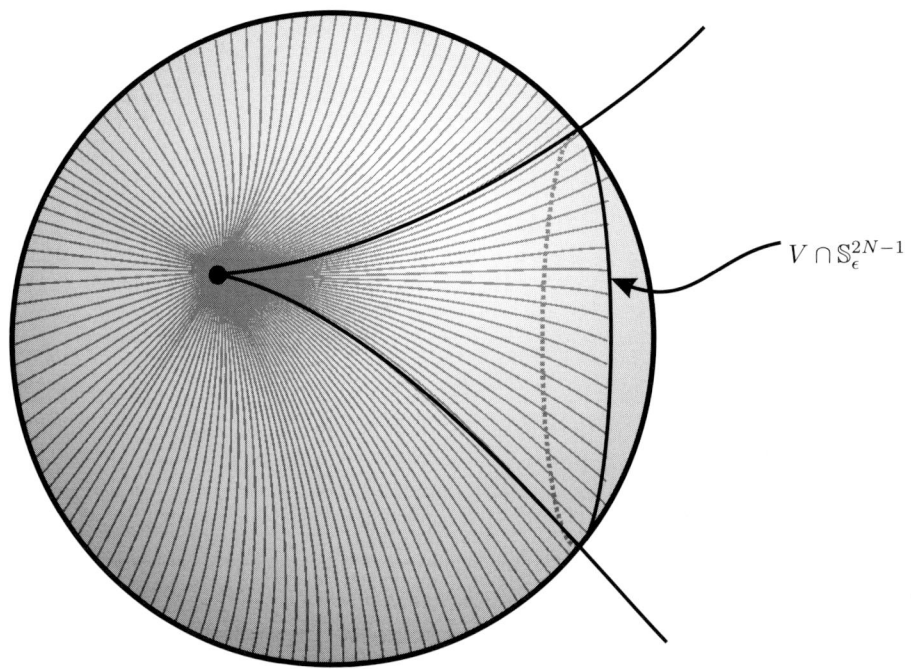

$V \cap \mathbb{S}_\epsilon^{2N-1}$

Figure 2: The conical structure

$\varepsilon > 0$ sufficiently small the function d_V has no critical points on $V \cap \mathbb{B}_\varepsilon$; therefore V^* meets transversally all sufficiently small spheres around the origin in $\mathbb{R}^N$. The gradient vector field of d_V is now everywhere transversal to the spheres around 0, and it can be assumed to be integrable. Hence it defines a 1-parameter family of local diffeomorphisms of V^* taking each link into "smaller" links. Thus, near 0, V is the cone over the intersection $V \cap \mathbb{S}_\varepsilon$ of V with a small sphere around the origin and the diffeomorphism type of the intersection $V \cap \mathbb{S}_\varepsilon$ is independent of ε for all sufficiently small spheres.

A more refined argument, due to A. Durfee [66] and based on the "Curve Selection Lemma" of [168], shows that the diffeomorphism type of the manifold $K := V \cap \mathbb{S}_\varepsilon$ is also independent of the choice of the embedding of V in $\mathbb{R}^N$. In other words, up to diffeomorphism K depends only on the analytic structure of V at 0 and not on the choice of the equations that define V nor on the sphere $\mathbb{S}_\varepsilon$, provided this is small enough.

This same result is also proved by Lê-Teissier in [136].

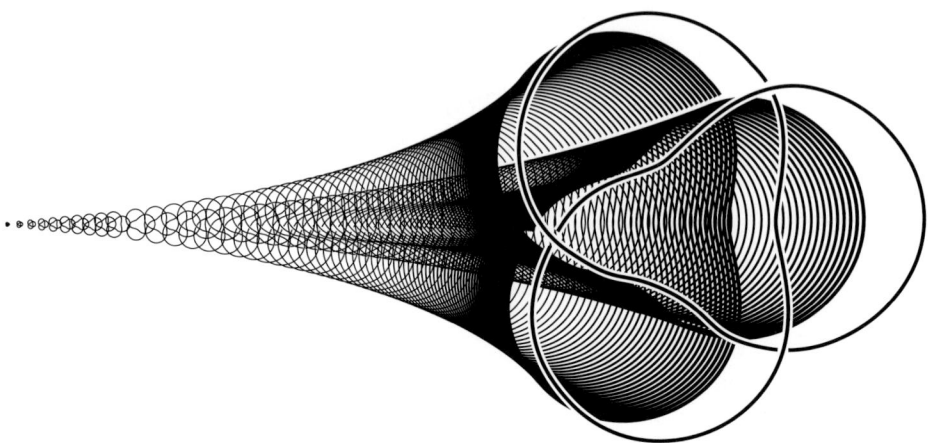

Figure 3: The link of the singularity determines the topological type.

2.1 Definition. The manifold $K := V \cap \mathbb{S}_\varepsilon$ is called *the link* of V at 0.

It is worth saying that the word "link" is being used in two different ways. In Definition 2.1 the name "link" comes from the notation used in PL-topology to denote the boundary of a regular neighborhood of a point in a triangulated manifold. But we also used the word "link" in Section 1 above to denote a disjoint union of circles in the 3-sphere. Both names are common in the literature. Thus given a polynomial map in two complex variables, as for instance $f(z_1, z_2) = z_1^p + z_2^q$, one has that *its link* (i.e., the intersection $K = f^{-1}(0) \cap \mathbb{S}_\varepsilon$) is a *link in the 3-sphere* $\mathbb{S}_\varepsilon$ (i.e., a disjoint union of circles embedded in the sphere). This is why some authors prefer to call the manifold K of 2.1 in some other way, as for instance "the boundary of the singularity". However we prefer to stick to the classical notation and call K the link of the singularity, just taking care that this does not lead to any confusion.

Notice however that we have not said anything yet about the embedded topological type of V. We claim that just as for the Pham-Brieskorn singularities of the previous section, one has in general that for $\varepsilon > 0$ sufficiently small, the pair $(\mathbb{B}_\varepsilon, V \cap \mathbb{B}_\varepsilon)$ is homeomorphic to the cone over the pair $(\mathbb{S}_\varepsilon, V \cap \mathbb{S}_\varepsilon)$. In other words, not only is $V \cap \mathbb{B}_\varepsilon$ the cone over $V \cap \mathbb{S}_\varepsilon$ (as we know already), but the pair $(\mathbb{B}_\varepsilon, V \cap \mathbb{B}_\varepsilon)$ is the cone over the pair $(\mathbb{S}_\varepsilon, V \cap \mathbb{S}_\varepsilon)$. To show this we only need to refine a little the above argument, following [168]: for each $x \in V^*$, let U_x be an open neighborhood of x in $\mathbb{B}_\varepsilon$, and let r_x be a vector field on U_x obtained by parallel extension of the gradient vector field ∇d_V on $U_x \cap V^*$. We may assume that the covering $\{U_x\}$ of V^* is locally finite. Now take the radial vector field ∇d on $\mathbb{B}_\varepsilon - V$, and glue all these vector fields by a partition of unity to obtain a vector field on $\mathbb{B}_\varepsilon - \{0\}$, which is everywhere transversal to the spheres around 0 and is tangent to V^*. Furthermore, we can choose this vector field to be integrable (see [168]). One has that the solutions of this vector field converge to 0 and provide a local family of diffeomorphisms that yield the following theorem.

2.2 Theorem (Milnor). *The homeomorphism type of V near 0 and the isotopy class of the embedding of V in $\mathbb{R}^N$ are determined by the intersection of V with a small sphere $\mathbb{S}_\varepsilon$ around the isolated singularity $0 \in V$. More precisely, for every sufficiently small ball $\mathbb{B}_\varepsilon$ around 0 one has a homeomorphism of pairs:*

$$(\mathbb{B}_\varepsilon, \mathbb{B}_\varepsilon \cap V) \cong \mathrm{Cone}\,(\mathbb{S}_\varepsilon, \mathbb{S}_\varepsilon \cap V),$$

where $\mathbb{S}_\varepsilon = \partial \mathbb{B}_\varepsilon$ is the boundary.

2.3 Remark. It is well known that Theorem 2.2 holds for non-isolated singularities. The main difference is that in this situation the link K will no longer be a manifold but a singular space. Actually, given an analytic germ (V, p), if we only care for the topology of V near p, then the fact that V is locally a cone over the link of p follows immediately from the work of Lojasiewicz in [144], where he proved that every analytic space can be triangulated. But one can say more, just as in [168]: given an embedding of (V, p) in some affine space $\mathbb{R}^N$, the embedded topological type of V is a cone over the intersection of V with a small sphere around p. The proof of this result is along the same lines as that of Theorem 2.2 above; this can be found for instance in [99, 48]. The proof uses Whitney stratifications (see for instance [136]). The idea of the proof is that given a real analytic germ $(V, 0)$, we can choose the representative of $(V, 0)$ small enough so that there exists a Whitney stratification $\{S_\alpha\}_{\alpha \in A}$ of the ball $\mathbb{B}_\varepsilon$ with a finite number of strata, compatible with V (i.e., V is the union of strata), and such that each stratum meets transversally all the spheres in $\mathbb{B}_\varepsilon$ with centre at 0. Each stratum S_α is a real analytic submanifold of $\mathbb{B}_\varepsilon$, so one can construct a vector field on S_α just as we did for V^* above, by considering the gradient vector field of the restriction to S_α of the function distance to 0 (squared). The Whitney conditions guarantee that we can put all these vector fields together, by a partition of unity, in order to obtain a vector field on all of $\mathbb{B}_\varepsilon$ (for $\varepsilon > 0$ sufficiently small), which is transversal to all small spheres around the origin, pointing outwards, and it is *stratified*, i.e., at each point $x \in V$ it is tangent to the stratum S_α that contains x. In this way we obtain a 1-parameter family of homeomorphisms taking the pair $(\mathbb{S}_\varepsilon, \mathbb{S}_\varepsilon \cap V)$ into the "smaller" links $(\mathbb{S}_{\varepsilon'}, \mathbb{S}_{\varepsilon'} \cap V)$ for $\varepsilon > \varepsilon' > 0$, and these homeomorphisms preserve each stratum S_α.

I.3   Ehresmann's fibration lemma

The content of this section is for smooth mappings between manifolds and is included in this text for completeness, since it motivates the fibration theorem for real singularities of Milnor that we shall prove in the next section. This result was originally proved in [74] (see also [1], for instance).

3.1 Theorem (Ehresmann's fibration lemma). *Let M and N be ($\mathbb{C}^\infty$ for simplicity) oriented, differentiable manifolds of dimensions $n + k$ and k respectively. Assume M is closed (=compact without boundary) and let*

$$f : M \to N$$

be a proper differentiable map which is a surjection everywhere, i.e., the Jacobian matrix $Df(x) = \left(\left(\frac{\partial f_i}{\partial x_j}(x) \right) \right)$ has rank k for each $x \in M$. Then f is the projection map of a (locally trivial) fibre bundle.

Proof. The fact that f is a submersion everywhere implies that all points in M are regular points of f, so the Implicit Function Theorem implies that all the fibres $f^{-1}(y)$, $y \in N$, are smooth submanifolds of M of codimension k, i.e., dimension n. So the proof of Theorem 3.1 amounts to proving the local triviality of the projection. In other words we must show that given any $y \in N$ there exists an open disc $\mathbb{D}_y \subset N$ such that $f^{-1}(\mathbb{D}_y)$ is a product $\mathbb{D}_y \times f^{-1}(y)$.

At each point $x \in M$ we have a splitting of the tangent bundle of M as

$$T_x M \cong T_x(f^{-1}(y)) \oplus \nu_x(f^{-1}(y)) \, , \, y = f(x) \, ,$$

where $\nu(f^{-1}(y))$ is the normal bundle of the fibre of f containing x, for some Riemannian metric. The fact that $Df(x)$ has rank n implies that $\nu_x(f^{-1}(y))$ is carried by $Df(x)$ isomorphically into the tangent bundle of N at y. Therefore the normal bundle of $f^{-1}(y)$ is trivial and it is isomorphic to the pull-back by f of the tangent space $T_y N$.

Let $\{a_1, \ldots, a_k\}$ be a basis of $T_y N$. We can always extend this to a set of k linearly independent integrable vector fields on a small disc $\mathbb{D}_y$ of y in N, which we still denote by $\{a_1, \ldots, a_k\}$. These define k local flows around y which can be used to parametrize $\mathbb{D}_y$, since the vector fields $\{a_1, \ldots, a_k\}$ are linearly independent. By the arguments above one has that $\{a_1, \ldots, a_k\}$ lift to a trivialization $\{\alpha_1, \ldots, \alpha_k\}$ of the normal bundle $\nu(f^{-1}(y))$, which extends to a set of n integrable vector fields on a tubular neighborhood $\mathcal{N}_y$ of the fibre $f^{-1}(y)$, linearly independent everywhere and orthogonal to all the fibres of f. By the compactness of the fibres of f, there exists a time $t > 0$ such that all these flows on $\mathcal{N}_y$ are defined for all times t' with $|t'| < t$ and they do not have fixed points. Moreover, we can choose the flows so that they parametrize the neighborhood $\mathcal{N}_y$. By re-scaling the time, if necessary, we can make this parametrization of $\mathcal{N}_y$ compatible with the one on the base $\mathbb{D}_y$. Hence the neighborhood $\mathcal{N}_y$ is of the form $\mathbb{D}_y \times f^{-1}(y)$, proving Theorem 3.1. $\quad\square$

I.4   Milnor's fibration theorem for real singularities

We now consider a real analytic map,

$$f = (f_1, \ldots, f_k) : (\mathbb{B}_r \subset \mathbb{R}^{n+k}, 0) \longrightarrow (\mathbb{R}^k, 0) \, ,$$

where $\mathbb{B}_r$ is a (sufficiently) small ball around 0 in $\mathbb{R}^{n+k}$.

4.1 Definition. We say that f satisfies *the Milnor condition* at $0 \in \mathbb{R}^{n+k}$ if its Jacobian matrix $Df(x)$ has rank k for all $x \neq 0$ in $\mathbb{B}_r$.

In other words this means that f is a surjection at every point in $\mathbb{B}_r$, except (possibly) at the origin. As Milnor himself points out in his book, this is a very stringent hypothesis when $k > 1$. We shall discuss this point later in the text, in Chapter VII.

Notice that if f satisfies the Milnor condition at 0, then by the first Thom-Mather transversality theorem (see [1]), if $V = f^{-1}(0)$ is transversal to a given sphere $\mathbb{S}_\varepsilon$ around 0, then the fibres $f^{-1}(t)$ will also be transversal to $\mathbb{S}_\varepsilon$ for all $t \in \mathbb{R}^k$ sufficiently near 0.

The following result of Milnor is a straightforward generalization of Ehresmann's fibration lemma, and its proof is just a refinement of the proof of Theorem 3.1 so we only sketch it. For Milnor this was an intermediate step for his fibration theorem for real singularities, which we state below as Theorem 4.3. However we state it as a theorem because this is the version of Milnor's fibration theorem which is more common in the literature and it also generalizes more naturally to other settings (c.f. the last section in this chapter and Chapter VII below).

4.2 Theorem. *Let $f : (\mathbb{B}_r \subset \mathbb{R}^{n+k}, 0) \longrightarrow (\mathbb{R}^k, 0)$ be real analytic and assume it satisfies the Milnor condition at 0. Let $\varepsilon > 0$ be sufficiently small so that $\mathbb{S}_\varepsilon$ and all the spheres around 0 in $\mathbb{R}^{n+k}$ of radius less than ε meet $V = f^{-1}(0)$ transversally. Let $\delta > 0$ be small enough with respect to ε so that all fibres $f^{-1}(t)$ with $\|t\| \leq \delta$ meet $\mathbb{S}_\varepsilon$ transversally. Let $\mathbb{S}_\delta^{k-1}$ be the sphere in $\mathbb{R}^k$ of radius δ and centre at 0. Then the map*

$$f : f^{-1}(\mathbb{S}_\delta^{k-1}) \cap \mathbb{B}_\varepsilon \longrightarrow \mathbb{S}_\delta^{k-1},$$

where $\mathbb{B}_\varepsilon \subset \mathbb{R}^{n+k}$ is the (open or closed) ball bounded by $\mathbb{S}_\varepsilon$, is the projection of a (locally trivial) fibre bundle.

Proof. By hypothesis the derivative of f is surjective at each point away from 0. For $x \in V^* = V - \{0\}$ the kernel of Df spans the tangent bundle of the fibres of f while the pull-back of $T_t \mathbb{D}_\delta$ spans the normal bundle $\nu(f^{-1}(t))$ of $f^{-1}(t)$, where $\mathbb{D}_\delta$ is the disc in $\mathbb{R}^k$ bounded by $\mathbb{S}_\delta^{k-1}$. Hence a choice of a basis for $T_t \mathbb{D}_\delta$ lifts to a trivialization of $\nu(f^{-1}(t))$. Furthermore, since the fibres of f are transversal to the boundary sphere $\mathbb{S}_\varepsilon$, the sections $\{s_1, \ldots, s_k\}$ that define this trivialization can be chosen to be tangent to $\mathbb{S}_\varepsilon$. Each section s_i generates a local flow ψ_λ^i, $\lambda \in \mathbb{R}$, and by compactness we can choose a time t_0 for which all these flows are well defined on all the fibres of f. This implies that each fibre of f has a neighborhood which is a product, proving Theorem 4.2. $\qquad \square$

It is worth saying that Theorem 4.2 could have been stated (with the same proof) as giving a fibre bundle over all of $\mathbb{D}_\delta - \{0\}$. Let us set $N(\varepsilon, \delta) := f^{-1}(\mathbb{D}_\delta) \cap \mathbb{B}_\varepsilon$ with $\mathbb{B}_\varepsilon$ being, from now on, the closed ball. This manifold $N(\varepsilon, \delta)$ is usually referred to as *a Milnor tube* for f. The boundary of $N(\varepsilon, \delta)$ is the union of the manifold $f^{-1}(\mathbb{S}_\delta^{k-1}) \cap \mathbb{B}_\varepsilon$ in Theorem 4.2, together with the manifold

$$T_{\varepsilon,\delta} := N(\varepsilon,\delta) \cap \mathbb{S}_\varepsilon \,,$$

which is a regular neighborhood of $K = V \cap \mathbb{S}_\varepsilon$ in the boundary sphere $\mathbb{S}_\varepsilon$.

4.3 Theorem. *Let f be as in Theorem 4.2. Then, with the above hypothesis and notation, the fibre bundle*

$$f : N(\varepsilon,\delta) - V \longrightarrow \mathbb{D}_\delta - \{0\}$$

induces a fibre bundle projection:

$$\phi : \mathbb{S}_\varepsilon - Int\,(T_{\varepsilon,\delta}) \longrightarrow \mathbb{S}_\delta^{k-1} = \partial \mathbb{D}_\delta$$

which, restricted to the boundary $\partial T_{\varepsilon,\delta} = f^{-1}(\mathbb{S}_\delta^{k-1}) \cap \mathbb{S}_\varepsilon$, is the map f.

This theorem is a consequence of Theorem 4.2 and the following lemma (which is Theorem 1 in [167]).

4.4 Lemma. *The boundary of the tube $N(\varepsilon,\delta)$ is homeomorphic to $\mathbb{S}_\varepsilon$ under a homeomorphism that leaves $T_{\varepsilon,\delta}$ pointwise fixed.*

In fact the same proof shows that $N(\varepsilon,\delta)$ is homeomorphic to the ball $\mathbb{B}_\varepsilon$.

Proof. We follow the arguments of Milnor in [167], just replacing the hard part in his proof by Corollary 3.4 in [168], which is a consequence of the Curve Selection Lemma. Let $f_1, \ldots, f_k$ be the components of f and define a function:

$$r(x) = f_1^2(x) + \cdots + f_k^2(x) \,.$$

Let ∇r be its gradient. The level surfaces $r^{-1}(s)$, $s \in \mathbb{R}^+$ are the tubes $f^{-1}(|t|)$ for $|t|^2 = s$, and the vector field ∇r is transversal to these tubes (away from V). Consider also the vector field $\nabla(\iota)(x) = 2x$, which is the restriction to $\mathbb{D}_\varepsilon - V$ of the gradient of the function $x \overset{\iota}{\mapsto} x^2$. Both maps r and ι are polynomial and ≥ 0. Hence Corollary 3.4 of [168] implies that on $\mathbb{B}_\varepsilon - V$ the vector fields ∇r and $\nabla(\iota)$ are either linearly independent or one is a positive multiple of the other. Now define a vector field on $\mathbb{B}_\varepsilon - V$ by:

$$v(x) = \|x\| \cdot \nabla r + \|\nabla r\| \cdot x \,,$$

which bisects the angle between $\nabla r(x)$ and x. This is a smooth vector field on $\mathbb{B}_\varepsilon - V$ and by the Schwartz inequality ($\|v_1\|\|v_2\| \geq v_1 \cdot v_2$, with equality holding only when the two vectors are colinear) it satisfies:

$$v \cdot x > 0 \,, \qquad v \cdot \nabla r > 0. \qquad\qquad (*)$$

Now, given any point $x_o \in \mathbb{B}_\varepsilon - V$, let $\gamma(t)$ be the solution through x_o of the differential equation $dx/dt = v(x)$ in $\mathbb{B}_\varepsilon - V$. As we move along the path $\gamma(t)$, by (∗) the distance to 0 increases strictly, and so does the function r, until $\gamma(t)$ intersects the boundary sphere. We may thus define a homeomorphism

$$h : \partial N(\varepsilon, \delta) - \mathrm{Int}(T_{\varepsilon, \delta}) \to \mathbb{S}_\varepsilon - \mathrm{Int}(T_{\varepsilon, \delta})$$

as follows: given $x \in \partial N(\varepsilon, \delta) - \mathrm{Int}(T_{\varepsilon, \delta})$, take the path $\gamma(t)$ passing through x and follow it until it meets the sphere at a point $\gamma(t_1)$, then define $h(x) = \gamma(t_1)$. It is clear that this homeomorphism is the identity on the boundary of $T_{\varepsilon, \delta}$, so we can extend it to the interior as the identity. $\qquad\square$

I.5 Open book decompositions and fibred knots

The two concepts in the title of this section were introduced after Milnor published his book on singularities, so he does not mention these at all. However these important concepts are very useful in many ways, so I include them here because they help us to grasp better what Milnor's theorem for complex singularities is telling us, and what is the difference with the general statement in the real case.

The concept of "open book decompositions" was introduced by E. Winkeln-kemper in [264] and we refer to his appendix in [205] for a clear and updated account on the subject (see also [210]). In [247] Tamura defines the equivalent notion of "spinnable structures" on manifolds. Here we only give the basic notions that we use in the sequel.

5.1 Definition. An *open book decomposition* of a smooth n-manifold M consists of a codimension 2 submanifold N, called *the binding*, embedded in M with trivial normal bundle, together with a fibre bundle decomposition of its complement:

$$\pi : M - N \to \mathbb{S}^1 \,,$$

satisfying that on a tubular neighborhood of N, diffeomorphic to $N \times \mathbb{D}^2$, the restriction of π to $N \times (\mathbb{D}^2 - \{0\})$ is the map $(x, y) \mapsto y/\|y\|$. The fibres of π are called *the pages* of the open book.

It follows from the definition that the pages are all diffeomorphic and each page F can be compactified by attaching the binding N as its boundary, thus getting a compact manifold with non-empty boundary. Also, since the base of the fibration is the circle $\mathbb{S}^1$, one can lift a never-zero vector field on $\mathbb{S}^1$ to an integrable vector field on $M - N$ which is transversal to the fibres. Using the flow lines of this vector field one can define a "first return map" on the fibres, which is well defined up to isotopy. This diffeomorphism is known as *the monodromy* of the fibration. Since all the pages have the same binding N as boundary, it follows that h extends as the identity on N.

This brings us back to the original definition of open books in [264]. Winkeln-kemper defined open books as follows: start with a compact manifold $\overline{F}$ with non-empty boundary $\partial\overline{F}$, together with a diffeomorphism h of $\overline{F}$ which is the identity on $\partial\overline{F}$. Now form the mapping cylinder $\overline{F}_h$ of h, which is a manifold with boundary

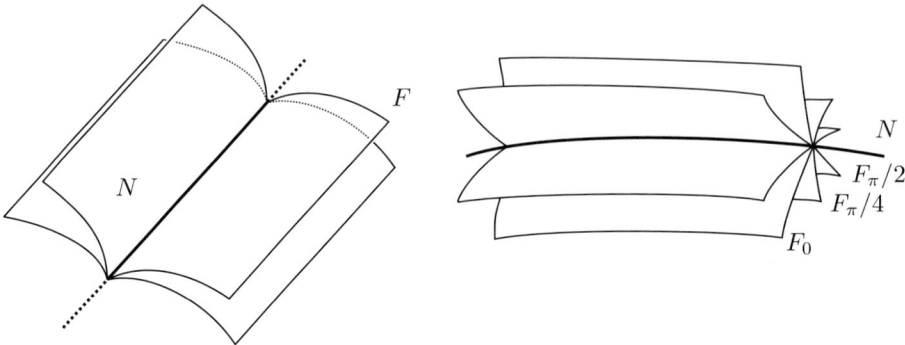

Figure 4: An open book.

$(\partial \overline{F}) \times \mathbb{S}^1$. Identifying (x, t) with (x, t') for each $x \in \partial \overline{F}$ and $t, t' \in \mathbb{S}^1$, we obtain a closed, differentiable manifold M. The fibres $\overline{F} \times t$ are the pages of the open book, and their common boundary $N = \partial \overline{F} \times t$ is the binding. Notice that in this definition the pages are already compact manifolds with boundary, and their interiors are the pages in Definition 5.1.

The concept of "fibred knots and links" actually arises from Milnor's fibration theorem for complex singularities. This was introduced by A. Durfee and B. Lawson in a paper [67] that generalizes the beautiful construction by Lawson, using Milnor's theorem, of codimension 1 foliations on odd-dimensional spheres.

We recall that a classical knot is an embedded circle in $\mathbb{S}^3$, and a classical link is a disjoint union of knots in the 3-sphere. These notions have been generalized as follows.

5.2 Definition. Let M be a smooth, closed, connected manifold. A *knot* in M means a smooth codimension 2 closed, connected submanifold N of M. If N has several connected components then it is called *a link* in M.

5.3 Definition. A knot (or link) $N \subset M$ is *fibred* if it is the binding of an open book decomposition of M.

I.6   On Milnor's fibration theorem for complex singularities

Consider now a complex analytic map,

$$f : (\mathbb{B}_r \subset \mathbb{C}^{n+1}, 0) \longrightarrow (\mathbb{C}, 0)$$

with a critical point at $0 \in \mathbb{C}^{n+1}$, which may be a non-isolated critical point of f. One has (Milnor's fibration theorem for complex singularities):

6.1 Theorem. *Let $\varepsilon > 0$ be sufficiently small and set $V = f^{-1}(0)$ and $K = V \cap \mathbb{S}_\varepsilon$. Then the map $\phi : \mathbb{S}_\varepsilon - K \to \mathbb{S}^1$ given by*

$$\phi(z) = \frac{f(z)}{|f(z)|}$$

is the projection map of a (locally trivial) fibred bundle. Furthermore, if V has an isolated singularity at 0, then $K \subset \mathbb{S}_\varepsilon$ is a fibred link, defining an open book decomposition of the sphere $\mathbb{S}_\varepsilon \cong \mathbb{S}^{2n+1}$.

In fact Milnor also showed that whenever $n > 1$, the link K is connected, so it is a knot in the sphere. [Milnor actually proved that K is $(n-2)$-connected, i.e., its homotopy groups vanish up to dimension $n-2$.]

6.2 Definition. A knot (or link if $n = 1$) defined by a holomorphic function as in Theorem 6.1 is called *an algebraic knot (or link)*.

The notion of "algebraic knot" was introduced by Lê Dũng Tráng in [125], and we refer to [127, 128, 62, 63] for basics about these knots. The proof of Theorem 6.1 is written in a very clear and elegant way in Milnor's book, so I will content myself by making a few comments about it. It is worth saying that Milnor's proof was written for polynomial maps, but all his arguments go through for holomorphic maps in general. We also refer the reader to [201] where Milnor's proof is adapted to meromorphic functions f/g and the key ideas are carefully explained. The point is to show first that the map ϕ in Theorem 6.1 has no critical points and so its fibres are all codimension 1 smooth submanifolds of the sphere $\mathbb{S}_\varepsilon$, and then construct a tangent vector field on $\mathbb{S}_\varepsilon - K$ which is transversal to the fibres of ϕ and whose solutions move at constant speed with respect to the argument of the complex number $\phi(z)$; this shows the local product structure around each fibre.

It is of course natural to compare Theorem 6.1 with the corresponding theorem for real singularities. In fact one is tempted to try to prove Theorem 6.1 by "pushing Theorem 4.3 further", and I believe this was the original approach followed by Milnor (see [167]). Two first difficulties, which are not serious problems, are that Theorem 4.3 requires an isolated critical point of f and it only gives a fibration away from a tubular neighborhood of the link. As pointed out by Milnor in [167, Chapter 11], this latter difficulty can be easily overcome and the fibration in Theorem 4.3 can be extended to the whole complement of the link with some extra work. The first difficulty can mean a problem, because the statement in general is false for real maps with non-isolated critical points. In fact it is false (in general) even for holomorphic maps which define complete intersection germs with non-isolated singularities. This was explained to me by Lê Dũng Tráng, who gave me the following example, which arises from Whitney's umbrella and is known as "Lê's example" (unpublished). Consider the function $f : \mathbb{C}^3 \to \mathbb{C}^2$ given by $f(x, y, z) = (x^2 - y^2 z, y)$. Its critical set Σ consists of the points where the Jacobian matrix Df has rank less than 2. One has:

$$Df(x, y, z) = \begin{pmatrix} 2x & -2yz & y^2 \\ 0 & 1 & 0 \end{pmatrix}.$$

Thus Σ consists of the z-axis $\{x = y = 0\}$. Notice $\Sigma = f^{-1}(0)$, and there are no more critical points of f. However the fibre over the point $(t^2, 0) \, t \in \mathbb{R}$ consists of the parallel lines $(\pm t, 0, z)$, while the fibre over a point of the form (t^2, λ) , $\lambda \neq 0)$, is the quadric $\{x^2 - \lambda^2 z = t^2\}$. Hence there cannot possibly be a local Milnor fibration. However this problem cannot occur for the situation envisaged in Theorem 6.1, that is when f is holomorphic and the target is 1-dimensional, essentially because in this case one has a theorem due to Hironaka [100], saying that f satisfies the "Thom a_f-condition" (see [136]). We can then follow the technique of Lê Dũng Tráng in [129] to show that one has a local Milnor fibration whenever one has the a_f-condition (see [201]).

There is one more difficulty for proving Theorem 6.1 using Theorem 4.3, but this again is a technical difficulty that can be overcome, and this is essentially what Milnor does in Lemma 5.9 in [168]. Indeed Theorem 4.3 is proved by using the fibration of Theorem 4.2, which for the case envisaged in Theorem 6.1 takes the form:

$$f : f^{-1}(\mathbb{S}^1_\delta) \cap \mathbb{B}_\varepsilon \longrightarrow \mathbb{S}^1_\delta \, ,$$

and then "inflating" the tube $f^{-1}(\mathbb{S}^{2n+1}_\delta) \cap \mathbb{B}_\varepsilon$ by means of a vector field $v(z)$ so that the tube becomes the complement in $\mathbb{S}^{2n+1}$ of an open tubular neighborhood of the link K. In general one does not have much control of the behaviour of this vector field and therefore one cannot say anything in general about the map ϕ in Theorem 4.3 nor its relation with the original map f. Still, in the complex case Milnor manages to construct a vector field in his Lemma 5.9, that we denote here by $\hat{v}$, making the same job as the vector field v in the proof of Theorem 4.3, with the additional feature that the flow lines of $\hat{v}$ carry the tube $f^{-1}(\mathbb{S}^{2n+1}_\delta) \cap \mathbb{B}_\varepsilon$ into the sphere, in such a way that the argument of the complex number $f(z)$ is constant along each flow line. Hence Milnor can assure that the fibres of the fibration given by Theorem 4.2 are taken into the fibres of Theorem 6.1.

There is finally one more difficulty, and this is a serious one: the behavior near the link K. There are real analytic singularities given by maps $\mathbb{R}^{2n} \to \mathbb{R}^2$ with a fibration as in Theorem 4.2 but which do not define an open book decomposition of the sphere $\mathbb{S}^{2n-1}$ (see [201]). Hence, in order to actually control the behavior of the fibres near the link, Milnor has to use completely different arguments (Section 4 of his book), which eventually are sufficient to complete the proof of Theorem 6.1 without using Theorem 4.2. Therefore the fibration Theorem 4.2 is given in [168] just as "an alternative" way to look at the fibration. Still, this alternative way has several advantages on its own and, as mentioned earlier, this is probably the most common version of Milnor's theorem.

6.3 Remarks. Milnor's fibration Theorem 4.3 was generalized by Hamm in [95] to germs of complex analytic, isolated complete intersection singularities; we refer to Looijenga's book [146] for a clear account on this subject. One may also consider the germ of an arbitrary complex analytic variety (V, p) whose singular set may have dimension more than 0, and the main theorem of [129] says that in this situation one has a topological locally trivial Milnor fibration as in Theorem 4.3 above.

It is worth remarking that Milnor's theorem also proves that one has a fibration as in Theorem 4.3 (essentially the same proof as above) even when the critical point of f is not isolated, and these two fibrations are equivalent. By the theorem of Andreotti-Frankel [6], this implies that the fibres of ϕ have the homotopy type of a CW-complex of middle dimension, and they are always parallelizable manifolds. (Milnor proves these facts directly, not using [6], because he needs essentially the same arguments to show that the link is highly connected.) These results have been generalized recently in [269, 201] for meromorphic functions:

6.4 Theorem. *Let $f, g : (\mathbb{C}^{n+1}, 0) \to (\mathbb{C}, 0)$ be holomorphic maps, so that all points in a punctured neighborhood of $0 \in \mathbb{C}$ are regular points of the meromorphic map*

$$f/g : (\mathbb{C}^{n+1}, 0) \to (\mathbb{C}, 0) \,.$$

Let $L_{fg} = (fg)^{-1}(0) \cap \mathbb{S}_\varepsilon$ be the link of fg. Then the map:

$$\phi := \frac{f\bar{g}}{|f\bar{g}|} = \frac{f/g}{|f/g|} : \mathbb{S}_\varepsilon \setminus L_{fg} \longrightarrow \mathbb{S}^1 \subset \mathbb{C}$$

is the projection of a $\mathbb{C}^\infty$ (locally trivial) fibre bundle, whose fibres F_θ are diffeomorphic to the complex manifolds $(f/g)^{-1}(t) \cap \overset{\circ}{\mathbb{D}}_\varepsilon$, where $t \in \mathbb{C}$ is a regular value of the meromorphic map f/g and $\overset{\circ}{\mathbb{D}}_\varepsilon$ is the interior of the disc in $\mathbb{C}^{n+1}$ whose boundary is $\mathbb{S}_\varepsilon$. Hence each fibre is a parallelizable manifold with the homotopy type of a CW-complex of dimension n.

We remark that this theorem is stated in [201] in terms of the function $f\bar{g}$; as explained there, both formulations are essentially equivalent.

I.7 The join of Pham and the topology of the Milnor fibre. The Milnor number

In his pioneer article [198], F. Pham studied the topology of the complex manifold $V_{(a_0,\dots,a_n)} \subset \mathbb{C}^{n+1}$ defined by:

$$z_0^{a_0} + \cdots + z_n^{a_n} = 1 \,,$$

where $n > 0$ and the a_i are integers ≥ 2. It is easy to see that this manifold is diffeomorphic to the Milnor fibre of the complex singularity defined by $f(z) = 0$, where f is the Pham-Brieskorn polynomial $f(z) = z_0^{a_0} + \cdots + z_n^{a_n}$. To explain Pham's results, let G_a denote the finite cyclic group of ath roots of unity. Given the integers $\{a_0, \dots, a_n\}$, denote by $J = J_{(a_0,\dots,a_n)}$ the join:

$$J = G_{a_0} * G_{a_1} * \cdots * G_{a_n} \subset \mathbb{C}^{n+1} \,,$$

which consists (see [164]) of all linear combinations

$$(t_0 \, \omega_0, \ldots, t_n \, \omega_n)$$

with the t_i real numbers ≥ 0 such that $t_0 + \cdots + t_n = 1$ and $\omega_j \in G_{a_j}$. Note that J can be identified with the subset $\mathcal{P} = \mathcal{P}_{(a_0,\ldots,a_n)}$ defined by:

$$\mathcal{P} = \{ z \in V_{(a_0,\ldots,a_n)} \mid z_j^{a_j} \in \mathbb{R} \text{ and } z_j^{a_j} \geq 0 \,, \text{ for all } j = 0,\ldots,n \} \,.$$

To see this, notice that $\mathcal{P}$ can also be described by the conditions:

$$z_j = u_j |z_j| \,, \; u_j \in G_{a_j} \,, \; t_j = |z_j|^{a_j}, \text{ for all } j = 0,\ldots,n.$$

Hence $\mathcal{P}$ is contained in the manifold $V_{(a_0,\ldots,a_n)}$. The set $\mathcal{P}$ is known as *the join of Pham* of the polynomial f. It is not hard to see that $V_{(a_0,\ldots,a_n)}$ has $\mathcal{P}$ as a *deformation retract* and therefore its homotopy type is that of $\mathcal{P}$. In fact, given a point $z \in V_{(a_0,\ldots,a_n)}$, first deform each coordinate z_j along a path in $\mathbb{C}$ chosen so that the trajectory described by $z_j^{a_j}$ is the straight line to the nearest point on the real axis, that we denote by $\hat{z}_j$. This carries z into a vector $\hat{z} = (\hat{z}_0,\ldots,\hat{z}_n)$ satisfying $\hat{z}_j^{a_j} \in \mathbb{R}$ for each j, and it is clear that this deformation leaves $V_{(a_0,\ldots,a_n)}$ invariant. Now, whenever one has that $\hat{z}_j^{a_j} < 0$, move $\hat{z}_j$ along a straight line to $0 \in \mathbb{C}$, moving the other components accordingly so that the deformation leaves $V_{(a_0,\ldots,a_n)}$ invariant. Hence the point $\hat{z} = (\hat{z}_0,\ldots,\hat{z}_n)$ moves along a straight line towards a point $\check{z} = (\check{z}_0,\ldots,\check{z}_n) \in V_{(a_0,\ldots,a_n)}$ whose coordinates are all ≥ 0 and one has that each coordinate $\check{z}_j$ is necessarily of the form $t_j \, \omega_j$ for some $t_j \geq 0$ and some $\omega_j \in G_{a_j}$. This gives a deformation of $V_{(a_0,\ldots,a_n)}$ into $\mathcal{P}$ that leaves this set invariant, so the join $\mathcal{P}$ *is a deformation retract* of $V_{(a_0,\ldots,a_n)}$. It is now an exercise to show that $\mathcal{P}$ *has the homotopy type of a wedge (or bouquet) of spheres* of real dimension n. Moreover, the number of spheres in this wedge is $(a_0 - 1) \cdot (a_1 - 1) \cdots (a_n - 1)$. Thus we have obtained:

7.1 Theorem (Pham). *The variety*

$$V_{(a_0,\ldots,a_n)} := \{ z \in \mathbb{C}^{n+1} \mid z_0^{a_0} + \cdots + z_n^{a_n} = 1 \}$$

has the set $\mathcal{P}$ as a deformation retract. Thus $V_{(a_0,\ldots,a_n)}$ has the homotopy type of a bouquet $\bigvee \mathbb{S}^n$ of spheres of dimension n, the number of spheres in this wedge being $[(a_0 - 1) \cdot (a_1 - 1) \cdots (a_n - 1)]$.

Let us now return to the situation envisaged in Section 6, of a complex analytic map $f : (\mathbb{B}_r \subset \mathbb{C}^{n+1}, 0) \longrightarrow (\mathbb{C}, 0)$ with an isolated critical point at $0 \in \mathbb{C}^{n+1}$. One has Milnor's fibration for f:

$$f : f^{-1}(\mathbb{S}_\delta^1) \cap \mathbb{B}_\varepsilon \longrightarrow \mathbb{S}_\delta^1 \,.$$

It is easy to see that for the Pham-Brieskorn polynomials $f(z) = z_0^{a_0} + \cdots + z_n^{a_n}$ one can take $\delta = 1$ and $\mathbb{B}_\varepsilon$ as the unit ball, so that the above manifold $V_{(a_0,\ldots,a_n)}$

is the Milnor fibre F. By Theorem 7.1, in this case F has the homotopy type of a bouquet of spheres of dimension n. More generally one has:

7.2 Theorem (Milnor). *Let* $f : (\mathbb{B}_r \subset \mathbb{C}^{n+1}, 0) \longrightarrow (\mathbb{C}, 0)$ *be complex analytic, with an isolated critical point at* $0 \in \mathbb{C}^{n+1}$. *Then the fibre* $F = f^{-1}(t) \cap \mathbb{B}_\varepsilon$ *is a parallelizable manifold with the homotopy type of a bouquet* $\bigvee \mathbb{S}^n$ *of spheres of middle dimension* n. *The number* $\mu(f)$ *of spheres in this bouquet is* ≥ 0 *and it is* 0 *iff* $0 \in \mathbb{C}^{n+1}$ *is a regular point of* f.

Notice that the manifold F can be equivalently defined by $F = \phi^{-1}(e^{i\theta})$ for some $e^{i\theta} \in \mathbb{S}^1 \subset \mathbb{C}$, where $\phi : \mathbb{S}_\varepsilon - K \to \mathbb{S}^1$ is given by $\phi(z) = f(z)/|f(z)|$. The proof of Theorem 7.2 in [168] is by Morse theory and it has several steps. The first one shows that F has the homology of a CW-complex of dimension n. But if we think of F as being defined by $f^{-1}(t) \cap \mathbb{B}_\varepsilon$, then this is an immediate consequence of the theorem of Andreotti-Frankel in [6], saying that the homology of every Stein space vanishes above the middle dimension. This fact implies, in particular, that F cannot have any compact component. Since it is a hypersurface in $\mathbb{C}^{n+1}$, its normal bundle is trivial and canonically trivialized by the gradient vector field $\overline{\nabla} f(z) = (\overline{\frac{\partial f}{\partial z_0}}, \ldots, \overline{\frac{\partial f}{\partial z_n}})$. We recall that a smooth manifold is parallelizable if its tangent bundle is trivial, and a rather simple and elegant argument in [113] shows that a manifold with no compact component is parallelizable iff it can be embedded in some Euclidean space with trivial normal bundle. Hence F is parallelizable.

7.3 Definition. The manifold F is called *the Milnor fibre* of f at 0. The number $\mu = \mu(f)$ of spheres in the bouquet given by Theorem 7.2 is called *the Milnor number* of f at 0.

It is worth saying that the Milnor number has several different interpretations, each pointing towards different properties of this invariant. The following theorem summarizes some of these properties, and it is all in [168].

7.4 Theorem (Milnor). *The number* μ *has the following interpretations:*

(i) *it equals the number of vanishing cycles on* $F = f^{-1}(t) \cap \mathbb{B}_\varepsilon$;

(ii) *it equals the multiplicity of the map-germ at* $0 \in \mathbb{C}^{n+1}$ *defined by the partial derivatives of* f. *That is,* μ *is the degree of the map* $z \mapsto \frac{(f_0(z), \ldots, f_n(z))}{\|(f_0(z), \ldots, f_n(z))\|}$ *from a small sphere* $\mathbb{S}_\varepsilon$ *centred at* 0 *into the unit sphere in* $\mathbb{C}^{n+1}$, *where* $f_i = \frac{\partial f}{\partial z_i}$;

(iii) *it is the dimension of the vector space obtained by taking the quotient of the local ring* $\mathcal{O}_0(\mathbb{C}^{n+1})$ *of holomorphic functions at* $0 \in \mathbb{C}^{n+1}$ *by the Jacobian ideal of* f:

$$\mu = \dim \mathcal{O}_0(\mathbb{C}^{n+1})/((f_0(z), \ldots, f_n(z))$$

(iv) *it is determined by the Euler characteristic of* F:

$$\mu = (-1)^{n+1}(1 - \chi(F)).$$

These are all immediate consequences of Theorem 7.2. In fact we know that near the singular point the variety V is locally a cone, so it has the homotopy of a

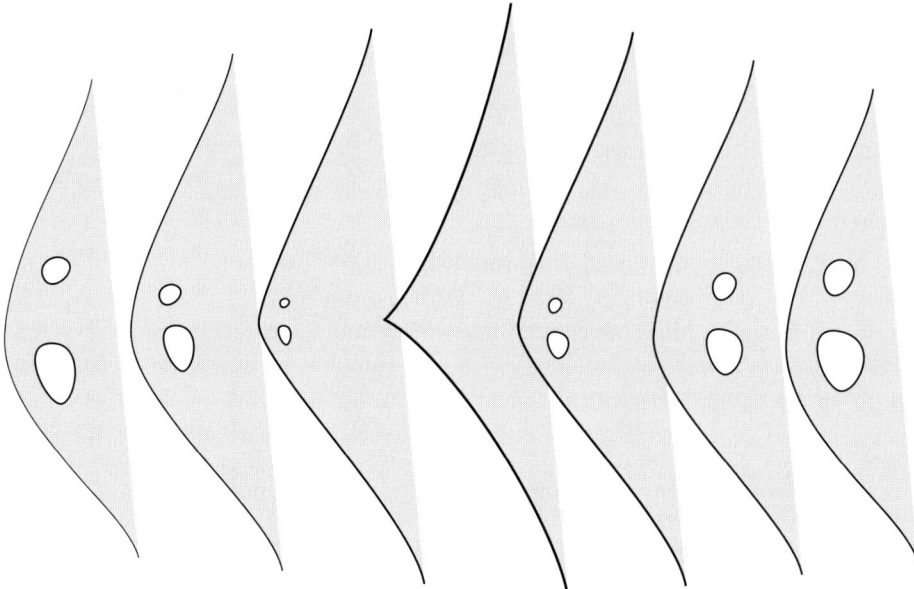

Figure 5: The Milnor fibration.

point and its homology vanishes in dimensions > 0. On the other hand Theorem 7.2 tells us that for each $t \neq 0$ with $|t|$ sufficiently small, the homology of $f^{-1}(t) \cap \mathbb{B}_\varepsilon$ in the middle dimension is a free Abelian group in μ generators. As the parameter t tends to 0 the non-singular fibre degenerates into $V \cap \mathbb{B}_\varepsilon$ and the homology of F disappears. To be precise, the ball $\mathbb{B}_\varepsilon$ has $V \cap \mathbb{B}_\varepsilon$ as a deformation retract (see the proof of Lemma 4.4 above); let r be such a retraction and let ι denote the inclusion $f^{-1}(t) \cap \mathbb{B}_\varepsilon \hookrightarrow \mathbb{B}_\varepsilon$. The composition $r \circ \iota$ gives a map $f^{-1}(t) \cap \mathbb{B}_\varepsilon \longrightarrow V \cap \mathbb{B}_\varepsilon$, which induces a morphism in homology (called a "specialization map"). This morphism carries $H_n(F)$ into $H_n(V \cap \mathbb{B}_\varepsilon) \cong 0$, so the μ generators of $H_n(F)$ vanish. Thus we call them *vanishing cycles*.

7.5 Remarks on Milnor numbers and vanishing cycles. For many purposes it is important to have preferred sets of vanishing cycles, as for instance to be able to control the bilinear pairing in $H_n(F)$ given by the intersection product, and to compute its signature (c.f. §8 below and IV.2). For the Pham-Brieskorn polynomials, in some cases the join of Pham can be used to construct a preferred set of vanishing cycles. In [131] Lê constructs an equivalent polyhedron for all singularities as above (and in the more general setting of his fibration theorem that we explain below). However the problem of finding a "distinguished basis" of vanishing cycles is indeed an issue and this can be used to obtain deep information about the singularities. We refer to [41, 43, 68, 85, 126] for very interesting results on this subject.

It is worth noting that the remarkable feature about Hamm's fibration theorem (for holomorphic map-germs which define isolated complete intersection singularities, see [95, 146]), is that the topology of the fibres behaves as in the case of hypersurfaces, in the sense that the fibres have the homotopy type of a bouquet of spheres of middle dimension. Thus one has a well-defined "Milnor number" for isolated complete intersection singularities. For curve singularities a notion of the Milnor number was introduced by R.-O. Buchweitz and G.-M. Greuel in [47]. For a smoothable curve singularity $(C, 0)$ (i.e., when there is a manifold "playing the role of a Milnor fibre") it is equal to $1 - \chi(\widetilde{C})$ for a smoothing $\widetilde{C}$ of the singularity, $\chi(\cdot)$ is the Euler-Poincaré characteristic. However, there exist surface singularities which have smoothings with different Euler-Poincaré characteristics ([204]). This does not permit us to generalize the notion of the Milnor number to higher dimensions so that for smoothable singularities it has the usual expression in terms of the Euler characteristic. In [72] is introduced a notion of Milnor number for every isolated singularity in an analytic variety. This is defined as the difference of two indices associated to every holomorphic 1-form on an isolated singularity germ, but it does not depend on the choice of the 1-form. One of these indices corresponds to the "1" in the difference $(1 - \chi(F))$ in (iv) above, the other corresponds to $\chi(F)$. The first index is the so-called "radial index" defined in [69, 70, 71, 35, 3, 233]. The second index is inspired in the homological index of Gomez-Mont in [90] and the index for 1-forms defined in [69, 70, 71] in analogy with the so-called GSV-index of [88, 232, 3]. In the case when the ambient variety V is a complete intersection germ in $\mathbb{C}^{n+k}$, this homological index coincides with the number of zeroes (counted with multiplicities) of an extension of the 1-form to a Milnor fibre of V.

There are also several possible generalizations of the concept of "Milnor number" for map-germs defined on singular spaces. The first one comes from Lê's fibration theorem in [129], by defining $\mu(f)$ to be $(-1)^{n+1}(1 - \chi(F))$, where n is the dimension of the ambient space and F is the Lê-Milnor fibre of f. A second definition comes from [90, 173] for functions on curve singularities, and from [108] for functions on isolated complete intersection germs in general, of any dimension. There is a third generalization coming from [36] by means of the Nash blow-up. These different generalizations of the Milnor number are compared in [234].

The concepts of Milnor number and vanishing cycles have also been extended to varieties with non-isolated singularities. On the one hand the vanishing cycles have given rise to the so-called "vanishing homology" when the singular set has dimension more than 0. We refer to [236, 237, 252, 253, 154, 155, 156, 158, 243, 221] for outstanding results in this direction and for very general "bouquet theorems" for the Milnor fibre. On the other hand, Parusinski defined in [195] a "generalized Milnor number" associated to each connected component of the singular set of a hypersurface in a complex manifold. Parusinski's Milnor number was extended in [35] to local complete intersections, and this generalization is inspired in the interpretation (iv) of the Milnor number in Theorem 7.4, together with the above comments on indices of vector fields on singular varieties. In fact, using that language one can prove ([232]) that the Milnor number of an isolated complete intersection singularity is obtained, up to sign, by localizing at the singularity two different characteristic classes associated to singular varieties: one is the top-dimensional Schwartz-MacPherson class, the other is the top-dimensional Fulton-Johnson class. When the singular set S has dimension $s > 0$, one can also localize other Schwartz-MacPherson and Fulton-Johnson classes (see [34, 35] for more about these classes). The differences amongst them define the so-called *Milnor classes* $\mu_i(S_j) \in H_{2i}(S_j; \mathbb{Z})$, where the S_j are the connected components of the singular set S and $i = 0, \ldots, s_j = \dim S_j$. We refer to [5, 34, 37, 35, 196, 221, 266] for different viewpoints of the Milnor classes and characteristic classes of singular varieties; [221] relates these with the vanishing cycles.

I.8 Exotic spheres and the topology of the link

We recall that a complex analytic variety V is (analytically) normal at a point x if every bounded holomorphic function on a punctured neighborhood of x extends to a holomorphic function at x (see for instance [179]). Normality at a point implies that the singular set is locally of codimension at least 2 (for a hypersurface or complete intersection germ this condition is actually equivalent to normality).

D. Mumford proved in [176] that a 2-dimensional normal complex analytic variety is a topological manifold iff it is a complex manifold. More precisely, Mumford proved that if x is a normal point in a complex surface V and the link K of x is simply connected, then K is the 3-sphere $\mathbb{S}^3$ and V is a smooth complex manifold at x. This remarkable theorem motivated a lot of research on the topic by various authors, most notably by E. Brieskorn [40, 39], F. Hirzebruch [102, 103] and J. Milnor [168, 167, 166], and it was shown to be completely false in higher dimensions. In fact Brieskorn showed in [40] that for every $k > 1$ odd, the link of 0 in the algebraic hypersurface

$$z_0^2 + z_1^2 + \cdots + z_{n-1}^2 + z_n^3 = 0$$

which is a generalized trefoil knot, is homeomorphic to the usual sphere $\mathbb{S}^{2n+1}$. It was thus natural to ask whether this link is diffeomorphic to $\mathbb{S}^{2n+1}$ or if it is an "exotic" sphere.

Let us recall that a smooth n-dimensional closed oriented manifold σ^n is a topological sphere if it is homeomorphic to the standard unit sphere $\mathbb{S}^n$. Smale's theorem [238] implies that for $n \geq 5$ the manifold σ^n is a topological sphere iff it is a homotopy sphere, i.e., if it has the same homotopy groups as the n-sphere; σ^n is said to be an *exotic sphere* if it is homeomorphic but not diffeomorphic to $\mathbb{S}^n$. The first exotic sphere was found in dimension 7 by Milnor [165] in 1956, and it was later shown by Kervaire and Milnor in [113] that there are no exotic spheres in dimensions less than 7 and $\neq 3, 4$. In dimensions 3,4 the existence of exotic spheres is the famous Poincaré's conjecture and the methods of [113] break down.

Two differentiable structures on spheres of dimension ≥ 5 are said to be equivalent if there is an orientation preserving diffeomorphism among them. It is proved in [113] that for $n \geq 5$ these equivalence classes form a finite Abelian group Θ_n, with the connected sum as operation; the identity element is the standard $\mathbb{S}^n$. This group contains a "preferred subgroup", usually denoted $bP_{n+1} \subset \Theta_n$, of those homotopy spheres that bound a parallelizable manifold. This group is zero for n even and for $n \neq 3$ odd it is a finite cyclic group which has finite index in Θ_n. This cyclic group has order 1 or 2 for $n \equiv 1 \,(\mathrm{mod}\,4)$, but for $n \equiv 3 \,(\mathrm{mod}\,4)$ its order $|bP_{4m}|$ grows more than exponentially:

$$|bP_{4m}| = \left[2^{2m-2}(2^{2m-1} - 1)\right] \cdot \left[\text{numerator of} \left(\frac{4\mathcal{B}_m}{m}\right)\right],$$

where the $\mathcal{B}_m$ are the Bernoulli numbers. Thus for instance (see [113]), for $n = 7, 11, 15$ or 19 there are, respectively, $|bP_{n+1}| = 28, 992, 8128$ and 130816 non-equivalent differentiable structures on the n-sphere, that bound a parallelizable manifold.

We observe that the link K^{2n-1} of an isolated complex hypersurface singularity has a canonical differentiable structure, being the intersection $K = (V - 0) \cap \mathbb{S}_\varepsilon$ of two smooth submanifolds of $\mathbb{C}^{n+1}$, and with this differentiable structure it is the boundary of the Milnor fibre, which is a parallelizable manifold. Hence if one such link K is a topological sphere, then it necessarily represents an element in bP_{2n}.

Notice that the link K^{2n-1} of an isolated complex hypersurface singularity $(V, 0)$ in $\mathbb{C}^{n+1}$ is always $(n-2)$-connected, by [168]. Thus, if $n > 2$ then its homology vanishes in dimensions $i = 1, \ldots, n-2$. Hence, by the Poincaré duality isomorphism, its homology also vanishes in dimensions $n+i$, $i = 1, \ldots, n-2$; so the only possibly non-zero groups are in dimensions $i = n, n-1$ and of course $i = 0, 2n-1$ where it is the integers (or the corresponding ring of coefficients). If $H_{n-1}(K)$ vanishes then $H_n(K)$ also vanishes and K is a homology sphere. If $n \geq 2$, then K is simply connected by [168] and therefore it is a homotopy sphere whenever it is a homology sphere. Furthermore, if $n \geq 3$ then Smale's theorem [238] on Poincaré's Conjecture in dimensions ≥ 5 implies that K is actually homeomorphic to $\mathbb{S}^{2n-1}$.

Summarizing, for $n = 1, 2$ there are no exotic spheres appearing as the link of an isolated complex singularity. For hypersurfaces of dimension $n \geq 3$ the link is a topological sphere iff $H_{n-1}(K)$ vanishes (c.f. Lemma 8.2 in [168]). In order to have a criterion to decide when this group vanishes, recall one has the monodromy h of the corresponding Milnor fibration, also called *the characteristic homeomorphism* of the fibre. This is the first return map on the fibres, defined by lifting to $\mathbb{S}^{2n+1} - K$ a vector field on the base $\mathbb{S}^1$. Fixing a fibre F_0, this map induces a homomorphism

$$h_* : H_n(F_0) \to H_n(F_0),$$

known as *the monodromy representation* of the fundamental group $\pi_1(\mathbb{S}^1)$ of the base. Using this one can define *the characteristic polynomial* of the fibration:

$$\Delta(t) := \det(tId_* - h_*),$$

which is a polynomial of the form:

$$t^\mu + b_1 t^{\mu-1} + \cdots + b_{\mu-1}t \pm 1,$$

with integer coefficients, since $H_n(F)$ is free Abelian of rank μ. This is an n-dimensional generalization of the Alexander polynomial of a knot. The following theorem was proved in [40, Satz 1.(i)] for the Pham-Brieskorn polynomials and by Milnor [168, Theorem 8.5] for isolated hypersurface singularities in general:

8.1 Theorem. *For $n \geq 3$, K is a topological sphere if and only if the integer*

$$\Delta(1) = \det(I_* - h_*) \quad is \quad \pm 1.$$

To prove this result one uses the Wang sequence associated to the Milnor fibration (see p. 67 in [168]):

$$H_n(F_0) \xrightarrow{h_* - I_*} H_n(F_0) \longrightarrow H_n(\mathbb{S}_\varepsilon - K) \longrightarrow 0 \ .$$

This exact sequence implies that $\Delta(1) = \pm 1$ if and only if $H_n(\mathbb{S}_\varepsilon - K)$ vanishes. The result then follows from the isomorphisms below, which are given by the Alexander and Poincaré duality isomorphisms, respectively:

$$H_n(\mathbb{S}_\varepsilon - K) \cong H^n(K) \cong H_{n-1}(K) \qquad \qquad \Box$$

For instance consider the link $K_{(a_0,\dots,a_n)}$ of the Brieskorn singularity:

$$z_0^{a_0} + z_1^{a_1} + \cdots + z_n^{a_n} = 0\,.$$

It follows from the previous discussion about Pham's work (also from [39, 168]) that the nth homology group $H_n(F)$ of the Milnor fibre is free Abelian of rank:

$$\mu = (a_0 - 1)(a_1 - 1) \cdots (a_n - 1)\,.$$

It is shown in [39] that the roots of the characteristic polynomial

$$h_* : H_n(F; \mathbb{C}) \to H_n(F; \mathbb{C})$$

are the products $(\omega_0 \, \omega_1 \cdots \omega^{n+1})$, where each ω_j runs over the a_j^{th} roots of unity other than 1. Hence the characteristic polynomial is given by:

$$\Delta(t) = \prod (t - \omega_0 \, \omega_1 \cdots \omega_n)\,.$$

As an example consider the case:

$$z_0^2 + z_1^2 + \cdots + z_{n-1}^2 + z_n^3 = 0\,,$$

so that $a_0 = \cdots = a_{n-1} = 2$ and $a_n = 3$. The roots of h_* are:

$$\omega_0 = \cdots = \omega_{n-1} = -1 \quad , \quad \omega_n = \left(\frac{-1 \pm \sqrt{-3}}{2} \right)\,.$$

Hence:

$$\Delta(t) = t^2 - t + 1 \qquad \text{for } n \text{ odd};$$
$$\Delta(t) = t^2 + t + 1 \qquad \text{for } n \text{ even}.$$

Thus $K_{(2,\cdots,2,3)}$ is a topological sphere when n is odd, i.e., when the link has dimension $2n - 1 = 1, 5, 9, \ldots$, as claimed earlier. In dimensions 1 and 5 there are no exotic spheres, so K has to be the usual sphere. However in dimension 9 there is one exotic sphere and this is represented by the generalized trefoil knot K, but in order to decide this one has to use rather sophisticated arguments. Indeed it was shown in [113] that the elements in bP_{2n-1} that bound a parallelizable manifold F which is $(n-1)$-connected are distinguished by (see IV.1 below for the definition of this invariant):

(i) the signature $\sigma(F) \in \mathbb{Z}$ of the intersection pairing in $H_n(F)$, if $n \neq 2$ is even;

or

(ii) the Arf-Kervaire invariant $c(F) \in \mathbb{Z}_2$, if n is odd.

In the case $n = 2m$ one has that the intersection pairing in $H_n(F)$ is even and therefore its signature $\sigma(F)$ must be divisible by 8; if g_m represents the generator of bP_{4m} then (by [113]) one has that K represents the element

$$K \approx \left[\frac{\sigma(F)}{8}\right] \cdot g_m.$$

When n is odd a remarkable theorem of J. Levine [137] says the Kervaire invariant $c(F)$ is 0 if $\Delta(-1) \equiv \pm 1 \mod 8$, and it is 1 if $\Delta(-1) \equiv \pm 3 \mod 8$. So the differentiable structure on K is determined by the characteristic polynomial $\Delta(t)$. As an example of these results Brieskorn shows in [39, p. 11] that the link of the n-dimensional singularity $V_{(2,\ldots,2,d)}$, with $n, d \geq 3$ odd numbers, is a topological sphere and the structure on the link is exotic iff $d \equiv \pm 3 \mod 8$.

Also using these results Hirzebruch was able to prove in [103, p. 20-21] that the link of the singularity

$$z_0^3 + z_1^{6k-1} + z_2^2 + \cdots + z_{2m}^2 = 0, \qquad k \geq 1; m \geq 2$$

is a topological sphere which represents the element:

$$(-1)^m k g_m \in bP_{4m}.$$

As an example Hirzebruch remarks that for $m = 2$ one has that bP_8 is the whole group Θ_7 and one gets the 28 classes of differentiable structures on $\mathbb{S}^7$ by taking $k = 1, \ldots, 28$. Similarly $bP_{12} = \Theta_{11}$ and $k = 1, \ldots, 992$ gives all the differentiable structures in $\mathbb{S}^{11}$. More generally, Brieskorn proved in [39, Korollar 2]:

8.2 Theorem (Brieskorn). *Every homotopy sphere of dimension $m = 2n - 1 > 6$ that bounds a parallelizable manifold is the link $K_{(a_0,\ldots,a_n)}$ for some hypersurface singularity of the form*

$$z_0^{a_0} + z_1^{a_1} + \cdots + z_n^{a_n} = 0.$$

Chapter II

Motions in Plane Geometry and the 3-dimensional Brieskorn Manifolds

This chapter is about the theorems of Klein (1884) and Milnor (1975), relating certain groups of isometries in plane geometry to the Brieskorn varieties $V_{(p,q,r)} = \{z_1^p + z_2^q + z_3^r = 0\}$. There are three cases, according as the triple $p^{-1} + q^{-1} + r^{-1}$ is greater than, equal to or smaller than 1. The theorems say that in either case the link $M_{(p,q,r)}$ is diffeomorphic to a space of orbits of the form $\Gamma \backslash G$, where G is a 3-dimensional Lie group and Γ is a discrete subgroup of G. In the first case, when $p^{-1} + q^{-1} + r^{-1} > 1$, the group G is $SU(2)$, the universal cover of $SO(3)$, the group of rotations of the 2-sphere; $SU(2)$ is isomorphic to the group of unit quaternions, so it is diffeomorphic to the 3-sphere, and its discrete subgroups are either cyclic or the lifting to $SU(2)$ of a dihedral group or a group of symmetries of a platonic solid. When $p^{-1} + q^{-1} + r^{-1} < 1$ the group G is $\widetilde{SL}(2, \mathbb{R})$, the universal cover of $PSL(2, \mathbb{R})$, the group of orientation preserving isometries of the hyperbolic plane $\mathbb{H}^2$, and the subgroups in question come from the celebrated Schwarz triangle groups of isometries in hyperbolic plane geometry. The case $p^{-1} + q^{-1} + r^{-1} = 1$ is somehow different and we discuss that difference in Chapter III.

We start this chapter with an exposition of the groups of symmetries of the platonic solids from the classical point of view and we relate them (in §2) to the Schwarz triangle groups; for this we explain briefly the groups of isometries in the three classical plane geometries: Euclidean, spherical and hyperbolic. Then we discuss the Lie group structure on the sphere $\mathbb{S}^3$, both as unit quaternions or as the group $SU(2)$ of unitary transformations, and relate this to the group $SO(3)$ of rotations of the 2-sphere. In Section 4 we state Klein's theorem for the finite subgroups of $SU(2)$ and we give Milnor's proof of a refinement of Klein's theorem, which expresses beautifully the relation between the finite subgroups of $SO(3)$ and the Brieskorn singularities in the spherical case. We then speak about the group $PSL(2, \mathbb{R})$ and its universal cover $\widetilde{SL}(2, \mathbb{R})$, giving Milnor's interpretation of the latter as "labeled biholomorphic maps" of the upper half-plane. This is then used to explain Milnor's theorem for the Brieskorn singularities in the hyperbolic case.

II.1 Groups of motions in the 2-sphere. The polyhedral groups

In this section we study and describe the finite subgroups of $SO(3)$, the group of orientation preserving isometries of the 2-sphere, endowed with the metric induced from $\mathbb{R}^3$. This is a well-understood subject that has fascinated generations of mathematicians since long ago. In fact the classification of the simplest of such groups, those coming from Euclidean plane geometry, is usually attributed to Leonardo da Vinci. It seems that J.F.C. Hessel (1830) was the first to enumerate the finite 3-dimensional groups of symmetries, which were also studied later by A. Cayley, F. Klein, H. Poincaré and many others. More recently, the contributions to the subject by H.S.M. Coxeter in the 20th century are remarkable. The literature in this topic is vast and we could just refer to it. However we prefer to say a few things here, for the sake of completeness, to motivate similar constructions for the group $PSL(2, \mathbb{R})$, and for the joy of discussing a little about this important and very beautiful piece of mathematics. This also helps to set up the notation which can be confusing sometimes, since the groups in question can be considered in $SO(3)$, or lifted to the trivial double cover $O(3)$, or extended to the universal cover $SU(2)$. Those wanting to go deeper into the subject can look at [56, 55], or also at [171] where Milnor presents these groups from his own viewpoint, which is very elegant and closer to what we need for this text (see also [118]).

Consider first the *orthogonal group* $O(3)$ of linear transformations of $\mathbb{R}^3$ that preserve the usual metric in $\mathbb{R}^3$, given by the standard quadratic form $(1, 1, 1)$. This is the group generated by reflections on all 2-planes through the origin and it can be identified with the space of orthonormal 3-frames in $\mathbb{R}^3$; its elements are matrices with determinant ± 1. It has a preferred subgroup $SO(3)$ consisting of those frames which induce the usual orientation in $\mathbb{R}^3$, i.e., matrices with determinant 1 and the corresponding linear transformations of $\mathbb{R}^3$ preserve the orientation. This is the subgroup of $O(3)$ consisting of words of even length, because each reflection reverses the orientation. Since the composition of two reflections is a rotation, it follows that $SO(3)$ is *the group of rotations* of $\mathbb{S}^3$.

The group $SO(3)$ acts transitively on the unit 2-sphere $\mathbb{S}^2 \subset \mathbb{R}^3$. The stabilizer (or isotropy) subgroup of each point is $\mathbb{S}^1 \cong SO(2) \subset SO(3)$. This means that given any pair of points $x_1, x_2 \in \mathbb{S}^2$, and unit tangent vectors $v_1 \in T_{x_1}\mathbb{S}^2$, $v_2 \in T_{x_2}\mathbb{S}^2$, there is exactly one element in $SO(3)$ taking x_1 into x_2 and v_1 into v_2. Therefore $SO(3)$ can be identified with the unit tangent bundle of $\mathbb{S}^2$. Hence, topologically, $SO(3)$ is an $\mathbb{S}^1$-bundle over $\mathbb{S}^2$ with Euler class 2. One has an identification $\mathbb{S}^2 \cong SO(3)/SO(2)$. It is worth saying that every compact Lie group has an essentially unique bi-invariant metric (i.e., invariant under left and right multiplication by elements in the group), and the usual round metric on $\mathbb{S}^2$ comes from the bi-invariant metric on $SO(3)$ via the identification $\mathbb{S}^2 \cong SO(3)/SO(2)$.

We are interested in the finite subgroups of $SO(3)$. These are the cyclic groups of finite order, the (finite) dihedral groups and the polyhedral groups of symmetries of a tetrahedron, an octahedron and an icosahedron. Let us briefly introduce these groups (we refer to Klein's beautiful book [114] for more on the subject).

a) The cyclic groups C_r of any finite order $r > 0$. These are embedded in $SO(3)$ as subgroups of $SO(2) \cong \mathbb{S}^1$, the group of rotations of $\mathbb{R}^2$ (which is identified with the xy-plane in $\mathbb{R}^3$); C_r can be regarded as the group of symmetries of a regular polygon P_r in $\mathbb{R}^2$ with r-edges.

b) The dihedral groups D_r of order $2r > 1$. These are obtained from the cyclic groups by adding a generator of order 2. This generator can be represented by a rotation α in $\mathbb{R}^3$ by an angle of 180° around the x-axis, taking the z-axis into itself but with reversed orientation. Restricted to the xy-plane this is a reflection, but it is a rotation in $\mathbb{R}^3$. Its order is twice the order of the corresponding cyclic group.

There is a geometric way to "visualize" this group by looking at its action on the 2-sphere; this is convenient for later discussions. Start with a regular polygon $\widehat{P}_r$ (including its interior) in the xy-plane with r-edges, and let P_r denote its boundary. Let us join the r vertices of P_r with the centre 0 by r straight lines, getting a triangulation of $\widehat{P}_r$ with r triangles. Now project $\widehat{P}_r$ radially from the centre to get the unit disc $\mathbb{D}^2$ in the xy-plane, whose boundary $\mathbb{S}^1$ corresponds to P_r. The disc $\mathbb{D}^2$ inherits a triangulation from that of P_r. Next, make a stereographic projection from the south pole $S = (0, 0, -1)$. This carries $\mathbb{D}^2$ into the upper half of the sphere $\mathbb{S}^2$, which is covered by r congruent and symmetric triangles that we denote $\{T_1, \ldots, T_r\}$. The equator $\mathbb{S}^1 = \partial \mathbb{D}^2$ is pointwise fixed by this projection and the origin 0 is mapped to the north pole $(0, 0, 1)$. It is clear that each triangle is bounded by geodesic arcs in the sphere (for the usual round metric), with two right angles at the vertices where the edges meet the equator $\mathbb{S}^1$. All the r triangles meet at the north pole, so each has an angle $2\pi/r$ there. Thus the angles of each T_i are $(\frac{\pi}{2}, \frac{\pi}{2}, \frac{2\pi}{r})$. We notice that the action of C_r on $\mathbb{S}^2$ permutes these triangles. The generator α carries them into the southern hemisphere of the sphere, which gets equipped with a triangulation σ. All together we have $2r$ triangles in σ covering the sphere and bounded by geodesics. It is an exercise to show that either of these triangles works as the *fundamental domain* for the action of D_r. That is, if we pick any one of them and denote it $\widehat{T}$, then the $2r$ images of $\widehat{T}$ by the elements of D_r cover the whole sphere and provide the above triangulation.

We remark that later in this section we relate these groups with the "triangle groups" (p, q, r), and for this it will be important to look at the angles $(\frac{\pi}{2}, \frac{\pi}{2}, \frac{2\pi}{r})$ of the triangles in σ. To construct the triangle groups it is however convenient to consider the full group of reflections on the edges of a given triangle, thus getting subgroups of $O(3)$. The order of these groups is twice the order of the corresponding groups of rotations. Thus, in order to get a triangle which is a fundamental domain for the action of the full triangle group, we must double the number of triangles in the triangulation σ above. This is achieved by splitting each triangle T_i in two equal halves by joining the north and south poles by r geodesic arcs in $\mathbb{S}^2$ passing through the middle points of the r edges of P_r, which are contained in the equator. Hence the triangles will have angles $(\frac{\pi}{2}, \frac{\pi}{2}, \frac{\pi}{r})$, corresponding to the triangle group $(2, 2, r)$ as we indicate below.

c) The tetrahedral group. This is the group of rotations in $\mathbb{R}^3$ that preserve a regular tetrahedron $\mathcal{T}$, which is a pyramid with triangular base. It is a group of order 12. Notice that each symmetry of $\mathcal{T}$ must carry vertices into vertices, middle points of the faces into middle points of the faces and the same for the edges. We may construct the 12 elements of this group as follows. Let $(v_1, \ldots, v_4)$ be the vertices of $\mathcal{T}$ and let $(f_1, \ldots, f_4)$ be the faces, numbered in such a way that f_i is the face opposite to the vertex v_i. For each i, let $m_i \in f_i$ be its barycentre (the middle point), and let l_i be the line determined by v_i and m_i. Then rotating by an angle $2\pi/3$ along l_i as axis we get a symmetry of $\mathcal{T}$, that we denote ρ_i; it has order 3. Doing this for each of the four vertices we get the 12 elements of $\mathcal{T}$ (a combination of all these rotations produces the identity in the group). To get a fundamental domain for the corresponding action on the sphere is rather easy: take a triangulation of the faces by joining the point m_i by lines with each of the three vertices in the face f_i. We get a triangulation of $\mathcal{T}$ with 12 triangles. Now make a radial projection from the centre 0, mapping $\mathcal{T}$ into $\mathbb{S}^2$. The vertices $(v_1, \ldots, v_4)$ are fixed points of this projection. We get a triangulation σ of the sphere with 12 triangles. Anyone of these serves as fundamental domain. We may compute the angles of these triangles as follows. Each triangle $\widehat{T}$ in σ has two vertices amongst $(v_1, \ldots, v_4)$ and another vertex at the barycentre of one of the faces $(f_1, \ldots, f_4)$. At each vertex of σ corresponding to a vertex v_i of $\mathcal{T}$ we have six triangles of σ, two for each face of $\mathcal{T}$ containing the given vertex. Hence the angle of each triangle at this point is $\pi/3$. At a vertex of σ corresponding to the barycentre of a face f_j we have three vertices of σ, so the corresponding angles are $2\pi/3$. Thus the angles of each T_i in σ are $(\frac{\pi}{3}, \frac{\pi}{3}, \frac{2\pi}{3})$.

Observe that we have considered only symmetries of the tetrahedron given by rotations. There are also symmetries given by reflections. For instance, given a face f_i of $\mathcal{T}$, take the middle point m_{ij} of an edge e_{ij} of f_i, the vertex v_j in f_i opposite to the edge e_{ij}, and the vertex v_i of $\mathcal{T}$ opposite to the face f_i. The points v_i, v_j and m_{ij} determine a plane $\mathcal{P}$ which cuts the tetrahedron in two equal halves. The reflection in $\mathcal{P}$ determines a symmetry of $\mathcal{T}$. Starting with each vertex of $\mathcal{T}$ we get three symmetries in this way, corresponding to the edges in the opposite face of $\mathcal{T}$. This gives 12 more symmetries of $\mathcal{T}$, which together with the previous 12 rotations gives the full group of symmetries of the tetrahedron, with 24 elements. In this case the fundamental domain for the action on the sphere is obtained by dividing in two each triangle of the above triangulation σ in an appropriate way. For this, take the full barycentric triangulation of the faces of

$\mathcal{T}$. That is, for each face f_i take its barycentre m_i and join it by lines with the three vertices of f_i as before (this gives the triangulation considered above), and also join the barycentre m_i with the middle points m_{ij} of the three edges of f_i. This corresponds to intersecting $\mathcal{T}$ with its planes of symmetry. Now we have a triangulation of $\mathcal{T}$ with 24 symmetric triangles T_i. Map these to the sphere by radial projection and get a triangulation of $\mathbb{S}^2$ with 24 congruent and symmetric triangles. Let us measure the angles of these triangles. At the vertices which are also vertices of $\mathcal{T}$ the angle is still $\pi/3$; each triangle has now one such vertex. At the vertices which are middle points of the faces we now have six adjacent triangles, since each of the previous three triangles was divided in two. Thus the angle is also $\pi/3$. Finally, at the vertices corresponding to the middle points of the edges we have four adjacent triangles, so the angle is $\pi/2$. Hence the triangle $T \subset \mathbb{S}^2$ which serves as fundamental domain for the action of the full group Σ^* of reflections of the tetrahedral group has angles $(\frac{\pi}{2}, \frac{\pi}{3}, \frac{\pi}{3})$. This corresponds to the triangle group $(2,3,3)$.

It remains to prove that the triangle T is indeed a fundamental domain for the action of Σ^*. This is more easily seen by looking directly at the action on the tetrahedron: it is clear that the rotation ρ_i above permutes the three triangles $\widehat{T}$ (which are double triangles) in the face f_i, and it also permutes the three faces adjacent to the vertex v_i. Thus, iterating this rotation ρ_i and combining it with a rotation ρ_j around some other vertex, we see that the images of the face f_i cover all of $\mathcal{T}$ and the image of a single triangle $\widehat{T}$ is carried to the 12 triangles of this triangulation. This proves that $\widehat{T}$ is a fundamental domain for the action of $\Sigma \subset \Sigma^*$, the group of rotations of $\mathcal{T}$. To complete the proof we notice that by construction, $\widehat{T}$ is divided in two symmetric triangles T_i by a plane of symmetry of $\mathcal{T}$, and the reflection in that plane permutes these two triangles. Hence T is a fundamental domain for the action of Σ^*.

d) The octahedral group. An octahedron $\mathcal{O}$ is the polyhedron whose vertices are the middle points of the faces of a cube $\mathcal{C}$. It has six vertices (one for each face of $\mathcal{C}$), 12 edges and eight (triangular) faces. The middle points of these eight faces of $\mathcal{O}$ are vertices of a smaller cube. Hence the cube and the octahedron are dual polyhedra and their groups of symmetries are equal.

The full group of symmetries of the cube has 48 elements. This is a subgroup of $O(3)$. The octahedron $\mathcal{O}$ has nine planes of symmetry, i.e., planes in $\mathbb{R}^3$ with respect to which the reflection maps $\mathcal{O}$ into itself. Six of these planes are the following (see Figure 6). Each vertex, say v_1 has four adjacent vertices, say $v_2, \ldots, v_5$ and an opposite one v_6, and the four adjacent vertices determine a quadrangle with edges $e_{12}, e_{13}, e_{14}, e_{15}$. Suppose e_{12} and e_{14} are opposite edges in the quadrangle, and similarly for e_{13} and e_{15}. Let m_{12} denote the middle point of the edge e_{12}. Then the points v_1, v_6, m_{12} determine a plane of symmetry of $\mathcal{O}$ (which passes by m_{14}), and the points v_1, v_6 and m_{13} determine another plane of symmetry. There are three pairs of opposite vertices and each such pair determines two planes as above, so we have six planes. These divide each triangular face of $\mathcal{O}$ into six al-

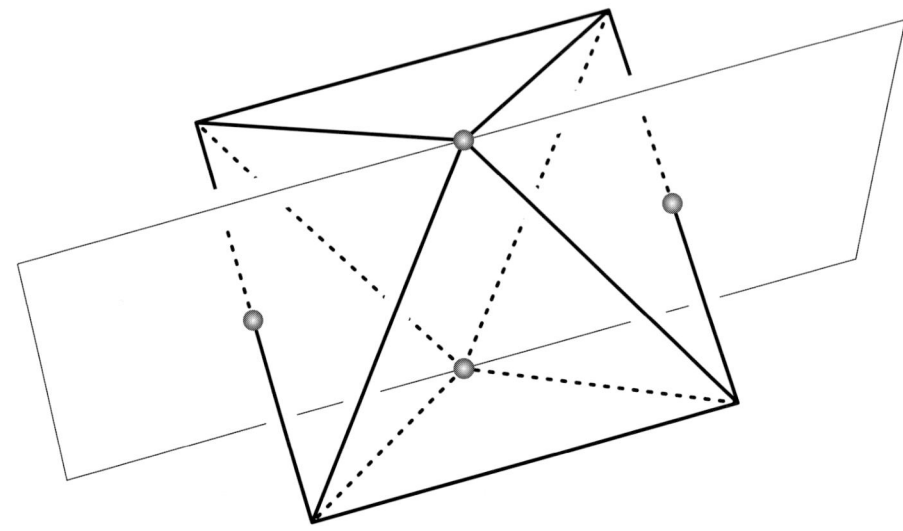

Figure 6: An octahedron and one of its planes of symmetry.

ternately congruent and symmetric triangles that have the middle point of the face as common vertex; they meet at each vertex of $\mathcal{O}$ in sets of eight and in sets of four at each middle point of the edges. Hence, if we put $\mathcal{O}$ with its vertices in the unit 2-sphere $\mathbb{S}^2$ and map $\mathcal{O}$ into $\mathbb{S}^2$ by radial projection from the origin, we get a triangulation of the sphere by triangles whose angles are $\pi/2$ (for the vertices which come from middle points of the edges), $\pi/3$ (for the vertices that correspond to middle points of the faces), and $\pi/4$ (for the vertices which are also vertices of $\mathcal{O}$). So this corresponds to the triangle group $(2, 3, 4)$. Any one of these 48 triangles T serves as fundamental domain for the action of the full octahedral group Σ^* in $\mathbb{S}^2$. To get a fundamental domain for the corresponding group Σ of rotations, which has 24 elements, we must take a "double" triangle $\widehat{T}$ as before.

We leave it as an exercise to show that these triangles T and $\widehat{T}$ are indeed fundamental domains for the actions of Σ^* and Σ, respectively.

e) The icosahedral (and dodecahedral) group. Just as the octahedron and the cube are dual polyhedra with the same groups of symmetries, so too the icosahedron and the dodecahedron are also dual: the middle points of the 12 pentagonal faces of the dodecahedron determine the 12 vertices of the icosahedron, and the middle points of the 20 triangular faces of the latter are the vertices of the former. Each has 30 edges.

Here is a method for actually making a dodecahedron (see Figure 7 below): start with a pentagon P_0 and put it flat over a table. Now attach five pentagons $P_1, \ldots, P_5$ to it by gluing each of them to one of the sides of P_0. And now glue each pentagon $P_1, \ldots, P_5$ to its two neighbors by the edge next to each of them.

In order to actually do this physically, you will have to fold them by the edge in common with P_0, as if these edges were the hinges of a door. The result looks like a bowl with boundary a regular polygon with 10 edges (the two free edges of each pentagon $P_1, \ldots, P_5$). Now make a second bowl like this and glue them together along their boundary. The result is a dodecahedron $\mathcal{D}$.

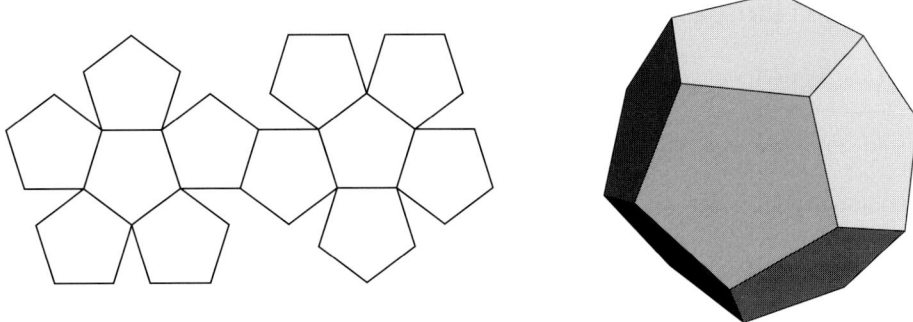

Figure 7: Making a dodecahedron.

Let us look at some of its planes of symmetries: we need 30 such planes to get the triangulation that gives the fundamental domain for the full group of symmetries of $\mathcal{D}$. Notice that each face has exactly one opposite face, so everything we do in one side gets automatically done in the other side. That is, we only look at one of our two bowls used to construct $\mathcal{D}$; this has six faces in it. Let us set $\mathcal{D}$ in $\mathbb{R}^3$ so that its vertices are in $\mathbb{S}^2$. Fix a face f_1 of the dodecahedron. Given one of its vertices v, take the middle point m of the opposite edge in that face. The points v and m determine a unique plane in $\mathbb{R}^3$ passing through the origin, and the reflection in this plane carries $\mathcal{D}$ into itself. Notice this plane cuts the face f_1 in two equal halves. The 3-sphere Doing this for the five vertices in f_1 we get a triangulation of this face with 10 congruent, symmetric triangles with a common vertex at the middle point of the face. Doing this for all the faces we get a triangulation of $\mathcal{D}$ with 120 triangles. Now map these to the sphere by radial projection. At each vertex in the sphere which is a vertex of $\mathcal{D}$, one has six triangles with that common vertex, hence they have angle $\pi/3$ there. As before, at each vertex coming from a middle point in an edge of $\mathcal{D}$ we have four triangles, so the angle is $\pi/2$. Finally, at each vertex coming from a middle point of the faces we have the 10 triangles, so the angle is $\pi/5$. Thus each of these 120 triangles has angles $(\frac{\pi}{2}, \frac{\pi}{3}, \frac{\pi}{5})$. This corresponds to the triangle group $(2, 3, 5)$. Let us show that anyone of these triangles serves as fundamental domain for the action of the full group of reflections on $\mathbb{S}^2$. As in the case of the tetrahedron, by construction it is enough to prove that the double triangle $\widehat{T}$ is a fundamental domain for the corresponding group of rotations. For this we observe that the action is transitive on the triangles $\widehat{T}$ contained in the same face (just take the appropriate rotation with axis the line determined by the middle point of the face and that of the opposite face). Furthermore, this rotation permutes the five pentagons adjacent to the given face, as well as with the other five pentagons adjacent to the opposite face. Hence the proof is complete by noticing that given any two opposite faces of $\mathcal{D}$, there is always a rotation permuting these faces.

II.2 Triangle groups and the classical plane geometries

In this section we describe the triangle groups, which were introduced by H.A. Schwarz and W. Dyck in the 1880s and are related to the polyhedral groups of Section 1. For this we consider the three classical 2-dimensional geometries: Euclidean, spherical and hyperbolic. Let us recall briefly a few facts about these geometries.

We recall that the group of Euclidean motions (isometries) of $\mathbb{R}^2$ is the affine group $E(2)$, isomorphic to $O(2) \times \mathbb{R}^2$, where $O(2)$ is generated by the reflections on lines through the origin in $\mathbb{R}^2$, and $\mathbb{R}^2$ is identified with its group of translations. Since translations are given by reflections on parallel lines, the group $E(2)$ is generated by the reflections on all possible lines in $\mathbb{R}^2$. In particular, given a triangle $T \subset \mathbb{R}^2$ with angles (α, β, γ), one has the *group of isometries* of $\mathbb{R}^2$ generated by the reflections on the edges of T. If two edges e_1, e_2 of T enclose the angle α at the vertex v_1, then the reflections on these two lines yield a rotation around v_1 by an angle of 2α, and similarly for the other angles. Hence if the angle is of the form π/p for some $p \in \mathbb{Z}$, $p > 1$, then repeating p-times the reflections on these two lines brings T back to its original position, while the $2p$-images of T fill out a neighborhood of the given vertex. We leave it as an exercise to think what happens when the angles are not of this form. (There are two remarkably different situations: when the angles are "rational", so that after a number of times the triangle comes back to its original place, and when the angles are "irrational".)

Now let us see what happens when we are on the 2-sphere. We know that its group of isometries is $O(3)$, which is generated by the reflections on all 2-planes through the origin in $\mathbb{R}^3$. Notice that each such plane meets $\mathbb{S}^2$ orthogonally in a circle of maximal length, i.e., an equator. These circles are *the geodesics* in $\mathbb{S}^2$, i.e., the curves that locally minimize the distance between points. Conversely, given any geodesic arc on the sphere, we know that this corresponds to a circle of maximal length; this circle is necessarily the intersection of $\mathbb{S}^2$ with a 2-plane through the origin in $\mathbb{R}^3$ and we can define a reflection on this plane. Hence, as in the Euclidean case, given a triangle T in $\mathbb{S}^2$, bounded by geodesics, we can look at the group of isometries of $\mathbb{S}^2$ generated by reflections on the edges of T. As before, if a given angle of T is of the form π/p for some $p \in \mathbb{Z}$, $p > 1$, then the reflections on the corresponding edges produce a rotation of an angle $2\pi/p$, and repeating this p-times brings T back to its original position, while its $2p$-images fill out a neighborhood of the rotation point.

Now we move to hyperbolic geometry. As models for $\mathbb{H}^2$, the 2-dimensional hyperbolic plane, one can take the Poincaré model of the unit disc, the Lobachevsky model of the upper half-plane, or whatever model one prefers. For this section I take the Poincaré model, so that $\mathbb{H}^2$ is the unit disc in $\mathbb{R}^2$ with the hyperbolic metric. Here is a way to define this metric. This elegant construction, which I learned from D.P. Sullivan, works in all dimensions and has the additional advantage that it gives for free the corresponding group of isometries. Let Möb$(3, \mathbb{R})$ denote the group of transformations of the extended plane

$$\widehat{\mathbb{R}}^2 = \mathbb{R}^2 \cup \{\infty\} \cong \mathbb{S}^2 \, ,$$

generated by inversions in all possible circles, including the reflections in straight lines, which are circles of infinite radius. Denote by Möb$(2, \mathbb{R})$ the subgroup of Möb$(3, \mathbb{R})$ generated by inversions that preserve the unit disc $\mathbb{D} = \mathbb{D}^2 \subset \mathbb{R}^2$. The generators of Möb$(2, \mathbb{R})$ are the inversions in circles (and lines) in $\widehat{\mathbb{R}}^2$ that meet orthogonally the boundary circle $\mathbb{S}^1 = \partial \mathbb{D}$. It is an exercise to see that Möb$(2, \mathbb{R})$ acts transitively on the interior $\overset{\circ}{\mathbb{D}}$ of $\mathbb{D}$. Furthermore, the isotropy subgroup at $0 \in \overset{\circ}{\mathbb{D}}$ is generated by the reflections on lines through the origin, so it is the group $O(2)$. Hence we can put the usual Riemannian metric on the tangent plane $T_0(\overset{\circ}{\mathbb{D}})$, tangent to $\overset{\circ}{\mathbb{D}}$ at 0, and spread this metric around using the group action. That is, given a point $x \in \overset{\circ}{\mathbb{D}}$, take an element g in Möb$(2, \mathbb{R})$ with $g(0) = x$; the derivative of g identifies the tangent planes $T_x(\overset{\circ}{\mathbb{D}}) \cong T_0(\overset{\circ}{\mathbb{D}})$ and the metric on the latter induces a metric in the former. This is well defined because the ambiguity lies in $O(2)$ and this group preserves the standard metric. Thus one has a well-defined Riemannian metric on $\overset{\circ}{\mathbb{D}}$ and it is an exercise to show that this metric is of constant negative curvature, so it gives the Poincaré model for hyperbolic geometry. By construction we see that the geodesics are circles in $\overset{\circ}{\mathbb{D}}$ that meet the boundary $\mathbb{S}^1$ orthogonally, and the isometries are compositions of inversions (that we shall call also "reflections") on these circles. By $\mathbb{H}^2$ we mean $\overset{\circ}{\mathbb{D}}$ with this metric.

Notice that Möb$(2, \mathbb{R})$ has the subgroup Möb$_+(2, \mathbb{R})$, of orientation preserving isometries of $\mathbb{H}^2$, as an index 2 subgroup, consisting of the words of even length. This group, also denoted $Iso_+(\mathbb{H}^2)$, is isomorphic to the famous group of Möbius transformations $z \mapsto \frac{az+b}{cz+d}$ with real coefficients and $ad - bc = 1$. It is also isomorphic to the group $PSL(2, \mathbb{R}) := SL(2, \mathbb{R})/\pm 1 = GL(2, \mathbb{R})/\mathbb{R}^*$. The previous discussion shows that as a manifold, Möb$(2, \mathbb{R})$ is diffeomorphic to $\mathbb{H}^2 \times SO(2)$, so it is an open solid torus $\mathbb{S}^1 \times \overset{\circ}{\mathbb{D}}$.

As before, given a triangle T in $\mathbb{H}^2$ bounded by geodesics, we can look at the group of isometries generated by the reflections (inversions) on the edges of T, and given an angle of T of the form π/p for some $p \in \mathbb{Z}$, $p > 1$, the reflections on the corresponding edges produce a rotation of an angle $2\pi/p$, and repeating this p-times brings T back to its original place.

From now on we essentially follow the exposition given in [171, §2]. Let us denote by P either the Euclidean plane (i.e., $\mathbb{R}^2$ with the Euclidean metric), the 2-sphere with its usual metric, or the hyperbolic plane. Given integers $p, q, r \geq 2$, there is a triangle $T = T_{(p,q,r)}$ in P with angles $\pi/p, \pi/q, \pi/r$. This triangle is Euclidean if $\frac{1}{p} + \frac{1}{q} + \frac{1}{r} = 1$, spherical if this sum is > 1 and hyperbolic if the sum is < 1. In the first case the area of T can be arbitrary, but in the spherical and hyperbolic cases the area is determined by the angles: it is $\left| \frac{\pi}{p} + \frac{\pi}{q} + \frac{\pi}{r} - \pi \right|$.

2.1 Definition.

(i) The *full (Schwarz) triangle group* $\Sigma^* = \Sigma^*_{(p,q,r)}$ is the group of isometries of P generated by reflections σ_1, σ_2, σ_3 in the three edges of $T_{(p,q,r)}$.

(ii) The *(Schwarz) triangle group* $\Sigma = \Sigma_{(p,q,r)}$ is the index 2 subgroup of $\Sigma^*_{(p,q,r)}$ consisting of orientation preserving maps.

It is customary to denote the triangle groups simply by (p, q, r), but this notation can be confusing in our case, so we use the notation of Definition 2.1.

From the previous discussion we know that in the spherical case the group Σ^* is a subgroup of $O(3)$ and $\Sigma \subset SO(3)$. In the hyperbolic case Σ^* is a subgroup of $\text{Möb}(2, \mathbb{R})$ and Σ is contained in $PSL(2, \mathbb{R})$. In the Euclidean case Σ^* lives in the affine group $E(2)$ and Σ is contained in $E^+(2)$, which is a semi-direct product $SO(2) \rtimes \mathbb{R}^2$.

The following theorem is due to Poincaré.

2.2 Theorem. *The group $\Sigma^*_{(p,q,r)}$ has a presentation with generators σ_1, σ_2, σ_3 and relations*

$$\sigma_1^2 = \sigma_2^2 = \sigma_3^2 = 1 \qquad and \qquad (\sigma_1 \sigma_2)^p = (\sigma_2 \sigma_3)^q = (\sigma_3 \sigma_1)^r = 1,$$

where it is understood that the edges e_1 and e_2 enclose the angle π/p, e_2 and e_3 enclose the angle π/q, and e_3, e_1 enclose the angle π/r. Furthermore, the triangle $T_{(p,q,r)}$ is a fundamental domain for the action of $\Sigma^*_{(p,q,r)}$ on P.

This theorem implies that the various images of the triangle $T_{(p,q,r)}$ cover the whole "plane" and they are pairwise disjoint, except for boundary points. These kinds of decompositions of the "plane" are usually called *tesselations*, as in Figure 8.

That $\Sigma^*_{(p,q,r)}$ is generated by σ_1, σ_2, σ_3 is by definition, and the above discussion shows that these generators satisfy the relations in Theorem 2.2, since each σ_i is a reflection (so that $\sigma_i = 1$) and the composition of any two of them produces a rotation (around the corresponding vertex) by twice the given angle. So what has to be proved in Theorem 2.2 is that there are not more relations among the generators and that the triangle $T_{(p,q,r)}$ is a fundamental domain, i.e., that its images under the action of Σ^* on P cover the whole "plane" and they are pairwise disjoint except for the boundaries of the triangles. We refer to [171, §2] for this.

An immediate consequence of Theorem 2.2 is:

2.3 Corollary. *In the hyperbolic and Euclidean cases $\frac{1}{p} + \frac{1}{q} + \frac{1}{r} \leq 1$ the group is infinite, while in the spherical case it is finite and its order is given by the area of the sphere divided by that of T:*

$$\left| \Sigma^*_{(p,q,r)} \right| = \frac{4}{(p^{-1} + q^{-1} + r^{-1}) - 1} \,.$$

Recall that the group $\Sigma_{(p,q,r)}$ of orientation preserving isometries of P consists of the elements of even length in $\Sigma^*_{(p,q,r)}$. Let us set (following [171]):

$$\tau_1 = \sigma_1\sigma_2 \,, \ \tau_2 = \sigma_2\sigma_3 \,, \ \tau_3 = \sigma_3\sigma_1 \,.$$

One has

$$\tau_1\tau_2\tau_3 = \sigma_1\sigma_2\sigma_2\sigma_3\sigma_3\sigma_1 = \sigma_1\sigma_2^2\sigma_3^2\sigma_1 = 1$$

and one has another consequence of Corollary 2.3 (see [171, 2.5]):

2.4 Corollary. *The group $\Sigma_{(p,q,r)}$ has a presentation with generators τ_1, τ_2, τ_3 and relations*

$$\tau_1^p = \tau_2^q = \tau_3^r = \tau_1\tau_2\tau_3 = 1 \,.$$

Let us now list the possibilities. We may assume that $2 \leq p \leq q \leq r$. In the Euclidean case $\frac{1}{p} + \frac{1}{q} + \frac{1}{r} = 1$ there are only three possibilities: the triple (p, q, r) must be $(2, 3, 6)$, $(2, 4, 4)$ or $(3, 3, 3)$. For the spherical case $\frac{1}{p} + \frac{1}{q} + \frac{1}{r} > 1$ the possibilities are: $(2, 2, r)$, $(2, 3, 3)$, $(2, 3, 4)$ or $(2, 3, 5)$. For all the other infinite triples we are in the hyperbolic case $\frac{1}{p} + \frac{1}{q} + \frac{1}{r} < 1$.

In the spherical case we already know from Section 1 which are the finite subgroups of $SO(3)$; we also know their orders and fundamental domains for their actions on the sphere. This allows us to deduce the following (well-known) theorem.

2.5 Theorem. *The finite subgroups of $SO(3)$ are the cyclic groups C_r of finite order and the triangle groups $\Sigma_{(p,q,r)}$ where (p, q, r) is one of the triples:*

(i) $(2, 2, r)$, *and $\Sigma_{(2,2,r)}$ is isomorphic to the dihedral group D_r of order $2r$.*

(ii) $(2, 3, 3)$, *and $\Sigma_{(2,3,3)}$ is isomorphic to the tetrahedral group of order 12.*

(iii) $(2, 3, 4)$, *and $\Sigma_{(2,3,4)}$ is isomorphic to the octahedral group of order 24.*

(iv) $(2, 3, 5)$, *and $\Sigma_{(2,3,5)}$ is isomorphic to the icosahedral group of order 60.*

Notice that the dihedral group can also be considered as the group of symmetries of a bi-prism over a regular polygon, so the spherical triangular groups are also called *polyhedral groups*.

In the hyperbolic case $\frac{1}{p} + \frac{1}{q} + \frac{1}{r} < 1$ the triangle groups are a special case of the so-called Fuchsian groups that we will describe in the next chapter. In the Euclidean case $\frac{1}{p} + \frac{1}{q} + \frac{1}{r} = 1$ the triangle groups, the analogous group of the quadrangle and $\mathbb{Z} \oplus \mathbb{Z}$, are the only discrete subgroups of $E^+(2)$ with compact quotient. We shall come back to this point in Chapter III.

Figure 8: Tesselations of $\mathbb{S}^2$ and $\mathbb{H}^2$ given by the triangle groups $\Sigma^*_{(2,3,5)}$ and $\Sigma^*_{(2,3,7)}$.

II.3 The 3-sphere as a Lie group and its finite subgroups

This section is about the geometry of quaternions and we refer to [56, Chapter 6] for a clear account of the subject.

Let us denote by $\mathcal{H}$ the space of quaternions. As a vector space $\mathcal{H}$ is $\mathbb{R}^4$ endowed with three additional complex structures given by the numbers

$$i = (0,1,0,0), \quad j = (0,0,1,0) \quad \text{and} \quad k = (0,0,0,1),$$

and defining relations:

$$i^2 = j^2 = k^2 = -1 \quad \text{and} \quad ij = k \,;\, jk = i \,;\, ki = j\,.$$

One has

$$ij = -ji \;;\; jk = -kj \;;\; ki = -ik\,.$$

This defines a multiplication of quaternions in the obvious way, by extending linearly the above operations. Every quaternion can be written uniquely as:

$$q = a_0 + a_1 i + a_2 j + a_3 k = (a_0 + a_1 i) + (a_2 + a_3 i)\, j$$

with $a_0, a_1, a_2, a_3 \in \mathbb{R}$.

This gives an identification $\mathcal{H} \cong \mathbb{C}^2$. In other words, every quaternion q can be written as a pair (z_1, z_2) of complex numbers, $q = z_1 + z_2 j$.

Given $q = a_0 + a_1 i + a_2 j + a_3 k$, we say that a_0 is its *real part*, while $a_1 i + a_2 j + a_3 k$ is a *pure quaternion*. Writing $q = a_0 + (a_1 i + a_2 j + a_3 k)$, one defines the quaternionic conjugation $q \mapsto \bar{q}$ by

$$\bar{q} = a_0 - (a_1 i + a_2 j + a_3 k)\,.$$

The norm $\|q\|$ of a quaternion is defined in the usual way, via the identification $\mathcal{H} \cong \mathbb{R}^4$, and just as for the complex numbers, it is an exercise to verify that quaternionic multiplication satisfies:

$$\|q\|^2 = q\,\bar{q} \quad \text{and} \quad \|q_1 q_2\| = \|q_1\|\,\|q_2\|.$$

Thus one has that multiplication in $\mathcal{H}$ preserves the unit quaternions and induces a Lie group structure on the unit sphere $\mathbb{S}^3$ in $\mathcal{H}$. In particular, if we identify $\mathcal{H}$ with $\mathbb{C}^2$ as above, one has an inclusion $\mathbb{C} \hookrightarrow \mathcal{H}$ given by $z \mapsto (z, 0)$; quaternionic multiplication restricts to the usual complex multiplication and one has the unit circle $\mathbb{S}^1$ as invariant set. This defines an inclusion $\mathbb{S}^1 \hookrightarrow \mathbb{S}^3$ as Lie groups.

We notice that we can identify a quaternion $q = z_1 + z_2 j$ with the complex matrix

$$A(q) = \begin{pmatrix} z_1 & z_2 \\ -\bar{z}_2 & \bar{z}_1 \end{pmatrix}.$$

With this identification one has that the norm of q is the determinant of $A(q)$, thus one has an induced map: $(\mathcal{H} - \{0\}) \overset{\psi}{\to} GL(2, \mathbb{C})$ into the 2×2 invertible matrices with complex coefficients. Furthermore, the column vectors of $A(q)$ satisfy:

$$\langle (z_1, -\bar{z}_2), (z_2, \bar{z}_1) \rangle = z_1 \bar{z}_2 - \bar{z}_2 z_1 = 0 \,,$$

where $\langle \cdot, \cdot \rangle$ is the usual Hermitian product in $\mathbb{C}$. This means that these column vectors an orthogonal complex 2-frame in $\mathbb{C}^2$; thus restricting ψ to the unit quaternions, one has that its image consists of all 2×2 unitary matrices with determinant 1. That is, we have a canonical identification of $\mathbb{S}^3$ with the *special unitary group* $SU(2)$, consisting of all matrices of the form $\begin{pmatrix} z_1 & z_2 \\ -\bar{z}_2 & \bar{z}_1 \end{pmatrix}$ with determinant 1. With this identification, the circle $\mathbb{S}^1 \subset \mathbb{C}$ gets embedded in $SU(2)$ as the group of diagonal matrices of the form $\begin{pmatrix} e^{i\theta} & 0 \\ 0 & e^{-i\theta} \end{pmatrix}$. The centre of $SU(2)$ is $\pm I$ and the projection

$$\mathbb{S}^3 \cong SU(2) \overset{\pi}{\longrightarrow} (SU(2)/ \pm I) \cong \mathbb{R}P^3$$

identifies pairs of opposite points in the sphere. This projection corresponds to the adjoint representation of $SU(2)$ in its Lie algebra, which can be identified with the tangent space at the identity $T_1\mathbb{S}^3 \cong \mathbb{R}^3$.

Let us describe with more care the projection $SU(2) \overset{\pi}{\to} \mathbb{R}P^3$. For this we recall that the sphere $\mathbb{S}^2$ can be identified with $\mathbb{C}P^1$, the projective space of complex lines through the origin in $\mathbb{C}^2$. Since $SU(2)$ acts linearly on $\mathbb{C}^2$, its action descends to an action on $\mathbb{S}^2$. This action is by holomorphic transformations, so it preserves the orientation of $\mathbb{S}^2$. Moreover, this action is also transitive, because given any two lines in $\mathbb{C}^2$ there is a linear map in $SU(2)$ taking one into the other. The isotropy subgroup consists of the unitary matrices that preserve the z_1-axis, so it is the circle $\mathbb{S}^1$, regarded as the diagonal matrices in $SU(2)$, which is in fact the group $U(1)$. One gets an identification $\mathbb{S}^2 \cong SU(2)/U(1)$ and the fibre bundle projection $SU(2) \cong \mathbb{S}^3 \to \mathbb{S}^2$ is precisely *the Hopf fibration*; the fibre over each point $x \in \mathbb{S}^2 \cong \mathbb{C}P^1$ is precisely the unit circle in the corresponding complex line. We observe that the metric on $\mathbb{S}^2$ is also the one induced by this identification. Thus one gets a homomorphism from $SU(2)$ into the group $SO(3)$ of orientation preserving isometries of $\mathbb{S}^2$, which is actually surjective. Moreover, it is two-to-one, since two matrices in $SU(2)$ induce the same map in $\mathbb{C}P^1$, the space of lines in $\mathbb{C}^2$, iff they differ by sign. This identifies $\mathbb{S}^3$ with the universal cover of $SO(3)$, being a 2-fold cover. Furthermore, since the projection $\mathbb{S}^3 \to SO(3)$ has kernel ± 1, it follows that as a manifold $SO(3)$ is the real projective space $\mathbb{R}P^3$.

The theorem below summarizes the previous discussion.

3.1 Theorem. *The Lie group $\mathbb{S}^3$ of unit quaternions is isomorphic to the special unitary group $SU(2)$. Its centre consists of the quaternions ± 1 and the projection $\pi : SU(2) \cong \mathbb{S}^3 \to SO(3)$ is a two-fold cover with kernel the centre ± 1 of $\mathbb{S}^3$. One* *has a commutative diagram:*

$$
\begin{array}{ccccc}
\mathbb{Z}_2 & \xrightarrow{\;\cong\;} & \mathbb{Z}_2 & \longrightarrow & 1 \\
\downarrow{\scriptstyle \iota} & & \downarrow{\scriptstyle \iota} & & \downarrow \\
\mathbb{S}^1 & \xrightarrow{\;i_1\;} & \mathbb{S}^3 \cong SU(2) & \xrightarrow{\;pr_1\;} & \mathbb{S}^2 \\
\downarrow{\scriptstyle \tilde{\pi}} & & \downarrow{\scriptstyle \pi} & & \downarrow{\scriptstyle Id} \\
\mathbb{R}P^1 \cong \mathbb{S}^1 & \xrightarrow{\;i_2\;} & \mathbb{R}P^3 \cong SO(3) & \xrightarrow{\;pr_2\;} & \mathbb{S}^2
\end{array}
$$

where the middle line is the Hopf fibration, the bottom line comes from identifying $SO(3)$ with the unit tangent bundle of $\mathbb{S}^2$, and the first two columns are two-fold covering projections.

We remark that we are making here an abuse of notation. The multiplicative group of the unit quaternions is usually denoted by $Sp(1)$ and called *the symplectic group*; this is the 3-sphere canonically, hence our notation.

Now we want to describe the finite subgroups of the 3-sphere. These were known since the 19th century. For instance A. Cayley gave the list of these groups and the list of all the unit quaternions in each (see for instance [223]).

Since one has a 2-fold covering $\mathbb{S}^3 \to SO(3)$, which is actually a group homomorphism with kernel ± 1, it follows that each finite subgroup Σ of $SO(3)$ lifts to a finite subgroup of $\mathbb{S}^3$ doubling its order. The cyclic groups of order r lift to cyclic groups of order $2r$ and the triangle subgroups $\Sigma_{(p,q,r)}$ lift to the so-called *binary polyhedral* groups:

(i) the binary dihedral group, of order 4r;

(ii) the binary tetrahedral group, of order 24;

(iii) the binary octahedral group, of order 48;

(iv) the binary icosahedral group, of order 120.

These are usually denoted $< p, q, r >$ but to be consistent with our notation in §2 we prefer to denote them $\Gamma_{(p,q,r)}$.

3.2 Remark. It is worth making clear that these binary polyhedral groups **are not** the corresponding full triangle groups Σ^* of §2 above, though their orders coincide. The binary groups live in $\mathbb{S}^3$ while the full triangle groups live in $O(3)$. The groups $\mathbb{S}^3$ and $O(3)$ are both double covers of $SO(3)$ and this is why the orders of the groups coincide. However $O(3)$ is a trivial double cover, isomorphic to $SO(3) \times \mathbb{Z}_2$, while $\mathbb{S}^3$ is simply connected and is therefore the universal cover of $SO(3)$. This means that $\mathbb{S}^3$ is also isomorphic to the group Spin(3) (see [124] and Chapter IV below).

3.3 Theorem. *The only finite subgroups of the 3-sphere $\mathbb{S}^3$ are the double covers of the finite subgroups of $SO(3)$ and the cyclic groups of odd order.*

The proof of this well-known result is now easy: if a subgroup $\Gamma \subset SU(2)$ contains the centre ± 1, then it is the lifting of a subgroup of $SO(3)$ and there is nothing to prove. Thus we only have to prove that if $\Gamma \subset SU(2)$ does not contain the centre, then it is a cyclic group of odd order. For this we observe:

(i) if a subgroup $\Gamma \subset SU(2)$ does not contain the centre, then the restriction to Γ of the projection $\mathbb{S}^3 \xrightarrow{\pi} SO(3)$ is an injective morphism and therefore Γ is isomorphic to its image $\Sigma \subset SO(3)$;

(ii) if g is an element in $\mathbb{S}^3$ such that $g^2 = 1$, then $g = \pm 1$. In fact, think of g as a unitary matrix $\begin{pmatrix} z_1 & z_2 \\ -\bar{z}_2 & \bar{z}_1 \end{pmatrix}$. Then one has:

$$g^2 = \begin{pmatrix} z_1^2 - |z_2|^2 & z_2(z_1 + \bar{z}_1) \\ -\bar{z}_2(z_1 + \bar{z}_1) & \bar{z}_1^2 - |z_2|^2 \end{pmatrix} = \begin{pmatrix} 1 & 0 \\ 0 & 1 \end{pmatrix} ;$$

the two equations given by the anti-diagonal imply that either $z_2 = 0$ or else z_1 is purely imaginary, say $z_1 = iy_1$ with $y \in \mathbb{R}$; if this happens, then $z_1^2 = -y_1^2 \leq 0$ and by hypothesis one has $z_1^2 - |z_2|^2 = 1$, which is not possible. Hence one must have $z_2 = 0$ and therefore $z_1 = \pm 1$.

Now, the inverse of an element in $\mathbb{S}^3$ is uniquely defined, so the elements in a subgroup $\Gamma \subset \mathbb{S}^3$, others than the identity, necessarily come in pairs. Hence if $-1 \notin \Gamma$, then the order of Γ is odd, by (ii) above. Thus (i) implies that Γ is isomorphic to a subgroup of $SO(3)$ of odd order, hence Γ is cyclic by Theorem 2.5.

II.4   Brieskorn manifolds and Klein's theorem

We recall that $SU(2)$ acts linearly on $\mathbb{C}^2$ with the origin 0 as the only fix point; the action is free away from 0. Also, this action, being unitary, preserves the usual Hermitian metric on $\mathbb{C}^2$ and therefore preserves all the spheres centred at 0. So each orbit, other than 0 itself, is a 3-sphere with centre at 0; the unit sphere is the orbit of $(1, 0) \in \mathbb{C}^2$. Each finite subgroup $\Gamma \subset SU(2)$ acts naturally on $\mathbb{C}^2$ and on the unit 3-sphere $\mathbb{S}^3 \subset \mathbb{C}^2$. The space of orbits $\Gamma \backslash \mathbb{C}^2$ is a 2-dimensional complex analytic variety V_Γ (c.f. [52, 102]) with an isolated singularity at the image of 0, that we denote by $\hat{0}$. In Chapter I we defined the link of an isolated singularity in an analytic space; this turns out to be the smooth boundary of a regular neighborhood of the singular point. Hence the link of $\hat{0}$ in V_Γ is $M_\Gamma = \Gamma \backslash \mathbb{S}^3$, a smooth compact 3-manifold whose universal cover is $\mathbb{S}^3$ and Γ is its fundamental group. Notice that Γ acts naturally on the 3-sphere by left (or right) multiplication in the group. The action we are taking here corresponds to left multiplication (for compatibility with the complex structure). Felix Klein [114] proved in the 19th century a remarkable theorem about these surfaces V_Γ. To state his result, let us introduce one of his ideas in this direction, which is to me a keystone in today's geometric invariant theory (c.f. [178]).

Let H^d denote the space of homogeneous polynomials of degree d in two complex variables x, y. This is a vector space of dimension $d+1$, with a basis given by the monomials $\{x^d, x^{d-1}y, \ldots, xy^{d-1}, y^d\}$. The action of $SU(2)$ on $\mathbb{C}^2$ defines a linear action of this group on H^d in the obvious way: given $g \in SU(2)$ and $h \in H^d$ define the action by $(g, h) \mapsto h_g$ where h_g is defined by $h_g(x, y) = h(g(x, y))$. In this way one has a representation of $SU(2)$ in H^d. [In fact these are all the irreducible representations of $SU(2)$.] Of course that if we are given a subgroup $\Gamma \subset SU(2)$, one can restrict this action to Γ.

4.1 Definition. An element $h \in H^d$ is Γ-invariant if $h(g(x, y)) = h(x, y)$ for all $g \in \Gamma$ and all $(x, y) \in \mathbb{C}^2$.

It is clear that the set of all Γ-invariant polynomials of degree d is a complex vector space H^d_Γ. Moreover, given $h_1 \in H^{d_1}_\Gamma$ and $h_2 \in H^{d_2}_\Gamma$ their product is in $H^{d_1+d_2}_\Gamma$. Hence the set of all Γ-invariant polynomials of all degrees forms a graded algebra, that we denote by H^*_Γ, which as a vector space is the direct sum of the H^d_Γ.

The theorem of Klein that we are talking about can be phrased as follows.

4.2 Theorem (Klein). *Let Γ be a finite subgroup of $SU(2)$. Then the algebra H^*_Γ of Γ-invariant polynomials is generated by three homogeneous polynomials, say h_1, h_2, h_3, of various degrees, which satisfy a (essentially single) polynomial equation*

$$f(h_1, h_2, h_3) = 0,$$

where f is weighted homogeneous. These three polynomials define a Γ-invariant function $H : \mathbb{C}^2 \to \mathbb{C}^3$, which induces a homeomorphism between the orbit space $V_\Gamma = \Gamma \backslash \mathbb{C}^2$ and the surface $V = f^{-1}(0)$. This map is actually a diffeomorphism from $V_\Gamma - \hat{0}$ into $V - 0$ and carries $M_\Gamma = (\Gamma \backslash SU(2))$ into the link of 0 in V, which is $V \cap \mathbb{S}^5$, where $\mathbb{S}^5$ is the unit sphere in $\mathbb{C}^3$.

In fact Klein finds the explicit polynomials that generate these algebras for each of the finite subgroups of $SU(2)$. For instance, if $\Gamma = \mathbb{Z}_r$ is cyclic of order r, then the generators of the algebra $H^*_{\mathbb{Z}_r}$ can be taken to be:

$$h_1 = x^r + y^r \quad , \quad h_2 = ix^r - iy^r \quad \text{and} \quad h_3 = 2ixy,$$

which clearly satisfy:

$$h_1^2 + h_2^2 + h_3^r = 0$$

and the surface $V_{\mathbb{Z}_r} = \mathbb{Z}_r \backslash \mathbb{C}^2$ is the hypersurface V in $\mathbb{C}^3$ defined by

$$V = \{ z_1^2 + z_2^2 + z_3^r = 0 \}.$$

So the link of V is the lens space $\mathbb{Z}_r \backslash \mathbb{S}^3$. In particular, for $r = 2$ this is $\mathbb{R}P^3 = SO(3)$.

We refer to Klein's book for the proof of Theorem 4.2. Here we follow [171] and prove a slight refinement of Klein's theorem, which leaves out a few cases but is more precise in its statement and has the advantage of motivating Milnor's proof of his theorem for the Brieskorn manifolds in the hyperbolic case, which we discuss in Section 6 below.

4.3 Theorem. *Let* $\Gamma = \Gamma_{(p,q,r)}$ *be a binary triangle subgroup of* $SU(2)$ *and let* $\Pi = \Pi_{(p,q,r)}$ *be its commutator,* $\Pi = [\Gamma, \Gamma]$. *Then the algebra* H^*_Π *of* Π-*invariant polynomials is generated by three polynomials* h_1, h_2, h_3 *of orders* $k/p,\ k/q,\ k/r$, *respectively, where* k *is the order of the triangle subgroup* $\Sigma_{(p,q,r)} \subset SO(3)$, $\Sigma_{(p,q,r)} = \Gamma_{(p,q,r)}/\pm 1$, *which satisfy the (essentially single) polynomial relation:*

$$h_1^p + h_2^q + h_3^r = 0.$$

The corresponding Π-*invariant map* $\mathbb{C}^2 \to \mathbb{C}^3$ *determines a homeomorphism between the space of orbits* $\Pi\backslash\mathbb{C}^2$ *and the Brieskorn variety:*

$$V_{(p,q,r)} = \{(z_1, z_2, z_3) \in \mathbb{C}^3 \mid z_1^p + z_2^q + z_3^r = 0\}.$$

This homeomorphism is actually a diffeomorphism away from the singular point in each surface, and it induces a diffeomorphism between the corresponding links.

The statement "essentially single" in Theorem 4.3 means that the ideal of all polynomial relations among these polynomials is generated by the aforementioned relation. Also, we recall that the order of $\Sigma_{(p,q,r)}$ is half the order of $\Gamma_{(p,q,r)}$, which equals the order of the full triangle group $\Sigma^*_{(p,q,r)}$.

Hence (2.3) yields that the order of $\Sigma_{(p,q,r)}$ is:

$$\left|\Sigma_{(p,q,r)}\right| = \frac{2}{(p^{-1} + q^{-1} + r^{-1}) - 1}.$$

Let us make some comments before coming to the proof of Theorem 4.3. First we remark that the equivalence in Theorem 4.3 is in fact an isomorphism of analytic spaces, but we will not discuss that here. In fact these singularities are the prototype of what is called a "taut singularity" (see [121]). These are, by definition, isolated surface singularities whose topology determines uniquely the analytic structure. This is not always so. For instance, it follows from [171, Theorem 7.3] that the Brieskorn manifolds $M_{(2,7,14)}$ and $M_{(3,4,12)}$ are both circle bundles with Euler class -1 over a Riemann surface of genus 3. Hence the corresponding surfaces are homeomorphic. However (c.f.[168, Th. 9.1]) the Milnor number of the first surface is 78, while for the latter this number is 66, so they are not analytically equivalent. Similar remarks apply to many other Brieskorn manifolds, as for instance $M_{(2,9,18)}$ and $M_{(3,5,15)}$.

It is an exercise to show that the binary icosahedral group $\Gamma_{(2,3,5)}$ is a perfect group, i.e., it equals its own commutator. The commutator of $\Gamma_{(2,3,4)}$ is $\Gamma_{(2,3,3)}$, the commutator of $\Gamma_{(2,3,3)}$ is $\Gamma_{(2,2,2)}$, which is the so-called quaternion group, consisting of the quaternions $\pm 1, \pm i, \pm j, \pm k$. The commutator of the dihedral group $\Gamma_{(2,2,r)}$ is the cyclic group $\mathbb{Z}_r$.

Hence Theorem 4.3 gives the equivalences:

(iv) $V_{(2,2,r)} \cong (\mathbb{Z}_r \backslash \mathbb{C}^2)$,

(iii) $V_{(2,3,3)} \cong (\Gamma_{(2,2,2)} \backslash \mathbb{C}^2)$,

(ii) $V_{(2,3,4)} \cong (\Gamma_{(2,3,3)} \backslash \mathbb{C}^2)$,

(i) $V_{(2,3,5)} \cong (\Gamma_{(2,3,5)} \backslash \mathbb{C}^2)$.

In this list there are missing the cases $(\Gamma_{(2,3,4)} \backslash \mathbb{C}^2)$ and $(\Gamma_{(2,2,r)} \backslash \mathbb{C}^2)$, $r > 2$. Klein's theorem is also true for these, and the corresponding singular surfaces are:

$$z_1^2 + z_2^3 + z_2 z_3^3 = 0$$

for the binary octahedral group $\Gamma_{(2,3,4)}$, and

$$z_1^2 + z_2^2 z_3 + z_3^{r+1} = 0$$

for the binary dihedral group with $4r$ elements. In the special case $r = 2$ this surface is equivalent to $V_{(2,3,3)}$.

We also notice that the Brieskorn manifold $M_{(2,3,5)} \cong \Gamma_{(2,3,5)} \backslash \mathbb{S}^3$ is a homology sphere, since the group $\Gamma_{(2,3,5)}$ is perfect. This is the famous Poincaré's homology sphere. Notice that examples like this may not occur in higher dimensions, as pointed out in §8 of the first chapter. The point is that when the link of a hypersurface singularity has dimension more than 3, then it is highly connected and therefore it is a homology sphere iff it is a homotopy sphere, and in this case Smale's h-cobordism theorem implies that the manifold is actually a topological sphere.

The singularities constructed above play a very special role in geometry and the theory of singularities, and they receive different names according to the property one is looking at. They are called, for instance, Klein singularities, Du Val singularities or rational double points. They are also the simple (or 0-modal) singularities in Arnold's notation (see for instance [9]). We refer to [65] for a (yet not exhaustive) list of 15 characterizations of these singularities. Klein's theorem gives two of them. These groups can also be regarded as Coxeter groups and there are very interesting results in this respect, see for instance [42, 45].

Proof of 4.3. This theorem is carefully proved in Milnor's article [171], so we only indicate here the main ideas. We recall that *a character* of a group Γ means a 1-dimensional unitary representation of Γ, i.e., a homomorphism

$$\chi : \Gamma \to U(1) \subset \mathbb{C}^* := \mathbb{C} - 0.$$

Given $\Gamma \subset SU(2)$ and a character χ, let $H_\Gamma^{d,\chi}$ denote the set of all homogeneous polynomials of degree d which are χ-invariant, i.e., they transform according to the rule:

$$h(g(x,y)) = \chi(g)h(x,y)$$

so they are "invariant up to multiplication by χ". [These polynomials were called "invariant forms" by Klein in [114].] Since χ is a homomorphism, the set $H_\Gamma^{d,\chi}$ is a complex vector space. Furthermore, given characters χ_1, χ_2, and polynomials $h_1 \in H_\Gamma^{d_1,\chi_1}$ and $h_2 \in H_\Gamma^{d_2,\chi_2}$, the product $h_1 \cdot h_2$ is in $H_\Gamma^{d_1+d_2,\chi_1\chi_2}$, where the product of the characters is defined in the obvious way. Hence these polynomials form a bigraded algebra that we denote $H_\Gamma^{*,*}$. Notice that $H_\Gamma^{*,1}$ is by definition the set of Γ-invariant polynomials.

Given $\Gamma \subset SU(2)$ and $h \in H_\Gamma^{d,\chi}$ for some character χ, for every $g = xyx^{-1}y^{-1}$ in the commutator Π of Γ, one has $\chi(g) = 1$. Thus one has an inclusion $H_\Gamma^{d,\chi} \subset H_\Pi^d$. On the other hand, Π is a normal subgroup of Γ, hence the quotient group Γ/Π acts linearly on H_Π^d, which is therefore a representation space for this group. Since the group Γ/Π is finite and Abelian, it follows (see [171, 4.1]) that the representation space H_Π^d splits as the direct sum of the $H_\Gamma^{d,\chi}$ as χ varies over the characters of Γ. Hence one has an isomorphism of algebras:

$$H_\Pi^* = \bigoplus H_\Gamma^{*,*}. \tag{4.4}$$

Now consider a polynomial $h \in H_\Gamma^{d,\chi}$, for some d and some χ. Since h is homogeneous of degree d, it follows that h must vanish over d lines (maybe not necessarily different) $\ell_1, \ldots, \ell_d$ through the origin in $\mathbb{C}^2$. These lines are necessarily permuted by the elements of Γ, and the linear equations of $\ell_1, \ldots, \ell_d$ determine h uniquely, up to multiplication by a constant. Conversely, given d points in $\mathbb{C}P^1$ which are permuted by the action of Γ, the corresponding lines in $\mathbb{C}^2$ determine a homogeneous polynomial h of degree d, which satisfies that for each $\gamma \in \Gamma$, the rotated polynomial $f(g(x,y))$ is a scalar multiple of $f(x,y)$. Thus we can define a character of Γ by:

$$\chi(g) = f(g(x,y))/f(x,y).$$

By construction one has that h is in $H_\Gamma^{d,\chi}$.

Now suppose Γ is a triangle subgroup $\Gamma_{(p,q,r)}$ of $SU(2)$. Then its action on $\mathbb{C}P^1 \cong \mathbb{S}^2$ factors through an action of the triangle group $\Sigma_{(p,q,r)} \subset SO(3)$, because the element -1 carries each line into itself.

Now consider the triangle $T_{(p,q,r)} \subset \mathbb{S}^2$ whose edges give rise to the full triangle group $\Sigma_{(p,q,r)}^*$, and let $\widehat{T}$ be the triangle obtained by reflecting $T_{(p,q,r)}$ in one of its edges. The generators of $\Sigma_{(p,q,r)}$ are rotations around the vertices v_1, v_2, v_3 of $T_{(p,q,r)}$ by angles $2\pi/p, 2\pi/q, 2\pi/r$, and $\widehat{T}$ is a fundamental domain for the action of $\Sigma_{(p,q,r)}$, as we already know from Corollary 2.4 or from the discussion in §1. One has three special orbits, which correspond to the three vertices v_1, v_2, v_3 of $\widehat{T}$, and all other orbits have as many points as there are copies of the triangle $\widehat{T}$ in $\mathbb{S}^2$, i.e., the order k of $\Sigma_{(p,q,r)}$. Since the stabilizers of the vertices are cyclic groups of orders p, q, r, respectively, it follows that the orbits of the three vertices have $k/p, k/q, k/r$ points. These three special orbits determine as above three polynomials $h_1 \in H_\Gamma^{\frac{k}{p},\chi_1}$, $h_2 \in H_\Gamma^{\frac{k}{q},\chi_2}$, $h_3 \in H_\Gamma^{\frac{k}{r},\chi_3}$. [It is worth saying that if we think of the triangle groups as being groups of symmetries of polyhedra, then the three special orbits correspond to: (i) the vertices of the polyhedron; (ii) the middle points of the edges; and (iii) the middle points of the faces.]

The claim now is that these polynomials h_1, h_2, h_3 generate the bigraded algebra $H_\Gamma^{*,*}$, so they generate H_Π^* and they satisfy the polynomial relation of Theorem 4.3 above. (This is also Lemma 4.3 in [171].) Since these polynomials are Π-invariant, they define a map

$$\phi : \Pi \backslash \mathbb{C}^2 \longrightarrow \mathbb{C}^3 ,$$

whose image is contained in the Brieskorn variety $V_{(p,q,r)}$.

We claim that this map is injective. For this we consider points (x_1, y_1) and (x_2, y_2) in $\mathbb{C}^2$ which are not in the same Π-orbit. Denote by $\{\pi_1, \dots \pi_m\}$ the elements of Π, and choose some polynomial $g(x, y)$ (maybe not homogeneous) which vanishes on (x_1, y_1) but does not vanish at any point in the Π-orbit of (x_2, y_2). Now define a Π-invariant polynomial h by:

$$h(x, y) = g(\pi_1(x, y)) \, g(\pi_2(x, y)) \, \cdots g(\pi_m(x, y)) .$$

By construction one has $h(x_1, y_1) \neq h(x_2, y_2)$. Now we express h as the sum of homogeneous polynomials of various degrees. Since h_1, h_2 and h_3 generate the bigraded algebra $H_\Gamma^{*,*}$, we must have that one of these satisfies that $h_i(x_1, y_1) \neq h_i(x_2, y_2)$. Hence the map ϕ is injective. The rest now follows from general arguments, using the local conical structure of $V_{(p,q,r)}$ and the fact that away from the isolated singular point in each surface, both $V_\Gamma = \Pi \backslash \mathbb{C}^2$ and $V_{(p,q,r)}$ are complex manifolds of the same dimension (c.f. [171, §4] for details).

II.5 The group $PSL(2, \mathbb{R})$ and its universal cover $\widetilde{SL}(2, \mathbb{R})$

Let us look now at the group $PSL(2, \mathbb{R})$ of orientation preserving isometries of the hyperbolic plane $\mathbb{H}^2$, introduced in §2 above. We recall that it acts on $\mathbb{H}^2$ with isotropy $SO(2)$, hence it is diffeomorphic to the open solid torus $\mathbb{S}^1 \times \mathbb{H}^2$. Given points $z_1, z_2 \in \mathbb{H}$ and unit tangent vectors v_1, v_2 based at z_1, z_2, there is one and only one element in $PSL(2, \mathbb{R})$ taking z_1 to z_2 and whose derivative carries v_1 into v_2.

For the rest of this chapter it is actually better to work with the Lobachevsky model for hyperbolic geometry, so we think of $\mathbb{H}^2$ as being the upper half-plane $\text{Im } z > 0$ of the complex plane $\mathbb{C}$. For simplicity we set $\mathbb{H} = \mathbb{H}^2$. We identify $PSL(2, \mathbb{R})$ with the group of all Möbius transformations $z \mapsto \frac{az+b}{cz+d}$, where $\begin{pmatrix} a & b \\ c & d \end{pmatrix}$ is an element in $SL(2, \mathbb{R})$ and two such matrices define the same Möbius transformations iff they differ by sign.

Notice that the action of $PSL(2, \mathbb{R})$ on the tangent bundle $T\mathbb{H} \cong \mathbb{H} \times \mathbb{C}$ is via the derivative:

$$g \cdot (z, w) = \left(g(z), \frac{dg}{dz}(z) \cdot w \right),$$

and this action is free away from the zero-section, because if an element $g \in PSL(2, \mathbb{R})$ fixes a point $z \in \mathbb{H}$, then it is locally a rotation (more precisely it is an "elliptic element" in $PSL(2, \mathbb{R})$, using the standard terminology). It follows that given any point $(z, v) \in T\mathbb{H}$ with $|v| \neq 0$, the orbit of this point is a copy of $PSL(2, \mathbb{R})$. In particular the orbit of the point $(i, 1)$ embeds $PSL(2, \mathbb{R})$ in $T\mathbb{H}$ as the unit tangent bundle of the hyperbolic plane.

Since $PSL(2, \mathbb{R})$ is topologically a solid torus $\mathbb{S}^1 \times \mathbb{H}^2$, its fundamental group is isomorphic to $\mathbb{Z}$ and its universal cover $\widetilde{SL}(2, \mathbb{R})$ is diffeomorphic to $\mathbb{R} \times \mathbb{H}^2 \cong \mathbb{R}^3$, but its Lie group structure is quite interesting. This group $\widetilde{SL}(2, \mathbb{R})$ is a central extension of $PSL(2, \mathbb{R})$ and one has a short exact sequence:

$$0 \longrightarrow \mathbb{Z} \longrightarrow \widetilde{SL}(2, \mathbb{R}) \longrightarrow PSL(2, \mathbb{R}) \longrightarrow 0,$$

where the kernel $\mathbb{Z}$ of the projection $\widetilde{SL}(2, \mathbb{R}) \longrightarrow PSL(2, \mathbb{R})$ is the centre of $\widetilde{SL}(2, \mathbb{R})$.

We need in the sequel a description of this group due to Milnor [171, §5], which is not standard. For this we need to introduce the *differential forms of fractional degree*, first considered by H. Petersson (1930).

It is classical in the theory of Riemann surfaces to consider Abelian differentials, which are expressions of the form $f(z)\, dz$ with f holomorphic, as well as quadratic differentials, i.e., expressions of the form $f(z)\, dz^2$. More generally, for any integer $k \geq 0$, a differential form of degree k on $\mathbb{H}$ (or on an open set of $\mathbb{C}$) can be defined as a complex-valued holomorphic function of two variables, of the form

$$\phi(z, w) = f(z)\, w^k,$$

where z varies over $\mathbb{H}$ and $w\ (= dz)$ varies over $\mathbb{C}$. These can be regarded as functions defined on the holomorphic tangent bundle $T\mathbb{H} \cong \mathbb{H} \times \mathbb{C}$. Given such a differential form $\phi(z, w)$ on $\mathbb{H}$ and an element $g \in PSL(2, \mathbb{R})$, *the pull-back $g^*\phi$ is the form*

$$g^*\phi(z, w) = \phi\left(g(z), \frac{dg}{dz}(z) \cdot w \right) = f(g(z)) \left(\frac{dg}{dz} \right)^k w^k.$$

This allows us to speak of differential forms (of any integral degree) *invariant* under the action of a certain subgroup of $PSL(2, \mathbb{R})$, these are called *automorphic forms* and we will come to them in §6 below.

We need to generalize these concepts (following [171]), replacing the integer k by an arbitrary rational number. For this we make the convention that the variable w is to vary over the universal covering group $\widetilde{\mathbb{C}}^*$ of the non-zero complex numbers, which is of course equivalent to the additive group of $\mathbb{C}$; the isomorphism

$$\widetilde{e} : \mathbb{C} \longrightarrow \widetilde{\mathbb{C}}^*$$

of complex Lie groups is obtained by lifting the exponential map $e(z) = e^{2\pi i z}$ from $\mathbb{C}$ to $\mathbb{C}^*$. The kernel of the projection is isomorphic to $\mathbb{Z}$ and it is generated by the image of $\widetilde{e}(1)$.

5.1 Definition. (Milnor) A *differential form of (fractional) degree* $\alpha \in \mathbb{Q}$ on the upper half-plane $\mathbb{H}$ is a holomorphic function on $\mathbb{H} \times \widetilde{\mathbb{C}}^*$ of the form:

$$\phi(z, w) \ = \ f(z)\, w^\alpha\,,$$

where f is a holomorphic function on $\mathbb{H}$ and w varies over the universal covering group of $\mathbb{C}^*$, the non-zero complex numbers. Here it is understood that w^α is to be evaluated in $\widetilde{\mathbb{C}}^*$ and then projected to $\mathbb{C}^*$ to be multiplied by $f(z)$.

5.2 Definition. (Milnor) A *labeled holomorphic map* on $\mathbb{H}$ means a holomorphic function $g : \mathbb{H} \to \mathbb{H}$ with nowhere-vanishing derivative, together with a lifting of the derivative from $\mathbb{C}^*$ to $\widetilde{\mathbb{C}}^*$ (a "labeling").

As noted by Milnor ([171, p. 201]) one has:

5.3 Proposition. *The set of all labeled biholomorphic maps from* $\mathbb{H}$ *to itself forms a group, which is isomorphic to* $\widetilde{SL}(2,\mathbb{R})$, *the universal covering group of* $PSL(2,\mathbb{R})$.

This group acts on $\mathbb{H} \times \widetilde{\mathbb{C}}^*$ by:

$$g \cdot (s, w) \ = \ \left(g(z) \,, \frac{\widetilde{dg}}{dz}(z) \cdot w \right),$$

where $\widetilde{SL}(2,\mathbb{R})$ is acting on $\mathbb{H}$ by projecting it to $SL(2,\mathbb{R})$. This action is free and therefore each orbit is a copy of $\widetilde{SL}(2,\mathbb{R})$ embedded in the complex manifold $\mathbb{H} \times \widetilde{\mathbb{C}}^*$.

II.6 Milnor's theorem for the 3-dimensional Brieskorn manifolds. The hyperbolic case

In this section we state the following theorem of Milnor [171] and give an outline of its proof, which is formally similar to the proof of Theorem 4.3. We refer to Milnor's article for the complete proof.

6.1 Theorem. *Let* $\Gamma = \Gamma_{(p,q,r)}$ *be a triangle subgroup of* $PSL(2,\mathbb{R})$, *let* $\widetilde{\Gamma}$ *be its lifting to the universal covering group* $\widetilde{SL}(2,\mathbb{R})$, *and let* $\widetilde{\Pi} \subset \widetilde{SL}(2,\mathbb{R})$ *be the commutator subgroup of* $\widetilde{\Gamma}$. *Then the orbit space* $(\widetilde{\Pi}\backslash\widetilde{SL}(2,\mathbb{R}))$ *is diffeomorphic to the 3-dimensional Brieskorn manifold* $M_{(p,q,r)}$.

Before coming to the proof of this theorem, let us remark that when we lift $\Gamma_{(p,q,r)} \subset PSL(2,\mathbb{R})$ to a subgroup $\widetilde{\Gamma}$ of $\widetilde{SL}(2,\mathbb{R})$, the lifting necessarily contains the centre $C \cong \mathbb{Z}$ of $\widetilde{SL}(2,\mathbb{R})$, which is the kernel of the projection into $PSL(2,\mathbb{R})$. Hence the quotient spaces $(\Gamma \backslash PSL(2,\mathbb{R}))$ and $(\widetilde{\Gamma} \backslash \widetilde{SL}(2,\mathbb{R}))$ are obviously diffeomorphic 3-manifolds, and they are closed (i.e., compact and with empty boundary) because $(\Gamma \backslash PSL(2,\mathbb{R}))$ is a Seifert manifold (see III.1 below) that fibres with three exceptional fibres over the orbifold $\Gamma \backslash \mathbb{H}$, which is the sphere $\mathbb{S}^2$ with three marked points. However, in Theorem 6.1 we are considering the commutator subgroup $\widetilde{\Pi}$, and there is no reason why this group should contain the centre C, and in fact it does not in general. Hence the Brieskorn manifold $M_{(p,q,r)}$ will not be in general of the form $(G \backslash PSL(2,\mathbb{R}))$ for a subgroup of $PSL(2,\mathbb{R})$ (c.f. Chapter III). An exception is when the triple (p,q,r) is $(2,3,7)$. The corresponding triangle group is perfect and its lifting $\widetilde{\Gamma}_{(2,3,7)} \subset \widetilde{SL}(2,\mathbb{R}))$ is also perfect. [In fact a simple computation shows that this is the only triangle subgroup of $PSL(2,\mathbb{R})$ which is perfect and its lifting to the universal cover is also perfect; it would be interesting to understand why this happens.]

Milnor's proof of Theorem 6.1 is along the same lines as his proof of Theorem 4.3, with automorphic forms replacing the homogeneous polynomials. We recall (see for instance [30]) that the classical automorphic functions were introduced by Poincaré in the 1880s as holomorphic (or meromorphic) functions on $\mathbb{H}$ satisfying certain conditions with respect to a discrete subgroup of Möbius transformations. For Theorem 6.1 we need to consider what Milnor calls *automorphic forms* of fractional degree. For this we consider differential forms of fractional degree as in the previous section. These are expressions of the form $\phi(z,w) = f(z)w^\alpha$, where $f(z)$ is a holomorphic function on $\mathbb{H}$, and w varies over the universal covering group of $\mathbb{C}^*$, the non-zero complex numbers (we recall that w^α is to be evaluated in $\widetilde{\mathbb{C}}^*$ and then projected to $\mathbb{C}^*$ to be multiplied by $f(z)$).

Given a differential $\phi(z,w) = f(z)w^\alpha$ on $\mathbb{H}$ and an element $g \in \widetilde{SL}(2,\mathbb{R})$, its pull-back $g^*\phi$ is:

$$g^*\phi(z,w) = \phi\left(g(z), \frac{\widetilde{dg}}{dz}(z) \cdot w\right).$$

6.2 Definition. Let Γ be a discrete subgroup of $\widetilde{SL}(2,\mathbb{R})$ and let $\chi : \Gamma \to U(1) \subset \mathbb{C}^*$ be a character. A differential $\phi(z,w) = f(z)w^\alpha$ on $\mathbb{H}$ is χ-*automorphic* if

$$g^*(\phi) = \chi(g) \cdot \phi$$

for every $g \in \Gamma$. If $\chi \equiv 1$ is the constant character, so that $g^*(\phi) = \phi$ for all $g \in \Gamma$, then $\phi(z,w)$ is said to be Γ-*automorphic*.

Notice that such a form vanishes at a point x (to order s) if g vanishes at x (to order s), and in this case g vanishes in the whole orbit of x, since the form is automorphic.

Let us denote by $A_{\tilde{\Gamma}}^{\alpha,\chi}$ the complex vector space of χ-*automorphic* forms of degree α. We denote simply by $A_{\tilde{\Gamma}}^{\alpha}$ those corresponding to the constant character 1, i.e., the $\tilde{\Gamma}$-*automorphic* forms. We have, just as for the homogeneous polynomials of Section 4, that each of these is a vector space. Taking all of them, of all degrees and for all characters, we get the bigraded algebras

$$A_{\tilde{\Gamma}}^{*} \quad \text{and} \quad A_{\tilde{\Gamma}}^{*,*}$$

of automorphic and χ-automorphic forms, respectively.

The first step for proving Theorem 6.1 is to show that given the extended triangle group $\tilde{\Gamma} = \tilde{\Gamma}_{(p,q,r)}$ and its commutator $\tilde{\Pi}$, one has that each vector space $A_{\tilde{\Pi}}^{\alpha}$ splits as direct sum of its subspaces $A_{\tilde{\Gamma}}^{\alpha,\chi}$ as χ varies over the characters of $\tilde{\Gamma}$. Thus one has that the graded algebra $A_{\tilde{\Pi}}^{*}$ of $\tilde{\Pi}$-automorphic forms is the direct sum of the bigraded algebras $A_{\tilde{\Gamma}}^{*,*}$. This is easy and similar to the previous assertion for polynomials and subgroups of $SU(2)$.

The aim is to show that the algebra $A_{\tilde{\Pi}}^{*}$ is generated by three automorphic forms which satisfy the relation given by the corresponding Brieskorn polynomial. For this Milnor shows (Lemma 3.1) that the group $\tilde{\Gamma}_{(p,q,r)}$ has a presentation with generators $\gamma_1, \gamma_2, \gamma_3$, which represent rotations through the three vertices of the triangle $T_{(p,q,r)}$ and satisfy the relations:

$$\gamma_1^p = \gamma_2^q = \gamma_3^r = \gamma_1 \gamma_2 \gamma_3.$$

Letting k be the rational number defined by,

$$k = \frac{1}{1 - p^{-1} - q^{-1} - r^{-1}},$$

so that π/k is the area of the triangle $T_{(p,q,r)}$, Milnor constructs a character χ_0 of $\tilde{\Gamma}$ defining it on the generators by:

$$\chi_0(\gamma_1) = e^{2\pi i k/p} \quad , \quad \chi_0(\gamma_2) = e^{2\pi i k/q} \quad , \quad \chi_0(\gamma_3) = e^{2\pi i k/r},$$

and shows that if an automorphic form $\phi \in A_{\tilde{\Gamma}}^{\alpha,\chi}$ does not vanish at any vertex of $T_{(p,q,r)}$, then its degree α is a multiple of k and the character χ is $\chi_0^{\alpha/k}$. [Recall that in the spherical case k is the order of the triangle group in question, hence the order of the generic orbits.]

To construct the generators of the algebra of Π-invariant automorphic forms, one may recall that in the spherical case this was done by looking at the orbit of each vertex of T; such orbit defines a set of lines through the origin in $\mathbb{C}^2$ which are Γ-invariant and give rise to a homogeneous polynomial h_i (well defined up to a constant) and a character χ_i, defined ad hoc so that h is χ_i-invariant. In the hyperbolic case the key is (Lemma 6.3 in [171]) proving that the space $A_{\tilde{\Gamma}}^{k,\chi_0}$ contains one and (up to a constant multiple) only one automorphic form which vanishes at any given point of $\mathbb{H}$ (and therefore at the orbit of the point, since it is automorphic). If the chosen point is a vertex of T, say v_1, then the form one gets must have a p-fold holomorphic root ϕ_1, and Milnor shows that this must be also an automorphic form for some character χ_1 satisfying $\chi_1^p = \chi_0$ (Lemma 5.3 in [171]). Hence, just as for polynomials, the orbit in $\mathbb{H}$ of each vertex defines an automorphic form of Γ, for some character.

In this way Milnor arrives at his main Lemma (6.4), that given the extended triangle group $\widetilde{\Gamma} = \widetilde{\Gamma}_{(p,q,r)}$ and its commutator $\widetilde{\Pi}$, the bigraded algebra $A_{\widetilde{\Gamma}}^{*,*}$ is isomorphic to the graded algebra $A_{\widetilde{\Pi}}^*$ and is generated by three forms

$$\phi_1 \in A_{\widetilde{\Gamma}}^{k/p,\chi_1} \quad , \quad \phi_2 \in A_{\widetilde{\Gamma}}^{k/q,\chi_2} \quad , \quad \phi_3 \in A_{\widetilde{\Gamma}}^{k/r,\chi_3} \,,$$

where the characters satisfy the relation

$$\chi_1^p = \chi_2^q = \chi_3^r = \chi_0 \,.$$

Furthermore, the automorphic form ϕ_i has a simple zero at each point of the $\widetilde{\Gamma}$-orbit of the vertex v_i and no other zeroes, and they satisfy the polynomial relation:

$$\phi_1^p + \phi_2^q + \phi_3^r = 0.$$

The three forms ϕ_1, ϕ_2, ϕ_3 are complex-valued holomorphic functions on $\mathbb{H} \times \widetilde{\mathbb{C}}^*$ and they are never simultaneously zero, so they define a holomorphic map:

$$\Phi = (\phi_1, \phi_2, \phi_3) : h \times \widetilde{\mathbb{C}}^* \longrightarrow \mathbb{C}^3$$

whose image is contained in the complex manifold $V_{(p,q,r)} - 0$, obtained by removing the singular point from the corresponding Brieskorn variety. As we know from Section 5, the group $\widetilde{SL}(2,\mathbb{R})$ acts freely on $\mathbb{H} \times \widetilde{\mathbb{C}}^*$ by holomorphic transformations. Hence $\mathbb{H} \times \widetilde{\mathbb{C}}^*$ is foliated by the orbits of $\widetilde{SL}(2,\mathbb{R})$, which are all copies of this group. Thus the quotient of $\mathbb{H} \times \widetilde{\mathbb{C}}^*$ by $\widetilde{\Pi}$ is a (non-compact) complex manifold of dimension 2, which is foliated by manifolds of the form $(\widetilde{\Pi} \backslash \widetilde{SL}(2,\mathbb{R}))$. One can show that Φ is actually a biholomorphism and induces a diffeomorphism between one of the orbits $(\widetilde{\Pi} \backslash \widetilde{SL}(2,\mathbb{R}))$ (anyone you prefer) and the Brieskorn manifold $M_{(p,q,r)}$. $\qquad\qquad\qquad\qquad\qquad\qquad\qquad\qquad\qquad\qquad\qquad\qquad\qquad\quad$ $\square$

II.7   Brieskorn-Hamm complete intersections. The theorem of Neumann

Consider now a Pham-Brieskorn polynomial in n complex variables,

$$f(z) = z_1^{a_1} + \cdots + z_n^{a_n} \ , a_i \geq 2 \ , n \geq 3 \,.$$

Given any n-tuple $\alpha_1 = \alpha_{11}, \ldots, \alpha_{1n}$ of non-zero complex numbers, the variety:

$$V_{\alpha_1} = \{ \alpha_{11} z_1^{a_1} + \cdots + \alpha_{1n} z_n^{a_n} = 0 \}$$

has an isolated singularity at 0 and it is equivalent to a Brieskorn variety. However the way this is placed in $\mathbb{C}^n$ depends on the coefficients α_{ij}. Suppose you have another vector of coefficients $\alpha_2 = \alpha_{21}, \ldots, \alpha_{2n}$, then one has another equivalent Brieskorn variety V_{α_2}. It is an exercise to show that if all the (2×2)-minors of the $(2 \times 2n)$ matrix whose rows are α_1, α_2 have non-zero determinant, then these two varieties intersect transversally away from 0 and define a complete intersection singularity, of codimension 2, with an isolated singularity at 0. More generally, H. Hamm [96] showed that given any $(n-2) \times n$ matrix $A = ((\alpha_{i,j}))$ the corresponding $n-2$ varieties define a 2-dimensional complete intersection V_A with an isolated singularity at $0 \in \mathbb{C}^n$ if and only if every $(n-2) \times (n-2)$ subdeterminant is non-zero. In this case the link,

$$M = V_A \cap \mathbb{S}^{2n-1},$$

is a smooth 3-manifold whose diffeomorphism type depends only on the n-tuple $a_1, \ldots, a_n$ [96]. Following Hamm and [183], we call V_A a *Brieskorn-Hamm complete intersection*. If $n = 3$ this is a Brieskorn variety as before.

Now consider a convex n-sided polygon $\Delta \subset \mathbb{H}$, bounded by geodesics, whose interior angles are $\pi/a_1, \ldots, \pi/a_n$, let Σ^* be the group generated by the reflections (inversions) $\sigma_1, \ldots, \sigma_n$ in the edges of Δ, and let $\Sigma \subset PSL(2, \mathbb{R})$ be the group of orientation preserving isometries of $\mathbb{H}$ generated by reflections on the edges of Δ. These groups are called *polygonal*, for obvious reasons. They were considered by W. Dyck (1882) and one can show, just as for the triangle groups, that Σ^* has a presentation with generators $\sigma_1, \ldots, \sigma_n$ and relations $\sigma_i^2 = (\sigma_i \sigma_{i+1})^{a_i} = 1$ for all i modulo n. The group Σ is generated by rotations around the vertices by twice the corresponding angles, and it has a presentation:

$$\langle \tau_1, \ldots, \tau_n \mid \tau_1^{a_1} = \cdots = \tau_n^{a_n} = \tau_1 \cdots \tau_n = 1 \rangle.$$

Following [183] we lift Σ to the universal covering $\widetilde{SL}(2, \mathbb{R})$. Its lifting $\Gamma = \Gamma_{(a_1, \ldots, a_n)}$ has a presentation

$$\langle \gamma_1, \ldots, \gamma_n, c \mid \gamma_1^{a_1} = \cdots = \gamma_n^{a_n} = c, \gamma_1 \cdots \gamma_n = c^{n-2} \rangle.$$

(Notice that if $n = 2$ nothing changes if we drop the generator c.) Now let $\Pi \subset \widetilde{SL}(2, \mathbb{R})$ be the commutator subgroup of Γ. Just as for the triangle groups, one has that both Γ and Π act (freely) on $\mathbb{H} \times \widetilde{\mathbb{C}}^*$ and one has the corresponding algebras of automorphic and χ-automorphic forms of fractional degree, for the various characters of Γ. Again, since $\Pi \backslash \Gamma$ is finite and Abelian, one has an isomorphism of algebras $A_\Pi^* \cong A_\Gamma^{*,*}$ and W. Neumann proved in [183] that these algebras are generated by n forms $\phi_j \in A_\Gamma^{k/a_j, \chi_j}$, where k is given by the area of Δ: $k^{-1} = (n-2) - \sum a_i^{-1}$ and the characters $\chi_1, \ldots, \chi_n$ are defined on the generators of Γ by:

$$\chi_j(\gamma_i) = e^{2\pi i k / a_i a_j} \qquad \text{for } j \neq i;$$

and

$$\chi_j(\gamma_i) = e^{2\pi i k / (a_i^2 + 1)/a_j} \quad \text{for } j = i.$$

Each form ϕ_j has a simple zero at the vertex v_j of Δ and is uniquely determined by its zeroes up to a multiple. These n forms satisfy polynomial relations:

$$\alpha_{i1} \, \phi_1^{a_1} + \cdots + \alpha_{in} \, \phi_n^{a_n} = 0 \, ; \quad i = 1, \ldots, n-2 \, ,$$

such that the matrix $A = ((\alpha_{ij}))$ has all $(n-2) \times (n-2)$ minors with non-zero determinant. In this way one has the Brieskorn-Hamm complete intersection V_A determined by the matrix A and the angles of the polyhedron Δ. We arrive at the following beautiful generalization of Milnor's theorem.

7.1 Theorem (Neumann). *These n forms define a map*

$$\Phi = (\phi_1, \ldots, \phi_n) : \mathbb{H} \times \widetilde{\mathbb{C}^*} \longrightarrow \mathbb{C}^n$$

which is Π-invariant and defines a biholomorphism between the space of orbits $\Pi \backslash (\mathbb{H} \times \widetilde{\mathbb{C}^*})$ and the non-singular part $V_A^* = V_A - 0$ of the Brieskorn-Hamm complete intersection

$$V_A = \bigcap_{i=1}^{n-2} \{ \, \alpha_{i1} \, z_1^{a_1} + \cdots + \alpha_{in} \, z_n^{a_n} = 0 \, \} \, .$$

This map induces a diffeomorphism between $\Pi \backslash \widetilde{SL}(2, \mathbb{R})$ and the link of 0 in V_A.

In Chapter III we speak more about this theorem and improvements of it made by Neumann himself and by I. Dolgachev.

II.8   Remarks

To complete the story about the relation between groups of isometries in plane geometry and Brieskorn singularities, it remains to consider:

(i) What about the Brieskorn singularities $\{ z_1^p + z_2^q + z_3^r = 0 \}$ when $\frac{1}{p} + \frac{1}{q} + \frac{1}{r} = 1$?

(ii) What about the subgroups of $PSL(2, \mathbb{R})$, other than the triangle groups? And what about the quotients of $PSL(2, \mathbb{R})$ by the triangle groups themselves? (not going to the universal cover and then taking the commutator subgroup there).

(iii) What about the groups of isometries of the Euclidean plane?

These three questions are answered in Chapter III. The answer to (i) was given by Milnor himself in [171], showing that these singularities correspond to quotients related to the nilpotent group N of real 3×3 upper triangular matrices. The Brieskorn singularities correspond to three of the discrete subgroups of N; this study was completed in [59] and [183].

A subgroup Γ of $PSL(2, \mathbb{R})$ which is discrete is called a *Fuchsian* group, and it is said to be cocompact if the orbit space $\Gamma \backslash PSL(2, \mathbb{R})$ is compact. These are the groups relevant for question (ii), since the link of an isolated complex surface singularity is a compact 3-manifold. Dolgachev gave in [57, 58] a beautiful construction of surface singularities using cocompact Fuchsian groups, which is very closely related to Milnor's work for the Brieskorn manifolds. This provides, for every cocompact Fuchsian group, an isolated surface singularity whose link is $\Gamma \backslash PSL(2, \mathbb{R})$. This was extended in [59] and [183] to all discrete subgroups of $\widetilde{SL}(2, \mathbb{R})$.

Question (iii) was considered in [231] and we discuss it in Chapter III below.

Chapter III

3-dimensional Lie Groups and Surface Singularities

The theorems of Klein, Milnor and Neumann discussed in Chapter II, together with very deep and important results by Hirzebruch, Dolgachev, Neumann himself and others that I have not mentioned yet, give us a very good understanding of the relation between 3-dimensional Lie groups and complex surface singularities; this is the topic that we explore in this chapter. Our basic references here are [207, 59, 194, 204, 104, 186, 231]. It turns out that up to isomorphism, there are six different 3-dimensional Lie groups that admit discrete subgroups with compact quotient (we call these *uniform subgroups*, a term frequently used in the literature). In five of these cases the orbit spaces one gets are Seifert manifolds, and these are strongly related with the so-called quasi-homogeneous surface singularities. The remaining case gives rise to the cusp-singularities, whose links are torus bundles over the circle. In this chapter we explain these relationships. We begin by describing a few well-known facts about quasi-homogeneous singularities and Seifert manifolds, then we describe the 3-dimensional Lie groups with uniform subgroups and their orbit spaces. Their relation with singularities is discussed case by case, including two cases which apparently do not come into the scene and are usually left aside when talking about complex singularities (these are the solvable group $\widetilde{E}^+(2)$ and the Abelian group $\mathbb{R}^3$). We show that there are several interesting properties in common for all six cases. We also discuss a relationship between the property of being Gorenstein for the complex surfaces in question and the Lie algebras of the corresponding groups.

III.1 Quasi-Homogeneous surface singularities

The quasi-homogeneous singularities play a very distinguished role in singularity theory. Their geometry and structure has been widely studied by many authors, most notably by H. Pinkham, P. Orlik and Ph. Wagreich, I. Dolgachev and W. Neumann. We refer to [256] for a beautiful introduction to the subject.

We recall that a polynomial h in n complex variables is *weighted homogeneous* if there exist non-zero integers $(q_1, \ldots, q_n)$ and a positive integer d such that:

$$h(t^{q_1} z_1, \ldots, t^{q_n} z_n) = t^d h(z_1, \ldots, z_n).$$

Equivalently, we demand that there exist non-zero rational numbers $(w_1, \ldots, w_n)$, called *the weights* of h, for which h is a sum of monomials $z_1^{\alpha_1} \cdots z_n^{\alpha_n}$ such that

$$\frac{\alpha_1}{w_1} + \cdots + \frac{\alpha_n}{w_n} = 1.$$

Given these weights we recover d as being the smallest integer such that for each $i = 1, \ldots, n$ there exists an integer q_i so that $q_i w_i = d$. These are the q_i. For example, if h is a Pham-Brieskorn polynomial,

$$h(z_1, \ldots, z_n) = z_1^{a_1} + \cdots + z_n^{a_n},$$

then the weights are obviously $(a_1, \ldots, a_n)$. For the polynomial

$$f(z_1, z_2) = z_1^{a_1} + z_1 z_2^{a_2}, \quad a_1 > 1, a_2 \geq 1,$$

the weights are: $w_1 = a_1$ and $w_2 = a_1 a_2 / (a_1 - 1)$. In this case one has $q_1 = a_2$, $q_2 = a_1 - 1$ and $d = a_1 a_2$. One has in this case:

$$f(t^{a_2} z_1, t^{a_1 - 1} z_2) = t^{a_1 a_2} f(z_1, z_2),$$

as it should be.

Now let $V \subset \mathbb{C}^n$ be an algebraic variety defined by homogeneous polynomials $h_1, \ldots, h_n$ with same weights. Let $q_1, \ldots, q_n$ be the corresponding exponents, defined as above. One can define an action of $\mathbb{C}^*$ on $\mathbb{C}^n$,

$$\sigma : \mathbb{C}^* \times \mathbb{C}^n \longrightarrow \mathbb{C}^n$$

by $\sigma(t, (z_1, \ldots, z_n)) = (t^{q_1} z_1, \ldots, t^{q_n} z_n)$. Notice this is exactly the action considered in I.1 above for the Brieskorn varieties. Just as in that case, if the exponents q_i are all > 0, then the restriction of this action to the positive real numbers $\mathbb{R}^+$ gives a real analytic flow, whose orbits are transversal to all the spheres in $\mathbb{C}^n$ centred at 0, and they converge to the origin. Hence every orbit contains 0 in its closure. Conversely, if a complex analytic variety $V \subset \mathbb{C}^n$ admits such a $\mathbb{C}^*$-action, then V is globally defined and it is (globally) a cone over its intersection with the unit sphere. Furthermore, by [194, 1.1.2] one also has that V is actually algebraic and the ideal of polynomials (in n variables) vanishing on V is generated by weighted homogeneous polynomials. In fact one has even more (see [194, 1.1.3]): if $V \subset \mathbb{C}^n$ is an algebraic variety and there is a $\mathbb{C}^*$-action on V, $\mathbb{C}^* \times V \to V$, given by a morphism of algebraic varieties, then:

(i) there is an embedding of V in some $\mathbb{C}^m$ and a $\mathbb{C}^*$-action on $\mathbb{C}^m$ for which V is an invariant set and the induced action on V coincides with the given one; and

(ii) one can choose the coordinates suitably, so that the action takes the form:

$$\sigma(t, (z_1, \ldots, z_m)) = (t^{q_1} z_1, \ldots, t^{q_m} z_m)$$

for some $q_i \in \mathbb{Z}$.

This means that the fact that an analytic germ $(V, 0)$ admits a $\mathbb{C}^*$-action is something intrinsic of the germ, independent of the equations used to define it; and V is automatically algebraic.

1.1 Definition. By a *quasi-homogeneous surface singularity* we mean a 2-dimensional complex analytic variety V in some affine space $\mathbb{C}^n$ with an isolated singularity at 0, together with a *good* $\mathbb{C}^*$-action, where good means that 0 is in the closure of every $\mathbb{C}^*$-orbit.

As before, by restricting the action to the positive real numbers $\mathbb{R}^+ \subset \mathbb{C}^*$ one gets a real analytic flow on V, with a fixed point at 0, whose orbits are rays emanating from the origin, transversal to all the spheres around 0, and the orbit space $(V - 0)/\mathbb{R}^+$ can be identified with the link M, which is automatically a Seifert manifold, with a Seifert decomposition given by the $\mathbb{S}^1$-orbits of the restriction to $\mathbb{S}^1$ of the $\mathbb{C}^*$-action.

We recall that a *Seifert manifold* can be defined to be a 3-manifold M, that we assume to be closed, connected and orientable, endowed with a foliation by oriented circles, such that every fibre has a neighborhood which is diffeomorphic to a torus $\mathbb{S}^1 \times \mathbb{D}^2$ and is a union of orbits. (See for instance [206, 193, 110] for details). We can actually assume that this foliation on M is given by an $\mathbb{S}^1$-action. One has that there are a finite number of special orbits $\mathcal{O}_1, \ldots, \mathcal{O}_r$ and all other orbits are principal. The principal orbits are those where the $\mathbb{S}^1$-action is free. Each such orbit has a neighborhood which is a solid torus $\mathbb{T}$, a union of principal orbits. The special (exceptional) orbits are where the action has nontrivial isotropy. It is easy to see that the isotropies are necessarily finite and cyclic, and therefore the quotient space of M by this action is a closed 2-dimensional manifold S; one has a projection $p : M \to S$, which is a fibre bundle away from the exceptional fibres. Thus we speak of the *Seifert fibres* of the corresponding Seifert fibration of M. To each exceptional Seifert fibre one attaches a pair (α, β) of normalized (or reduced) Seifert invariants $(0 < \beta < \alpha)$. These integers satisfy the equation $\alpha a + \beta b = 0$ in $H_1(N, \mathbb{Z})$, where N is a small tubular neighborhood of the corresponding exceptional fibre, saturated with Seifert fibres, b is an oriented Seifert fibre on ∂N and a is an oriented curve on ∂N such that the intersection $a \cdot b$ is $+1$ on ∂N, oriented as the boundary of N.

Seifert manifolds are characterized, up to equivariant diffeomorphism, by their *Seifert invariants* (see [193, 1.10]). These are the normalized orbit invariants (α_i, β_i), $i = 1, \ldots, r$, of the exceptional fibres and (in our case) two other invariants that we now describe. One is the genus of the 2-dimensional manifold S. It can happen that S is non-orientable, and this has to be taken into account, but that will not happen for the cases that concern us in this text. The other relevant invariant is usually denoted by e; when there are no special fibres, so that M is an $\mathbb{S}^1$-bundle over S, e is the usual Euler number (see [242]). In the general case one must take the special fibres into account. To define it we recall that if E is an oriented $\mathbb{S}^1$ bundle over an oriented 2-dimensional, compact, connected, manifold B, then its usual Euler class is the primary (*i.e., non-automatically zero*) obstruction for constructing a section of E. This class lives in $H^2(B; \mathbb{Z})$ and it becomes a number when we evaluate it on the orientation class of B. If B has non-empty boundary, then $H^2(B; \mathbb{Z}) \cong 0$, so the bundle is trivial. However, if we fix a choice of a trivialization of E over ∂B, *i.e.*, a section of $\tau : \partial B \to E$, then one has an *Euler class of E relative to τ*, $e(E; \tau) \in H^2(B, \partial B; \mathbb{Z}) \cong \mathbb{Z}$; evaluating $e(E; \tau)$ on the orientation cycle of the pair $(B, \partial B)$ we obtain an integer, which is by definition, the *Euler number* of E relative to τ. Now, given the oriented Seifert fibration $\pi : M \to S$, let us remove from S small, pairwise disjoint, open discs around the points corresponding to the special fibres, and denote by B what is left. Let E be $\pi^{-1}(B)$, which is M minus a union of open solid tori. This is an $\mathbb{S}^1$ bundle over B. On each boundary torus T_i one can can choose a unique (up to isotopy) oriented curve a which intersects each Seifert fibre in exactly one point and satisfies that $m = \alpha[a] + \beta[b]$, where m is a meridian of T_i, (α, β) are the corresponding reduced orbit invariants and $[b]$ is the homology class represented by one Seifert fibre. This curve a determines a section of $E|_{T_i}$. Doing this for each boundary torus we obtain a section of E over ∂B. The *Euler number $e = e(M)$* of the Seifert fibration $\pi : M \to S$ is defined to be the *Euler number* of E relative to the given trivialization over ∂B.

1.2 Definition. The normalized (or reduced) Seifert invariants of the Seifert manifold M are:

$$\{\, g;\, e;\, (\alpha_1, \beta_1), \ldots, (\alpha_r, \beta_r)\,\}.$$

Now, given the surface V with $\mathbb{C}^*$-action, let $V^* = V - \{0\}$. Consider the subgroup $\mathbb{R}^+ \subset \mathbb{C}^*$ of positive real numbers. If we divide V^* by the induced $\mathbb{R}^+$-action we obtain the link M, which is then naturally equipped with an $\mathbb{S}^1$-action. Thus M becomes a Seifert manifold.

The orbit space $S = V^*/\mathbb{C}^*$ is an oriented 2-dimensional closed manifold, canonically diffeomorphic to the orbit space $M/\mathbb{S}^1$. The projection $V^* \xrightarrow{\pi} S$ endows V^* with the structure of a *holomorphic Seifert bundle*. This means that V^* minus the exceptional orbits is indeed a holomorphic $\mathbb{C}^*$-bundle over S minus the corresponding points, and the structure of V^* around the exceptional orbits is described by the orbit invariants (α_i, β_i), which can be defined as follows.

Let $\mathcal{O}_i$ be the corresponding exceptional $\mathbb{S}^1$-orbit on M. The isotropy group of $\mathcal{O}_i$ is necessarily a finite cyclic subgroup σ_i of $\mathbb{S}^1$, and its order is the corresponding invariant α_i. Now, given a point x in $\mathcal{O}_i$, let H be a 2-plane in the tangent space $T_x M$ transversal to the line tangent to the orbit. Using a local chart, we identify a small disc $\hat{U} \subset H$ with a small disc U in M, which is transversal to $\mathcal{O}_i$. Now let W be the union of all the $\mathbb{S}^1$-orbits that meet U. This is a solid torus, and it is a neighborhood of $\mathcal{O}_i$ in M. By the *slice theorem* (see for instance [193, 256]), the $\mathbb{S}^1$-action gives a representation $\hat{\sigma}$ of the isotropy group σ_i in $GL(2, \mathbb{R})$, regarded as the automorphisms of H, such that $\hat{\sigma}(\gamma)(\hat{U}) = \hat{U}$ for all $\gamma \in \sigma$, and W is equivariantly diffeomorphic to $(\mathbb{S}^1 \times \hat{U})/\sigma_i$, where σ_i acts on $\mathbb{S}^1 \times \hat{U}$ by $\gamma \cdot (g, u) = (g\gamma^{-1}, \sigma_i(\gamma) u)$. We may identify the plane H with $\mathbb{C}$ and the group σ_i with the αth roots of unity, so that the slice representation is just complex multiplication by these roots of unity. Then, for each γ in σ_i and for each $x \in \hat{U}$ the slice representation takes the form $\hat{\sigma}(\gamma)(z) = \gamma^{\nu_i}(z)$, for some integer ν_i, $0 < \nu_i < \alpha_i$. The integers (α_i, ν_i) are the *orbit invariants* of $\mathcal{O}_i$ as used in [194, 256]. The corresponding invariant β_i is defined by

$$\nu_i \beta_i \equiv -1 \mod (\alpha_i);$$

so it is the inverse of $-\nu_i$ modulo α_i.

As an example (see [256]), consider the affine surface V defined by $z_1^{12} + z_2^8 - z_2 z_3^5 = 0$. Then V is invariant under the action $t \cdot (z_1, z_2, z_3) \mapsto (t^{10} z_1, t^{15} z_2, t^{21} z_3)$, so V is quasi-homogeneous and it has six exceptional orbits. One of these is

$$\mathcal{O}_1 = \{(z_1, z_2, z_3) \in V \cap \mathbb{S}^5 \mid z_1 = z_2^7 - z_3^5 = 0\},$$

whose isotropy group is $\mathbb{Z}_3$, so $\alpha_1 = 1$. To compute β_1 we notice that $(0, 1, 1)$ is a point in $\mathcal{O}_1$ and a plane transversal to this orbit is given by $z_2 = z_3 = 0$. Then the slice representation is $\gamma \cdot (z_1, 0, 0) = (\gamma^{10} z_1, 0, 0)$ for γ a cubic root of unity. Thus $\nu_1 \equiv 10 \equiv 1 \mod 3$. Therefore $\beta_1 = 2$.

The Euler number e also has a very nice interpretation, coming from [204, 187]. For this, as in Chapter II, we let P denote one of the standard simply connected complex manifolds of dimension 1: $\mathbb{CP}^1$, $\mathbb{C}$ or the upper half-plane $\mathcal{H}$, which serve as models for the classical plane geometries. Since the bundle V^* over S is holomorphic, it turns out that the surface S is automatically a Riemann surface with marked points corresponding to the exceptional orbits, where it has an orbifold structure. By the Riemann-Köbe uniformization theorem we know that choosing the appropriate "plane", one has that P is the universal cover of S (as an orbifold) and there exists $\Gamma \subset Aut(P)$, a discrete group of holomorphic automorphisms of P with quotient $S = P/\Gamma$. Then a well-known theorem of Selberg (see also [80]) says that there exists a normal subgroup $\hat{\Gamma}$ of Γ of finite index which acts freely on P. Thus one has a finite covering map $\hat{S} \overset{H}{\to} S$ with covering group $H = \Gamma/\hat{\Gamma}$. It was noted in [204] (see also [171, 6.6]) that there is a holomorphic $\mathbb{C}^*$-bundle L over $\hat{S}$ with an action of H (something like a "pull-back bundle") with $L/H \cong V^*$, and one has a commutative diagram:

$$\begin{array}{ccc} L & \overset{H}{\longrightarrow} & V^* \\ \pi \downarrow & & \pi \downarrow \\ \hat{S} & \overset{H}{\longrightarrow} & S. \end{array} \qquad (1.3)$$

We recall that the $\mathbb{C}^*$-bundle L over $\widehat{S}$ is classified topologically by its Chern class $c_1(L) \in H^2(\widehat{S}; \mathbb{Z})$, which evaluated on the fundamental cycle $[\widehat{S}]$ gives the Euler number of L defined as above (see also Chapter IV). It turns out that the number

$$e(V^* \to S) = (\text{Euler number of } L)/\text{order of } H \in \mathbb{Q} \qquad (1.4)$$

is independent of the choice of the normal subgroup $\widehat{\Gamma}$ and the bundle L. This number is called *the virtual degree*, or *rational Euler number* of the Seifert bundle $V^* \to S$, and it is related to the Euler number e in 1.10, which is an integer, by the formula:

$$e(V^* \to S) = -e - \sum_{i=1}^{m} \frac{\beta_i}{\alpha_i} \ .$$

The rational Euler number $e(V^* \to S)$ of the Seifert bundle generalizes the usual Euler number of $\mathbb{S}^1$-bundles; this was introduced in [187, 204] for Seifert manifolds M in general, and it was shown that the corresponding Seifert manifold occurs as the link of a surface singularity if and only if $e(M \to S) < 0$.

In [194], completing previous results by several authors, there is given an explicit method for computing the Seifert invariants of all quasi-homogeneous surface singularities from the the weights of the corresponding $\mathbb{C}^*$-action, and they use this information to determine the weighted graph of a canonical good resolution for these singularities (c.f. Chapter IV below).

1.5 Remark. It is well known that there is a very close relation between quasihomogeneous singularities, graded algebras of finite type and automorphy factors. The theorems of Klein, Milnor and Neumann in Chapter II are an example of this, and we refer to [256] for more on the subject. Given a quasihomogeneous surface V as above, an *automorphy factor* for V means a diagram similar to (1.3) but replacing $\widehat{S}$ by the universal cover P of S. More precisely, given the $\mathbb{C}^*$-bundle $V^* \xrightarrow{\pi} S$, then (by [204, 54, 59]) there exists a holomorphic line bundle L over P with an action of Γ that commutes with the $\mathbb{C}^*$-action, such that $L/\Gamma = V^*$ and one has a commutative diagram:

$$
\begin{array}{ccc}
L & \xrightarrow{\ \Gamma\ } & V^* \\
\pi \downarrow & & \downarrow \pi \\
P & \xrightarrow{\ \Gamma\ } & S.
\end{array}
\qquad (1.6)
$$

The triple (P, L, Γ) is called an *automorphy factor* for V. If P is $\mathbb{CP}^1 \cong \mathbb{S}^2$, and $\Gamma \subset SU(2)$, then the canonical automorphy factor is $(\mathbb{CP}^1, T^*(\mathbb{CP}^1), \Gamma)$, where $T^*(\mathbb{CP}^1)$ is the cotangent bundle. When P is the upper half-plane $\mathcal{H}$, and $\Gamma \subset PSL(2, \mathbb{R})$, the canonical automorphy factor is $(\mathcal{H}, \mathcal{H} \times \mathbb{C}, \Gamma)$, since the tangent bundle of $\mathcal{H}$ is trivial, $T(\mathcal{H}) \cong \mathcal{H} \times \mathbb{C}$; the action of Γ on $T(\mathcal{H})$ is via the derivative, as explained in II.5 above.

1.7 Remark. It is clear that if V is a quasi-homogeneous singular variety, then the $\mathbb{C}^*$-action on V defines a holomorphic vector field on V, whose orbits are everywhere transversal to the link. This vector field plays the role of a "radial" vector field. One can prove (see [29]) that every small perturbation of this vector field is still "equivalent" to it in some sense that can be made precise, and these types of vector fields form an open dense set in the space of all germs at 0 of holomorphic vector fields on V. This motivates two questions:

Question 1: If an isolated complex singularity $(V, 0)$ admits a holomorphic vector field which is singular only at 0 and the field of complex lines that it spans is everywhere transversal to the link, does it follow that V is quasi-homogeneous? If this is true, it is probably a consequence of the deep results of K. Saito in [216], where he gives several characterizations of quasi-homogeneous singularities.

Question 2: Given an isolated singularity germ $(V, 0)$, one can always make sense of the concept of "the local radial (or Schwartz) index" of germs of vector fields on V (see for instance [3, 35, 69]); the results of [29] imply that there is a smallest possible index attained by the holomorphic vector fields on V. If V is quasi-homogeneous, then this index is 1. What is this number for isolated singularity germs which are not quasi-homogeneous? An alternative way of re-phrasing this question is to characterize the generic germs of holomorphic vector fields on V with the "least complicated topology".

III.2   3-manifolds whose universal covering is a Lie group

In this section we describe the closed, oriented 3-dimensional manifolds which are diffeomorphic to homogeneous spaces of the form $\Gamma \backslash G$, where G is a 3-dimensional Lie group and Γ is a discrete subgroup of G with compact quotient; for short we call these *uniform subgroups*, as it is common in the literature. Our basic reference is the article of Raymond and Vasquez [207], which completes previous work by various people. In fact the problem of classifying the closed 3-manifolds of the form $\Gamma \backslash G$ was considered in [23, Ch. III], where the authors give the list of the 3-dimensional Lie groups that admit uniform subgroups; they also list the corresponding uniform subgroups, but their list is incomplete and erroneous in some cases. There is also a classification of the 3-dimensional Lie groups with uniform subgroups given in [172, §4]. It turns out that in all cases but one, the corresponding orbit spaces $\Gamma \backslash G$ admit circle actions and therefore are Seifert manifolds. Thus the classification in [207] is based on the classification given in [206] of circle actions on 3-manifolds. The remaining case is when G is the solvable group $E(1, 1)$ which had to be treated separately. Notice that it is enough to consider the simply-connected Lie groups, for given another one, we can always lift it to the universal covering and work there.

Up to isomorphism, there are six different 3-dimensional, simply-connected Lie groups that admit uniform subgroups. These are: (i) two semi-simple groups: $SU(2)$ (compact) and $\widetilde{SL}(2, \mathbb{R})$, both of them known to us from Chapter II; two solvable (not-nilpotent) groups: $\widetilde{E}^+(2)$, which is the universal covering of the group $E^+(2)$ of orientation preserving isometries of the euclidean plane, and the inhomogeneous Lorenz group $E(1,1)$; the Heisenberg group N of 3×3 upper triangular real matrices, which is nilpotent, and $\mathbb{R}^3$ regarded as an Abelian (additive) group. We recall briefly which are the uniform subgroups in each case, and which are the corresponding quotient manifolds. Except for the case $G = SU(2)$, in all other five cases the underlying manifold is $\mathbb{R}^3$, with different Lie group structures.

It is worth noting that these Lie groups give rise to six of the eight geometries which are relevant for 3-manifolds theory according to W. Thurston (see [186]). Let us describe the six types of manifolds of the form $\Gamma \backslash G$.

(i) If $G = SU(2)$, this was widely studied in the previous chapter. Its finite subgroups are either cyclic or the triangle groups $\Gamma_{(2,2,r)}$, $r \geq 2$, $\Gamma_{(2,3,3)}$, $\Gamma_{(2,3,4)}$, $\Gamma_{(2,3,5)}$. The quotient $\Gamma \backslash SU(2)$ is a Seifert manifold which fibres over $(\Gamma_{(p,q,r)} \backslash \mathbb{S}^2) \cong \mathbb{S}^2$ with either two or three exceptional fibres, according as the group Γ is cyclic or a triangular group. In the second case the special orbits are those of the vertices of the triangle and the corresponding Seifert invariants are:

$$\{0\,;-2\,;(p,p-1),\,(q,q-1),\,(r,r-1)\,\}.$$

When $\Gamma = \mathbb{Z}_n$ is cyclic, the manifold $\Gamma \backslash SU(2)$ is the lens space $L(n,1)$, which is also a Seifert manifold but it has many such representations, unlike all the other cases.

(ii) Let $G = \widetilde{SL}(2,\mathbb{R})$. We start with $PSL(2,\mathbb{R})$; its discrete subgroups are called Fuchsian groups, and they are said to be cocompact when the quotient $\Gamma \backslash G$ is compact, which is the relevant case for us. Examples of these are the triangle and the polygonal groups of Chapter II. The elements of $PSL(2,\mathbb{R})$ are of three types: hyperbolic, parabolic and elliptic; the elliptic elements are conjugate to rotations and are the only type of elements in $PSL(2,\mathbb{R})$ that have fixed points is $\mathbb{H}$: the hyperbolic elements have two fixed points on the real axis (which is not in $\mathbb{H}$), and the parabolic elements have one fixed point on the real axis. A Fuchsian group Γ can have at most finitely many orbits of fixed points in $\mathbb{H}$, and the isotropy of each orbit is cyclic of finite order. Let us denote by $\alpha_1, \ldots, \alpha_m$ the orders of the isotropy subgroup of the different orbits of fixed points. For instance, for the triangle groups these are precisely p, q, r. The group Γ acts on $\mathbb{H}$ and the quotient is a Riemann surface of genus $g \geq 0$, which has an orbifold structure with marked points $x_1, \ldots, x_m$, which correspond to the orbits with non-trivial isotropy. Thus one has a branched covering projection $\mathbb{H} \to \Gamma \backslash \mathbb{H}$, ramified at these m points, with branching indices the α_i. The set of integers $\{g; \alpha_1, \ldots, \alpha_m\}$ is called *the signature* of the Fuchsian group, and it characterizes the group up to conjugation by a quasi-conformal diffeomorphism (see for instance [28]).

For instance, if Γ is the fundamental group of a Riemann surface S of genus $g > 1$, then Γ is a subgroup of $PSL(2,\mathbb{R})$ that acts freely on $\mathbb{H}$ with quotient S; hence the signature is in this case $\{g; 0\}$. The projection $p : \mathbb{H} \to \Gamma \backslash \mathbb{H} = S$ is now a covering projection, without ramifications, and the orbit space $\Gamma\backslash G$ is the unit tangent bundle of S. For a triangle group the signature is $\{0; p, q, r\}$.

In general, there is a projection $\Gamma\backslash G \to \Gamma\backslash\mathbb{H}$ which is a Seifert fibration, with exceptional fibres over the m points $x_1, \ldots, x_m$ which have non-trivial isotropy. The orders α_i of these isotropy groups gives the orbit invariants α_i of Definition 1.2.

Now recall that $G = \widetilde{SL}(2, \mathbb{R})$ is a central extension of $PSL(2, \mathbb{R})$ and one has an exact sequence:

$$0 \to \mathbb{Z} \to \widetilde{SL}(2, \mathbb{R}) \xrightarrow{\pi_\infty} PSL(2, \mathbb{R}) \to 1 \,.$$

For every positive integer $r \geq 1$ one has an r-fold cyclic cover G_r of $PSL(2, \mathbb{R})$ and projections

$$G_\infty = \widetilde{SL}(2, \mathbb{R}) \longrightarrow G_r \xrightarrow{\pi_r} G_1 = PSL(2, \mathbb{R}) \longrightarrow 0 \,.$$

In fact one has central extensions,

$$0 \longrightarrow \mathbb{Z}/r \longrightarrow G_r \xrightarrow{\pi_r} G_1 \longrightarrow 1 \,,$$

$$0 \longrightarrow \mathbb{Z} \longrightarrow G_\infty \xrightarrow{p_r} G_r \longrightarrow 1 \,.$$

According to [207], each discrete cocompact subgroup $\Gamma_\infty \subset G_\infty$ has finite index, say $r = r(\Gamma)$, in $\pi_\infty^{-1}(\pi_\infty(\Gamma_\infty))$. For instance, if $\Gamma \subset PSL(2, \mathbb{R})$, then its lifting $\widetilde{\Gamma}$ to $\widetilde{SL}(2, \mathbb{R})$ obviously has index $r = 1$ in $\pi_\infty^{-1}(\pi_\infty(\Gamma_\infty))$; if we now start with a subgroup $\Gamma \subset SL(2, \mathbb{R})$, then its lifting $\widetilde{\Gamma}$ to $\widetilde{SL}(2, \mathbb{R})$ may have $r = 1$ or $r = 2$ according as Γ contains or not the kernel of the projection $SL(2, \mathbb{R}) \to PSL(2, \mathbb{R})$, respectively. One has that the projection Γ_r of Γ_∞ to the r-fold covering group G_r only meets the kernel of the projection $p_r : G_r \to G_1$ at the identity and one has

$$\Gamma_r\backslash G_r \cong \Gamma_\infty\backslash G_\infty \,.$$

Then, according to [207], the cocompact discrete subgroups of $\widetilde{SL}(2, \mathbb{R})$ and the diffeomorphism type of the quotient $M_\Gamma = \Gamma_\infty\backslash G_\infty$, are characterized by the corresponding integer r, together with the signature of its image $\Gamma_1 = \pi_\infty(\Gamma_\infty)$ in $PSL(2, \mathbb{R})$. As mentioned before, these quotients are all Seifert manifolds; their (normalized) Seifert invariants can be computed from the signature of Γ_1. These are (see [207]):

$$\{ g; e = 2g - 2; (\alpha_1, \alpha_1 - 1), \ldots, (\alpha_m, \alpha_m - 1) \} \,,$$

where g is the genus of $\Gamma_1\backslash\mathbb{H} = S$. The corresponding integer r was found in [231, 2.4] to be:

$$r = \frac{\chi(M_\Gamma \to S)}{e(M_\Gamma \to S)} \,,$$

where $e(M_\Gamma \to S) = \sum_{i=1}^m \frac{\beta_i}{\alpha_i}$ is defined in (1.4) above, and the *Euler characteristic* $\chi(M_\Gamma \to S)$ of the Seifert fibration is defined by:

$$\chi(M_\Gamma \to S) = \chi(S) - \sum_{i=1}^m \frac{\alpha_i - 1}{\alpha_i} \,.$$

The invariant $\chi(M_\Gamma \to S)$ appears in Pinkham's and Dolgachev's work. It is shown in [186, p. 250–251] that it distinguishes the possible geometries in the link $M = \Gamma \backslash G$ of a quasi-homogeneous singularity: the group G is $SU(2)$, $\widetilde{SL}(2,\mathbb{R})$ or the Heisenberg group N according as $\chi(M_\Gamma \to S)$ is > 0, < 0, or $= 0$, respectively (see the discussion below).

(iii) $G = N$, the Heisenberg group of real matrices of the form

$$\begin{pmatrix} 1 & x & t \\ 0 & 1 & y \\ 0 & 0 & 1 \end{pmatrix}.$$

It has one discrete subgroup Γ_k for each integer $k \geq 1$, which consists of the elements in N whose entries are integral multiples of k. The manifolds one gets are $\mathbb{S}^1$-bundles over the torus $\mathbb{T} = \mathbb{S}^1 \times \mathbb{S}^1$. Notice one has a natural projection of N into $\mathbb{R}^2$ regarded as the plane defined by the x, y coordinates, whose quotient by the action of Γ_k is the 2-torus. The fibre of this projection is $\mathbb{R}$, spanned by the t-coordinate, which maps into the $\mathbb{S}^1$-fibres of the $\mathbb{S}^1$-bundle over the torus. As noticed in [171, p. 222], the first integral homology group of the quotient $\Gamma_k \backslash N$ is isomorphic to $\mathbb{Z} \oplus \mathbb{Z} \oplus \mathbb{Z}_k$, so the Euler number of this bundle is $\pm k$, the sign depending on the choice of orientations. This happens because the group operation on N is given by:

$$[x, y, t] \cdot [x', y', t'] = [x + x', y + y', xy' + t + t'],$$

where we are writing $[x, y, t]$ for the corresponding matrix (c.f. the Abelian case $G = \mathbb{R}^3$ below).

iv) $G = E(1, 1)$. This is the inhomogeneous Lorentz group of affine transformations of $\mathbb{R}^2$ that preserve the quadratic form $x^2 - y^2$. This group and its discrete subgroups are beautifully explained in [104]. One has an extension

$$1 \to \mathbb{R}^2 \to G \to \mathbb{R} \to 1$$

with $[G, G] = \mathbb{R}^2$, which presents G as a semi-direct product of $\mathbb{R}^2$ and $\mathbb{R}$. Its discrete subgroups are all of the form $\Gamma = (\mathbb{Z} \times \mathbb{Z}) \rtimes \mathbb{Z} := \mathcal{M} \rtimes \mathcal{V}$. The quotient of $\mathbb{R}^2$ by the lattice $\mathcal{M}$ is a 2-torus $\mathbb{T}^2$, while the quotient of $\mathbb{R}$ by $\mathcal{V}$ is a circle. The quotient $\Gamma \backslash G$ is a torus bundle over $\mathbb{S}^1$ with infinite cyclic monodromy, where both eigenvalues are positive and distinct from 1; all such possibilities occur.

v) $G = \widetilde{E}^+(2)$, the universal cover of the affine group $E^+(2)$, which can be regarded as the group of real matrices:

$$\begin{pmatrix} \cos 2\pi\theta & -\sin 2\pi\theta & x_1 \\ \sin 2\pi\theta & \cos 2\pi\theta & x_2 \\ 0 & 0 & 1 \end{pmatrix}.$$

The group G is a semi-direct product $\mathbb{R}^2 \rtimes \mathbb{R}$, where $\mathbb{R}$ acts on $\mathbb{R}^2$ by the rotations induced from $SO(2) \subset E^+(2)$. It is a central extension of $E^+(2)$,

$$0 \to \mathbb{Z} \to \tilde{E}^+(2) \to E^+(2) \to 1.$$

We see from [207, 1.3] that every uniform subgroup of G is the lifting of one such subgroup in $E^+(2)$, and these are $\mathbb{Z}^2 = \mathbb{Z} \oplus \mathbb{Z}$, the triangle groups $(2,3,6)$, $(2,4,4)$, $(3,3,3)$ and the quadrangle group $(2,2,2,2)$. For $\Gamma = \mathbb{Z}^2$ the quotient $M_\Gamma = \Gamma \backslash G$ is the 3-torus $\mathbb{T}^3$, but regarded as a homogeneous space of $E^+(2)$, not equivalent to its structure as an Abelian Lie group considered below. In the other cases the corresponding quotient M_Γ is a Seifert manifold with Seifert invariants $\{0; -2; (\alpha_1, \alpha_1 - 1), \ldots, (\alpha_n, \alpha_n - 1)\}$, where n is either 3 or 4 and $(\alpha_1, \ldots, \alpha_n)$ is one of the above triples (if $n = 3$) or quadruple (if $n = 4$).

(vi) $G = \mathbb{R}^3$. In this case the only discrete subgroup with compact quotient is $\mathbb{Z}^3 = \mathbb{Z} \oplus \mathbb{Z} \oplus \mathbb{Z}$, up to equivalence, and the quotient is the 3-torus $\mathbb{T}^3$ regarded as an Abelian group with the product structure.

This classification of 3-manifolds which are quotients of 3-dimensional Lie groups by discrete subgroups motivates the following interesting question, asked of me some time ago by Etienne Ghys, whose answer I do not know:

2.1 Question. Which 3-manifolds arise as quotients $H \backslash G$, where G is a Lie group of dimension $n + 3$ and H is a subgroup of G of dimension n? That is, which 3-manifolds are homogeneous spaces?

Recently E. Ghys mentioned to me that he thinks this question has been answered already, but I could not find it in the literature. Of course, once we know the answer then a natural next problem is to decide which of these homogeneous 3-manifolds are links of complex surface singularities?

More generally one may study which closed 3-manifolds arise as orbit spaces of the form $M = \rho(H) \backslash X$, where X is a smooth Riemannian manifold, H is a connected Lie group and ρ is an action of H on X whose orbits are all of codimension 3 with the same isotropy everywhere. This is closely related to the problem of classifying cohomogeneity 3 actions, which is no doubt an interesting and hard problem (c.f. [174, 106, 98, 245]).

III.3 Lie groups and singularities I: quasi-homogeneous singularities

One has the following theorem of [59, 186], which culminates from the theorems of Klein, Milnor and Neumann described in Chapter II, as well as previous results by Dolgachev in [57, 58], see also [187].

3.1 Theorem. *Let $(V, 0)$ be the germ of a quasi-homogeneous surface singularity which is Gorenstein. Then its link M is diffeomorphic to an orbit space of the*

form $\Gamma\backslash G$, where G is one of the groups $SU(2)$, $\widetilde{SL}(2,\mathbb{R})$ or N, and Γ is a discrete
subgroup of G. Conversely, let G be one of these three Lie groups and $\Gamma \subset G$ a
discrete subgroup with compact quotient $M_\Gamma = \Gamma\backslash G$; then M_Γ is diffeomorphic to
the link of a normal, Gorenstein quasi-homogeneous surface singularity.

We recall that an isolated surface singularity germ $(V,0)$ is (analytically)
normal if every bounded holomorphic function on $V^* = V - 0$ extends to a holomorphic function at 0 (see for instance [179]). This implies that the singularity is
isolated. (For hypersurface or complete intersection germs these two conditions are
actually equivalent, i.e., normal iff the singular set has codimension more than 1.)
A normal surface singularity germ $(V,0)$ is *Gorenstein* if there exists a nowhere-
vanishing holomorphic 2-form on V^*. [Notice that the canonical bundle $\mathcal{K}$ of V^*,
i.e., the bundle of holomorphic 2-forms, is 1-dimensional. Hence the existence of
a nowhere-vanishing holomorphic 2-form on V^* is equivalent to saying that $\mathcal{K}$ is
holomorphically trivial. By normality, this means that the dualizing sheaf of V is
free at 0, which is the usual requirement for a singularity to be Gorenstein.]

It is worth saying that one cannot expect to have in this generality such
an explicit description of the equivalence between the orbit spaces $\Gamma\backslash G$ and the
corresponding singularities, as one does in Milnor's theorem for the Brieskorn
manifolds (or in its generalization by Neumann described in Chapter II), because
in the general case the structure of the local ring of these singularities can be quite
complicated, with many generators and relations. However one does have that
in all cases the corresponding local rings are given by appropriate automorphic
forms (see [59]). Moreover (see [183, 186]), their analytic structure may have a
large-dimensional deformation space, so one has to look also at the embedding
of the group Γ in the corresponding group G, similarly to classical Teichmüller
theory. This means in particular that the singularity one gets from a given discrete
subgroup $\Gamma \subset G$ does not depend only on the abstract group Γ, but also on the
way it is embedded in G.

We refer to the articles of Neumann and/or Dolgachev for the proof of this
theorem and for the interesting study in [186] of the deformations of both, the
analytic structure on the variety and the embedding of the group Γ in G.

Here we follow [231] and describe how to construct explicitly the surface singularities associated to the orbit spaces $\Gamma\backslash G$ in each of the three cases, and we show
that these singularities are Gorenstein by exhibiting explicit nowhere-vanishing
holomorphic 2-forms in all cases. This construction is essentially Dolgachev's and
the arguments below rely on his work, specially in [59].

3.2. In the case $G = SU(2)$ the corresponding singularities are those of Klein
studied in Chapter II. It is clear that these singularities admit a good $\mathbb{C}^*$-action,
and they are normal (since they are isolated hypersurface singularities, or also by
Grauert's contraction criterium that we explain in Chapter IV). The corresponding
2-form on $V_\Gamma = \Gamma\backslash\mathbb{C}^2$ is obtained by noticing that the holomorphic form $\Omega =$
$dz_1 \wedge dz_2$ on $\mathbb{C}^2$ is clearly nowhere-vanishing and invariant under the action of

all $SU(2)$, which acts on $\mathbb{C}^2$ by holomorphic transformations. Hence this form descends to the quotient V_Γ and is nowhere-vanishing on $V_\Gamma^* = V_\Gamma - 0$.

Alternatively, one may consider (say for the commutators of the binary triangle groups) the corresponding Brieskorn variety in $\mathbb{C}^3$,

$$V_{p,q,r} = \{\, z_1^p + z_2^q + z_3^r = 0 \,\},$$

whose link is $G = SU(2)$ divided by the commutator of the corresponding triangle group (one has analogous singularities for all the finite subgroups of $SU(2)$, see Chapter II). One has in $\mathbb{C}^3$ the form $\widetilde{\Omega} = dz_1 \wedge dz_2 \wedge dz_3$; this induces a *canonical* 2-form ω on $V_{p,q,r} - 0$, obtained by contracting $\widetilde{\Omega}$ with the gradient vector field ∇f, where f is the corresponding Pham-Brieskorn polynomial. In local coordinates ω is:

$$\omega = \frac{dz_1 \wedge dz_2}{\partial f/\partial z_3} = \frac{dz_2 \wedge dz_3}{\partial f/\partial z_1} = \frac{dz_3 \wedge dz_1}{\partial f/\partial z_2}. \tag{3.2.1}$$

It is easy to see that the 2-forms on the right-hand side of this equation, which are defined where the denominator is not zero, coincide when any two of them are defined and so they give a global 2-form, which coincides with the one given by contracting $\widetilde{\Omega}$ with ∇f. It is now an exercise (see [229]) to show that the pull-back of this 2-form w to $\mathbb{C}^2$ under Klein's map $\mathbb{C}^2 \to \mathbb{C}^3$ is a constant multiple of the form $\Omega = dz_1 \wedge dz_2$, and therefore these two forms on $V_\Gamma = \Gamma \backslash \mathbb{C}^2$ coincide up to a constant.

3.3. Let us describe now the case $G = \widetilde{SL}(2, \mathbb{R})$. We consider first the case $\Gamma \subset PSL(2, \mathbb{R})$. The corresponding singularities are known as *Dolgachev singularities* and were introduced in [57, 58]. We recall (Chapter II) that $PSL(2, \mathbb{R})$ acts on the tangent bundle $T\mathbb{H}$ of the hyperbolic plane by:

$$g \cdot (z, w) \mapsto \left(g(z), \frac{\widetilde{dg}}{dz} w \right).$$

This action is free away from the zero section, which is $\mathbb{H}$ and where we may have fixed points, all with finite cyclic isotropy. Thus we may embed $PSL(2, \mathbb{R})$ in $T\mathbb{H}$ as the boundary of a tubular neighborhood of $\mathbb{H}$ in $T\mathbb{H}$, regarded as a complex manifold. The orbit space $\widetilde{V}_\Gamma = \Gamma \backslash T\mathbb{H}$ is a complex analytic surface with (possibly) singularities contained in the Riemann surface $\Gamma \backslash \mathbb{H} \subset \widetilde{V}_\Gamma$, which correspond to the special orbits of the Γ-action on $\mathbb{H}$. Then Dolgachev showed that $\Gamma \backslash \mathbb{H}$ can be blown down to a point that we may denote by 0 (this follows essentially from Grauert's criterium, c.f. Chapter IV). The result is a complex analytic surface V_Γ with an isolated normal singularity at 0. It is clear from the construction that the link of V_Γ is $\Gamma \backslash PSL(2, \mathbb{R})$. One may exhibit an explicit 2-form for these singularities, thus proving that they are Gorenstein, but we will do this in general below for all the uniform subgroups of $\widetilde{SL}(2, \mathbb{R})$.

It is worth saying that the singularities one gets in this way may not be hypersurfaces nor complete intersections, and not even smoothable, as shown by

Pinkham. For instance, for the triangle subgroups of $PSL(2, \mathbb{R})$, Dolgachev showed that there are only 14 of them for which the corresponding singularity can be defined by a single equation; these correspond to the 14 exceptional unimodal singularities of Arnold (c.f. [257]).

   To construct the singularities corresponding to arbitrary uniform subgroups of $\widetilde{SL}(2, \mathbb{R})$ we adopt the notation introduced in [171, 231].

3.4 Definition. For every integer $r \geq 1$, an *r-labeled biholomorphic map* of $\mathbb{H}$ to itself means a biholomorphic map $f : \mathbb{H} \to \mathbb{H}$, together with a lift $\frac{\widetilde{df}}{dz}$ of its derivative to the r-fold cover of the multiplicative group $\mathbb{C}^*$.

   Taking $r = 1$ we recover the group $G_1 = PSL(2, \mathbb{R})$ as the group of 1-labeled biholomorphic maps of $\mathbb{H}$, while for $r = \infty$ we get the labeled biholomorphic maps of Milnor, which form the group $\widetilde{SL}(2, \mathbb{R})$. It is an exercise to see that the set of all r- labeled biholomorphic maps forms the r-fold cyclic covering group G_r of $PSL(2, \mathbb{R})$.

   If we denote by $\mathbb{C}_r \cong \mathbb{C}$ the r-fold cyclic cover of $\mathbb{C}$ branched at 0, we then have an action of G_r on $\mathbb{H} \times \mathbb{C}_r$, which embeds G_r as the orbit of $(i, 1)$, where 1 is a selected point in $\mathbb{C}_r$ lying over $1 \in \mathbb{C}$. Given $\Gamma \subset G_r$, we have an analytic surface $\Gamma \backslash \mathbb{H} \times \mathbb{C}_r$ with isolated singularities on $S_\Gamma = \Gamma \backslash (\mathbb{H} \times 0)$; the 3-manifold $\Gamma \backslash G_r = M_\Gamma$ is embedded in $\Gamma \backslash \mathbb{H} \times \mathbb{C}_r$ as the boundary of a tubular neighborhood of S_Γ. Observe that we are now having an automorphy factor: as in [57, 58, 59], we may now blow down S_Γ to a point (essentially by Grauert's criterium, c.f. Ch. IV) to obtain a complex analytic surface V_Γ with a normal singularity at a point that we may denote by 0, with an obvious $\mathbb{C}^*$-action. The link of V_Γ is $\Gamma \backslash G_r = M_\Gamma$.

   To show that this singularity is Gorenstein we observe, following [229], that one has on $\mathbb{H} \times \widetilde{\mathbb{C}}^*$ the holomorphic 2-form:

$$\Omega = \frac{dz \wedge dw}{w^2},$$

which is nowhere-vanishing and, we claim, is invariant under the action of $\widetilde{SL}(2, \mathbb{R})$. In fact, for every $g \in \widetilde{SL}(2, \mathbb{R})$ we have:

$$g^* \Omega(z, w) = \frac{g^* dz \wedge g^* dw}{\left(\frac{dg}{dz}(z)\right)^2 w^2} = \left(\frac{\left(\frac{dg}{dz}(z)\right)^2}{\left(\frac{dg}{dz}(z)\right)^2 w^2}\right) dz \wedge dw = \frac{dz \wedge dw}{w^2},$$

so Ω is invariant. Hence it descends to a nowhere-vanishing holomorphic form on $\Gamma \backslash \mathbb{H} \times (\mathbb{C}_r - 0)$, and so defines a nowhere-vanishing holomorphic 2-form on $V_\Gamma - 0$.

3.5 Remark. We notice that in the two cases above, $G = SU(2)$ or $G = \widetilde{SL}(2, \mathbb{R})$, one has free actions of G on a certain complex 2-manifold X, which is $\mathbb{C}^2 - 0$ or $\mathbb{H} \times \widetilde{\mathbb{C}}^*$. In both cases the G-orbits foliate X by copies of G, and one has a G-invariant nowhere-vanishing holomorphic form on all of X. We will see that one has a similar picture for all the six Lie groups in question.

3.6. Consider now the Heisenberg group N of real matrices of the form

$$\begin{pmatrix} 1 & x & t \\ 0 & 1 & y \\ 0 & 0 & 1 \end{pmatrix}.$$

For simplicity we denote such a matrix by $[x, y, t]$. The group operation on N is given by:

$$[x, y, t] \cdot [x', y', t'] = [x + x', y + y', xy' + t + t'].$$

We may define the smooth map $N \times \mathbb{C}^2 \xrightarrow{\ \rho\ } \mathbb{C}^2$ by:

$$\rho([x, y, t], (z_1, z_2)) = (z_1 + x + iy, e^{2\pi i t} z_2).$$

This is not a group action since:

$$\rho\Big([x, y, t], \rho([x', y', t'], (z_1, z_2))\Big) = (z_1 + x + x' + i(b + b'), e^{2\pi i(t+t')} z_2),$$

while

$$\rho\Big(([x, y, t] \cdot [x', y', t']), (z_1, z_2)\Big) = (z_1 + x + x' + i(b + b'), e^{2\pi i(t+t'+xy')} z_2), \quad (3.6.1)$$

so we get an extra factor $e^{2\pi i(xy')}$ in the second variable. Still we observe that ρ is an action restricted to every discrete subgroup of N, since in these cases $xy' \in \mathbb{Z}$.

Notice that we can embed $\mathbb{C}$ in N by $z \mapsto [x, y, 0]$, though this is not a group homomorphism. Still, the restriction to $\mathbb{C} \subset N$ of the map ρ is a group action of $\mathbb{C}$ on $\mathbb{C}^2$, acting by translations in the first variable. We may also define an embedding of $\mathbb{R}$ in N (this time as a subgroup) by $t \mapsto [0, 0, t]$. The map ρ gives an action of $\mathbb{R} \subset N$ on $\mathbb{C}^2$, it is multiplication in the second variable by $e^{2\pi i t}$, which are rotations. It follows that for each $(z_1, z_2) \in \mathbb{C}^2$ with $z_2 \neq 0$, the image of the map:

$$\rho_{(z_1, z_2)} : N \to \mathbb{C}^2$$

given by $\rho_{(z_1, z_2)}([x, y, t])$ is a cylinder $\mathbb{C} \times \mathbb{S}^1$. This defines a foliation of $\mathbb{C}^2$ by cylinders, with one special leaf $\mathcal{L}_0 = \{(z_1, 0)\}$ (where the foliation is singular), and every discrete subgroup Γ of N acts on each leaf of this foliation. By the previous comments we see that one also has an $\mathbb{S}^1$-action on each leaf of the foliation, given by the restriction of ρ to $\mathbb{R} \times \mathbb{C}^2$.

We now recall that, up to isomorphism, N has one discrete subgroup Γ_k for each integer $k > 0$, consisting of those triangular matrices as above, for which the coefficients x, y, t are all multiples of k. Given $\Gamma_k = \Gamma \subset N$ discrete, we can pass to the quotient $\widetilde{V}_\Gamma = \Gamma \backslash \mathbb{C}^2$. The singular leaf $\mathcal{L}_0$ becomes a 2-torus $\mathbb{T}^2 = \Gamma \backslash \mathcal{L}_0$, and all other leaves of $\mathcal{F}$ become copies of $M_\Gamma = \Gamma \backslash N \cong \Gamma \backslash (\mathbb{C} \times \mathbb{S}^1)$, so they are $\mathbb{S}^1$-bundles over the torus and the complex surface $\widetilde{V}_\Gamma$ is a holomorphic line bundle over the 2-torus. The Chern class of this bundle is $\pm k$, depending on the choices of orientations, due to the contribution of the $\{x, y\}$-variables in (3.6.1). This means that M_Γ is a circle bundle over the 2-torus with Euler class $\pm k$. Choosing the orientations so that the self-intersection of $\mathbb{T}^2$ is $-k$, we have that the contraction criterium of Grauert says that we can blow $\mathbb{T}^2$ down to a point to get a surface singularity V_Γ, whose minimal resolution (see Chapter IV) is $\widetilde{V}_\Gamma$.

For $k = -1, -2, -3$ the surfaces one gets in this way are homeomorphic to the Brieskorn varieties,

$$V_{(p,q,r)} = \{ z_1^p + z_2^q + z_3^r = 0 \} \,,$$

where (p, q, r) is one of the triples $(2, 3, 6)$, $(2, 4, 4)$ or $(3, 3, 3)$. This follows either by [171], by [194] or by [59, 186]. Milnor tells us that the corresponding Brieskorn manifolds are circle bundles over the torus, with Euler class equal to $-k$, where k is the greatest common divisor of (p, q, r); Orlik and Wagreich [194] tell us that the minimal resolutions of these singularities are holomorphic line bundles over the 2-torus, with the corresponding Chern classes.

We remark that Dolgachev in [59] gives a finer argument than we do here, and he proves the above diffeomorphisms directly.

We also remark that these singularities are *elliptic*, meaning by this that their geometric genus is 1 (see [122] and Chapter IV below). Moreover, by [122] they are also *minimally elliptic*, which amounts to ellipticity plus Gorenstein. To see that they are actually Gorenstein we observe, following [231], that for every $h = [x, y, t] \in N$ the map $\theta_h : \mathbb{C}^2 \to \mathbb{C}^2$ given by:

$$\theta_h(z_1, z_2) = \rho([x, y, t], (z_1, z_2)) = (z_1 + (x + iy) , e^{2\pi i t} z_2)$$

is holomorphic and leaves invariant the holomorphic 2-form

$$\Omega = \frac{dz_1 \wedge dz_2}{z_2}$$

since for every $g = [x, y, t] \in N$ one has:

$$g^* \Omega(z_1, z_2) = \frac{g^* dz_1 \wedge g^* dz_2}{e^{2\pi i t} z_2} = \frac{dz_1 \wedge e^{2\pi i t} dz_2}{e^{2\pi i t} z_2} = \Omega(z_1, z_2) \,.$$

Hence this form descends to $V_\Gamma - 0$ and is never-vanishing there. $\square$

III.4   Lie groups and singularities II: the cusps

This section is essentially taken from [104], where Hirzebruch constructs and studies the cusp singularities.

Let K be a totally real quadratic field over $\mathbb{Q}$, so it has two different embeddings in $\mathbb{R}$, that we denote by $x^{(1)}, x^{(2)}$, for $x \in K$. Let ϱ be the ring of algebraic integers in K, and let $x \mapsto x'$ be the non-trivial automorphism of K. The Hilbert modular group:

$$SL_2(\varrho) = \left\{ \begin{pmatrix} a & b \\ c & d \end{pmatrix} \,\middle|\, a, b, c, d \in \varrho \,, ad - bc = 1 \right\}$$

acts on $\mathbb{H} \times \mathbb{H}$, where $\mathbb{H}$ is the upper half-plane, by

$$\begin{pmatrix} a & b \\ c & d \end{pmatrix} (z_1, z_2) = \left(\frac{a z_1 + b}{c z_1 + d}, \frac{a' z_1 + b'}{c' z_1 + d'} \right).$$

The corresponding projective group $\widehat{P} = SL_2(\varrho)/\{\pm Id\}$ acts effectively on $\mathbb{H} \times \mathbb{H}$, and for each point $x \in \mathbb{H} \times \mathbb{H}$ the isotropy group $\widehat{P}_x \subset \widehat{P}$ is finite and cyclic. The orbit space $\widehat{P} \backslash (\mathbb{H} \times \mathbb{H})$ is a complex space, which can be compactified by adding to it a finite number of points, called the cusps. The resulting space is a compact algebraic surface with isolated singularities at these cusps, whose links are torus bundles over the circle, diffeomorphic to quotient spaces of the form $\Gamma \backslash E(1,1)$.

More generally, given K, let $\mathcal{M}$ be a $\mathbb{Z}$-module of rank 2 contained in K, and let $U_{\mathcal{M}}^+$ be the group of totally positive units in K that leave $\mathcal{M}$ invariant; we recall that an element $x \in K$ is totally positive (denoted $x \gg 0$) if each of its two embeddings in $\mathbb{R}$ is > 0. The elements of $U_{\mathcal{M}}^+$ are automatically algebraic integers since they leave $\mathcal{M}$ invariant (see [104]). The group $U_{\mathcal{M}}^+$ is free of rank 1, $U_{\mathcal{M}}^+ \cong \mathbb{Z}$.

We recall that the group $PSL(2, \mathbb{R})$ acts on $\mathbb{H}$ via the Möbius transformations; thus its product $PSL(2, \mathbb{R})^2 = PSL(2, \mathbb{R}) \times PSL(2, \mathbb{R})$ acts naturally on $\mathbb{H} \times \mathbb{H}$. An element of $PSL(2, \mathbb{R})$ is *parabolic* when it has exactly one fixed point in $\mathbb{C}P^1$, and in this case the fixed point is in $\mathbb{R}P^1 \cong \mathbb{R} \cup \infty$. Similarly, an element $\gamma = (\gamma^{(1)}, \gamma^{(2)})$ of $PSL(2, \mathbb{R})^2$ is called *parabolic* iff both $\gamma^{(1)}, \gamma^{(2)}$ are parabolic. Such an element has exactly one fixed point in $(\mathbb{C}P^1)^2$, which belongs to $(\mathbb{R}P^1)^2$.

A discrete subgroup $\Gamma \subset PSL(2, \mathbb{R})^2$ is said to be *irreducible* if it contains no element $\gamma = (\gamma^{(1)}, \gamma^{(2)})$ such that $\gamma^{(i)} = 1$ for some i and $\gamma^{(j)} \neq 1$ for some j. It is proved in [235] that an irreducible discrete subgroup $\Gamma \subset PSL(2, \mathbb{R})^2$ has at most finitely many distinct orbits of parabolic points. Let us look at the geometry near a parabolic point. According to [235, p. 45], for any parabolic fixed point x of an irreducible discrete subgroup $\Gamma \subset PSL(2, \mathbb{R})^2$ with $\Gamma \backslash \mathbb{H} \times \mathbb{H}$ of finite volume, the parabolic element $\rho \in PSL(2, \mathbb{R})^2$ with $\rho x = x$ can be chosen so that $\rho \Gamma_x \rho^{-1}$ is contained in $PGL(K) \subset PSL(2, \mathbb{R})^2$, where Γ_x is the isotropy of x, K is a suitable totally real field and $PGL(K)$ is the projectivization of

$$GL^+(2, K) = \left\{ \begin{pmatrix} a & b \\ c & d \end{pmatrix} \,\middle|\, a, b, c, d, \in K; \ ad - bc \gg 0 \right\}.$$

Then one has an exact sequence (see [104, 1.5 (15)]):

$$0 \to \mathcal{M} \to \rho \Gamma_x \rho^{-1} \to \mathcal{V} \to 1,$$

where $\mathcal{M}$ is a $\mathbb{Z}$-module of rank 2 contained in K and $\mathcal{V}$ is a non-trivial subgroup of $U_{\mathcal{M}}^+$ of finite index. The field K, the equivalence class of $\mathcal{M}$ and the group $\mathcal{V}$ are completely determined by the parabolic orbit and do not depend on the choice of ρ. In this case one has that $\rho \Gamma_x \rho^{-1}$ is a semi-direct product $\mathcal{M} \rtimes \mathcal{V}$ that we may denote by $\Gamma(\mathcal{M}, \mathcal{V})$. The parabolic orbit is called a cusp (similarly to the classical theory of Kleinian groups) of type $(\mathcal{M}, \mathcal{V})$, a name which is also used for the image point 0 of this orbit in the quotient space of orbits.

For any positive number d, the group $\Gamma(\mathcal{M}, \mathcal{V})$ acts freely on

$$W = \{(z, w) \in \mathbb{H} \times \mathbb{H} \mid \operatorname{Im}(z) \cdot \operatorname{Im}(w) \geq d\}.$$

The orbit space $\Gamma(\mathcal{M}, \mathcal{V}) \backslash W$ is a non-compact manifold with compact boundary

$$M = \Gamma(\mathcal{M}, \mathcal{V}) \backslash \partial W.$$

The space $\Gamma(\mathcal{M}, \mathcal{V}) \backslash W \subset \Gamma(\mathcal{M}, \mathcal{V}) \backslash \mathbb{H} \times \mathbb{H}$ can be compactified by adding to it the point 0, whose link is M. By [104, p. 202] the analytic surface $V_\Gamma = \Gamma(\mathcal{M}, \mathcal{V}) \backslash \overline{\mathbb{H} \times \mathbb{H}}$ has a normal singularity at 0.

We notice that the group $G = E(1, 1)$, homeomorphic to $\mathbb{R}^3$, can be identified with the group of affine transformations of $\mathbb{H} \times \mathbb{H}$ of the form:

$$(z, w) \longmapsto (t_1 z + a_1, \, t_2 w + a_2),$$

where $a_1, a_2, t_1, t_2 \in \mathbb{R}$ and $t_1 t_2 = 1$. Thus G acts freely on $\mathbb{H} \times \mathbb{H}$ and we can identify G with the orbit of a point, say (i, i). It follows that the link M can be identified with the quotient $\Gamma(\mathcal{M}, \mathcal{V}) \backslash E(1, 1)$ and it is therefore a torus bundle over the circle $\mathbb{S}^1 \cong \mathcal{V} \backslash \mathbb{R}$. Of course the various orbits of G fill out all of $V_\Gamma - 0$.

We notice too that the holomorphic 2-form

$$\Omega = \frac{dz \wedge dw}{z \, w}$$

on $\mathbb{H} \times \mathbb{H}$ is obviously invariant under the above action of $E(1, 1)$; it is also invariant under the action of the isotropy group of the cusp, so it descends to a never-zero holomorphic 2-form on a punctured neighborhood of ∞ in V_Γ.

III.5   Lie groups and singularities III: the Abelian and $E^+(2)$-cases

It is known ([186, 59]) that the only normal complex surface singularities whose link is of the form $\Gamma \backslash G$, with G a Lie group and Γ a uniform subgroup, are those described in Sections 3 and 4 above: the weighted homogeneous Gorenstein singularities and the cusp singularities. It is well known (see for instance [184] or [246]) that the 3-torus cannot be the link of a normal singularity in a complex surface; this observation extends to all quotients of $E^+(2)$ by uniform subgroups. Even so, there are interesting similarities among both of these cases and the previous ones, and this is what we discuss in this section, following [231]. We start with the Abelian case, which is somewhat similar to that of the nilpotent case N.

Let us recall from Chapter II that one has a canonical isomorphism $\tilde{e} : \mathbb{C} \to \widetilde{\mathbb{C}}^*$ of the additive (complex) Lie group $\mathbb{C}$ with the universal covering group $\widetilde{\mathbb{C}}^*$ of the non-zero complex numbers; this isomorphism is obtained by lifting to $\widetilde{\mathbb{C}}^*$ the exponential map $e(z) = e^{2\pi i z} : \mathbb{C} \to \mathbb{C}^*$. Given $G = \mathbb{R}^3$, regarded as a Lie group under the coordinate-wise addition, one has an action of $G \cong \mathbb{C} \times \mathbb{R}$ on $\mathbb{C} \times \widetilde{\mathbb{C}}^*$ given by:

$$((x, y), t), (z, w)) \mapsto (z + (x + iy), \widetilde{e}(t) \cdot w).$$

This foliates $\mathbb{C} \times \widetilde{\mathbb{C}}^*$ by copies of $\mathbb{R}^3$ embedded as the orbits of the action. The natural projection into $\mathbb{C} \times \mathbb{C}^*$ induces an action on $\mathbb{C} \times \mathbb{C}^*$ given by

$$((x, y), t), (z, w)) \mapsto (z + (x + iy), e(t) \cdot w),$$

which extends in the obvious way to an action on $\mathbb{C}^2$ whose orbits are all cylinders $\mathbb{C} \times \mathbb{S}^1$ that wrap around the "special leaf" $\mathbb{C} \times \{0\}$. This foliation $\mathcal{F}$ is invariant under the action of the whole group G, so in particular it is invariant under $\Gamma = \mathbb{Z}^3 = \mathbb{Z} \oplus \mathbb{Z} \oplus \mathbb{Z}$, which is the only uniform subgroup of G. Notice that the action of $\mathbb{Z}$ on the second component is already trivial since this is the kernel of the map $\widetilde{\mathbb{C}}^* \rightarrow \mathbb{C}^*$. The quotient $\widetilde{V}_\Gamma = \Gamma \backslash \mathbb{C}^2$ is a complex manifold which fibres as a trivial holomorphic line bundle over the torus $\mathbb{T}^2 = \mathbb{Z}^2 \backslash (\mathbb{C} \times \{0\})$. The complement of the torus in $\widetilde{V}_\Gamma$ is diffeomorphic to a cylinder $\mathbb{T}^3 \times \mathbb{R}$, where $\mathbb{T}^3 = \mathbb{S}^1 \times \mathbb{S}^1 \times \mathbb{S}^1$; these 3-torii are quotients of the leaves of the foliation $\mathcal{F}$ by the action of Γ.

Just as in the nilpotent case, one has the holomorphic 2-form,

$$\Omega = \frac{dz \wedge dw}{w}$$

on $\mathbb{C} \times \widetilde{\mathbb{C}}^*$, which is never-vanishing and G-invariant, so it descends to $\widetilde{V}_\Gamma - \mathbb{T}^2$.

Hence, so far this case is entirely analogous to the previous cases; the problem now is that $\mathbb{T}^2$ is embedded in $\widetilde{V}_\Gamma$ with self-intersection 0. Therefore Grauert's criterium does not apply (see IV.5.8 and IV.5.9) and the torus cannot be blown down complex analytically.

5.1 Question. Can the torus $\mathbb{T}^2$ be blown down real analytically?

By [4] we know that 5.1 is true up to homeomorphism. Thus, the 3-torus is the link of an isolated singularity in a real analytic 4-dimensional variety $(V, 0)$ which has a holomorphic structure (compatible with the real analytic structure at 0) and has a nowhere-vanishing holomorphic 2-form on $V - 0$. This seems to be related with an interesting class of singularities that come from the example below, taken from [50], which is the starting point for Chapter VI below.

Consider the linear vector field in $\mathbb{C}^3$ given by $F(z) = (\lambda_1 z_1, \lambda_2 z_2, \lambda_2 z_2)$, where the $\lambda_j = \mu_j + i\nu_j$ are non-zero complex numbers. Let V_F be the variety

$$V_F = \{z \in \mathbb{C}^3 \mid \langle F(z), z \rangle = \sum_{j=1}^3 \lambda_j |z_j|^2 = 0\}$$

of points where the vector field F is tangent to the sphere through the point and centre at 0. This variety is real analytic since it is defined by the real analytic equations:

$$\sum_{j=1}^{3} \mu_j \left(x_j^2 + y_j^2 \right) = 0 = \sum_{j=1}^{3} \nu_j \left(x_j^2 + y_j^2 \right)$$

where $z_j = x_j + iy_j$. It may consist of $0 \in \mathbb{C}^3$ alone, for instance when the μ_j are all > 0. However, if we assume that the eigenvalues λ_i are chosen so that $0 \in \mathbb{C}$ is in the interior of the convex hull of $\{\lambda_1, \lambda_2, \lambda_3\}$, then one has that:

(i) V_F has real codimension 2 in $\mathbb{C}^3$; and $0 \in \mathbb{C}^3$ is its only singular point;

(ii) the link M of V_F is the 3-torus $\mathbb{T}^3$ and there is a transitive action of $\mathbb{T}^3$ on M; and

(iii) at each point $z \in (V_F - 0)$ the complex line spanned by F is transversal to V_F.

The first two conditions tell us that the 3-torus is the link of the real analytic variety V_F, which is actually a complete intersection in $\mathbb{C}^3$, and the third condition yields that $V_F - 0$ is, canonically, a complex manifold (see Chapter VI), though it is not (in general) embedded as a complex submanifold of $\mathbb{C}^3$.

Let us consider now the case of $\widetilde{E}^+(2)$, or rather we look at $G = E^+(2)$, since all the uniform subgroups of the former come from the latter. This group acts on $\mathbb{C} \cong \mathbb{R}^2$ in the obvious way, and this action extends naturally to the tangent bundle $T\mathbb{C} = \mathbb{C} \times \mathbb{C}$ via the derivative:

$$g \cdot (z, w) \longmapsto \left(g(z), \frac{dg}{dz}(z) \cdot w \right) .$$

This action is free away from the line $z = 0$ and therefore the orbits foliate $\mathbb{C} \times \mathbb{C}$ by copies of $G = \widetilde{E}^+(2)$ that degenerate into the singular leaf of the foliation, given by $z = 0$. Hence, every discrete subgroup Γ of $G = E^+(2)$ acts freely on $T\mathbb{C}$, away from the zero section, but it has fixed points there (except for $\Gamma = \mathbb{Z}^2$ which acts freely also on this line). The quotient $V_\Gamma = \Gamma \backslash T\mathbb{C}$ is a complex analytic surface with zero, three, or four singular points, as the case may be, all contained in the divisor $S_\Gamma = \Gamma \backslash \mathbb{C}$, which is either a torus, if $\Gamma = \mathbb{Z}^2$, or $\mathbb{C}P^1$ with three or four marked points in the other cases. The action of Γ on $T\mathbb{C}$ preserves the leaves of the above foliation and descends to a foliation of $V_\Gamma - S_\Gamma$ by copies of $\Gamma \backslash E^+(2)$ embedded as boundaries of regular neighborhoods of S_Γ, which is the special leaf.

Again, as in the Abelian case, one cannot blow down this divisor complex analytically. In the case $\Gamma = \mathbb{Z}^2$ this follows because the divisor S_Γ, which is a 2-torus, is embedded with trivial normal bundle. In the other cases this can be seen as follows: at each singularity of V_Γ one has, locally, a quotient singularity of the form $\mathbb{Z}_n \backslash \mathbb{C}^2$ for an appropriate $n = 3, 4$. One can resolve this and get a divisor D in the resolution $\widetilde{V}_\Gamma$; D consists of a central curve isomorphic to S_Γ and one "branch" of rational curves for each singularity of V_Γ. Then S_Γ can be blown down iff D can be blown down, but an easy calculation shows that the intersection matrix of D is not negative-definite, so Grauert's criterium says we cannot blow it down.

Just as in the Abelian case, one may ask whether S_Γ can be blown down real analytically. By [4] this is always the case up to homeomorphism.

Finally we observe that the holomorphic form $\Omega = dz \wedge dw$ on $T\mathbb{C}$ is globally defined and G-invariant, so it descends to V_Γ minus the singular points, and it is never-zero.

III.6   A uniform picture of 3-dimensional Lie groups

Let us put together some of what we have said in the previous sections about 3-dimensional Lie groups and their uniform subgroups, in relation with surface singularities.

Let G be any one of the six 3-dimensional Lie groups with a uniform subgroup $\Gamma \subset G$. Then we know from the previous discussion that there is associated to G a complex 2-dimensional manifold, equipped with a canonical holomorphic 2-form and a foliation $\mathcal{F}$ of X defined, in all cases but one, by an action of G. There are several interesting properties common to all cases, and others that miss in one or two cases. It is convenient to recall first which are the specific manifolds, the 2-forms and the action of G on X in each case.

(i) For $G = SU(2)$, $X = (\mathbb{C}^2 - 0)$ and the form is $\Omega = dz_1 \wedge dz_2$. The action is the usual one by linear transformations on $\mathbb{C}^2$.

(ii) For $G = \widetilde{SL}(2, \mathbb{R})$, $X = (\mathbb{H} \times \widetilde{\mathbb{C}^*})$ and $\Omega = \frac{dz \wedge dw}{w^2}$. The action on $\mathbb{H}$ is via the Möbius transformations, and on $\mathbb{C}^*$ is by lifting the derivative to the universal cover of the multiplicative group $\mathbb{C}^*$.

(iii) For $G = N$, the Heisenberg group, $X = (\mathbb{C} \times \mathbb{C}^*)$ and $\Omega = \frac{dz_1 \wedge dz_2}{z_2}$. This is the only exception in which we do not have a group action, but a map $G \times X \to X$
$$([x, y, t], (z_1, z_2)) \mapsto (z_1 + (x + iy)\, e^{2\pi i t} z_2)$$
which is an action restricted to the uniform subgroups of G. I believe that one should be able to read from [59] all the information we need to fully unify this case with the others, but I have not been able to this this yet.

(iv) For $G = E(1, 1)$, $X = (\mathbb{H} \times \mathbb{H})$ and $\Omega = \frac{dz \wedge dw}{z\, w}$. The action of G on X is given by identifying this group with the group of affine transformations of X of the form:
$$(z, w) \longmapsto (t_1 z + a_1 ,\, t_2 w + a_2),$$
where $a_1, a_2, t_1, t_2 \in \mathbb{R}$ and $t_1 t_2 = 1$.

(v) For $G = E^+(2)$, $X = (\mathbb{C} \times \mathbb{C}^*)$ and $\Omega = dz \wedge dw$. Here X is regarded as the non-zero tangent vectors on $\mathbb{C}$. The action on $\mathbb{C}$ is the obvious one; on $\mathbb{C}^*$ it is multiplication by the derivative.

(vi) For $G = \mathbb{R}^3$, $X = (\mathbb{C} \times \mathbb{C}^*)$ and $\Omega = \frac{dz_1 \wedge dz_2}{z_2}$. The action is given by the same map as in (iii) above, which is now an action.

Having this in mind, let us recall some of the properties that these cases have in common.

(a) Except for the case $G = N$, the leaves of $\mathcal{F}$ are the orbits of the G-action, which is free in all cases.

(b) In all cases, every uniform discrete subgroup Γ of G acts on X preserving the leaves of $\mathcal{F}$; and the space of orbits $\Gamma \backslash \mathcal{L}$ is the homogeneous space $\Gamma \backslash G$.

(c) In all cases one has a canonical holomorphic trivialization of the canonical bundle $\mathcal{K}$ of X given by the 2-form Ω, which is G-invariant in all cases but one: $G = N$, where we have Γ-invariance for all uniform subgroups. This induces an $Sp(1)$ structure on the tangent bundle TX.

(d) The orbit space $\widetilde{V}_\Gamma^* = \Gamma \backslash X$ is a complex analytic manifold, foliated by copies of $\Gamma \subset G$, with a nowhere-vanishing holomorphic 2-form. For $G = SU(2)$, $\widetilde{SL}(2,\mathbb{R})$, $E(1,1)$ or N, one has that the complex manifold $\widetilde{V}_\Gamma^*$ is biholomorphic to $V - 0$, where V is a 2-dimensional surface with a normal singularity at 0. For $G = E^+(2)$ and $G = \mathbb{R}^3$, the complex manifold $\widetilde{V}_\Gamma^*$ is homeomorphic to $V - 0$, where V is a real algebraic variety with an isolated singularity at 0 (by [4, 4.1]). I believe that the 1-point compactification of $\widetilde{V}_\Gamma^*$ should be itself real analytic, but I do not know how to prove this (c.f. [161, Lemme 1]).

Summarizing, one has:

6.1 Theorem. *There are, up to isomorphism, six 3-dimensional, simply connected Lie groups with uniform subgroups $\Gamma \subset G$. Given any such group G and a uniform subgroup Γ, there is associated to G a canonical complex 2-dimensional manifold X, equipped with a canonical holomorphic 2-form and a foliation $\mathcal{F}$ of X defined, in all cases but one, by an action of G. The manifold X is in all cases given by an automorphy factor of some line bundle. The quotient $\widetilde{V}_\Gamma^* = \Gamma \backslash X$ is a complex manifold, foliated by copies of $\Gamma \backslash G$, with a canonical never-vanishing holomorphic 2-form. This manifold $\widetilde{V}_\Gamma^*$ is actually an open cylinder $\Gamma \backslash G \times (0,1)$, and one of its ends can be compactified by attaching to it a smooth divisor S_Γ, so that we get a complex analytic surface $\widetilde{V}_\Gamma = \widetilde{V}_\Gamma^* \cup S_\Gamma$ which may have isolated, normal singularities at S_Γ. In four of the six cases in question, this divisor can be blown down complex analytically and the result is a complex analytic surface V_Γ with a normal singularity P, whose link is the 3-manifold M_Γ. In the remaining two cases the divisor S_Γ can only be blown down real analytically, so the quotient V_Γ is (homeomorphic to) a 4-dimensional real analytic space with an isolated singularity P, whose link is M_Γ and which has a complex structure away from P.*

III.7 Lie algebras and the Gorenstein property

An essential feature of Lie groups is that they have their Lie algebras, which can be defined as being the vector space of all left invariant vector fields on the group, endowed with the multiplication given by the Lie bracket. These play an important role in the classifications given in [23, 172, 207] of 3-dimensional Lie groups and their uniform subgroups. On the other hand we know from [59, 186] and the previous discussion, that there is a close connection between closed 3-manifolds of the form $\Gamma\backslash G$ and complex surfaces which are normal and Gorenstein, i.e., they admit a nowhere-zero holomorphic 2-form. In this section we discuss an aspect of how the property of being Gorenstein relates to the Lie algebras of the corresponding groups. This goes back to [224, 229, 231].

Given an oriented Lie group G of dimension n, and a positively oriented basis $\{v_1, \ldots, v_n\}$ of its tangent space at the identity T_eG, multiplication on the left by an element $g \in G$ carries T_eG isomorphically into T_gG and defines a basis of this space by left translating the basis $\{v_1, \ldots, v_n\}$. In this way we obtain a trivialization $\mathcal{L}$ of the tangent bundle TG by left invariant vector fields, and every left invariant vector field is a linear combination of these. Of course this trivialization $\mathcal{F}$ of TG depends on the choice of basis for T_eG, but this dependence is determined just by a linear transformation at T_eG carrying one basis into the other.

If Γ is a discrete subgroup of G, then the orbit space $M_\Gamma = \Gamma\backslash G$ is a smooth manifold, in fact a homogeneous space, and the left-invariant vector fields descend to M_Γ. Thus the trivialization $\mathcal{L}$ of the tangent bundle TG determines a trivialization of the tangent bundle of M_Γ that we also denote $\mathcal{L}$. This trivialization $\mathcal{L}$ induces a metric on M_Γ with identical curvature properties to the left-invariant metric on G defined by $\mathcal{L}$. This type of metrics on homogeneous spaces plays an important role in geometry (see for instance [172]). One also has several other important geometric structures on M_Γ determined by $\mathcal{L}$, as for instance a spin structure with its Dirac operator (see [124] and Chapter IV below).

Let us consider now a complex 2-dimensional manifold X. The structure group of its tangent bundle TX is $GL(2, \mathbb{C})$, and we can always reduce it to $U(2)$ by endowing X with some Hermitian metric. If we try to reduce the structure group of TX further to $SU(2)$ we may run into problems, since this is not always possible. The obstruction for doing so is the first Chern class of TX (c.f. IV.1 below). To see this we notice that $U(2)$, being the structure group of TX, acts naturally on the space of differential forms (of all degrees) on X. In particular it acts on the forms of type $(2,0)$, i.e., on the sections of the bundle associated to TX whose fibre at $x \in X$ is $\bigwedge^{(2,0)} T_x^*X$; this is isomorphic to the bundle of holomorphic 2-forms on X, which is called the *canonical bundle* $\mathcal{K}$ of X. The action of $U(2)$ on $\bigwedge^{(2,0)} T_x^*X$ is by the determinant, and $SU(2)$ is precisely the subgroup of $U(2)$ which acts trivially on $\mathcal{K}$. Hence we can reduce the structure group of TX from $U(2)$ further to $SU(2)$ if and only if the bundle $\mathcal{K}$ is trivial, and the specific reductions of the structure group of TX to $SU(2)$ correspond to the specific trivializations of $\mathcal{K}$. Now, the canonical bundle $\mathcal{K}$ is 1-dimensional, so it is classified (up to a differentiable isomorphism) by its Chern class, which is the negative of the first Chern class of X (see [101, Th. 4.4.3]. Hence $c_1(X)$ vanishes iff $\mathcal{K}$ is trivial and this happens iff we can reduce the structure group of TX to $SU(2)$.

Now suppose the manifold X is endowed with a never-zero holomorphic 2-form, as in the previous sections. Then this form defines a specific reduction to $SU(2)$ of the structure group of TX. Since $SU(2)$ is isomorphic to the group of unit quaternions $Sp(1)$, this means that we have at each point of X multiplication of tangent vectors by the quaternions i, j, k.

Let us now return to the situation envisaged above. In each of the six cases in question we have a specific complex 2-manifold X equipped with a never-zero holomorphic 2-form. Thus we have, by the previous discussion, a specific reduction of the structure group of TX to $SU(2) \cong Sp(1)$. Moreover, the normal bundle of the foliation is trivial in all cases. Choosing a trivialization of this bundle one gets, via multiplication by i, j, k, a trivialization $\mathcal{L}$ of $T\mathcal{F}$, the bundle tangent to the foliation. If $G \neq N$, this trivialization of $T\mathcal{F}$ is left invariant under the G-action on X and so defines a basis of the Lie algebra of G.

In the nilpotent case we still have the holomorphic 2-form on X and therefore a multiplication by the quaternions i, j, k, which induces a canonical trivialization of the tangent bundle of each leaf of the foliation $\mathcal{F}$. However, there is not a group action of G preserving the leaves, but only an action restricted to each discrete subgroup, so I do not know how to relate these vector fields with the Lie algebra of G.

It is worth noting that this same construction applies to the link of every 2-dimensional isolated hypersurface (or complete intersection) singularity [224]. In fact, if V is the germ of an affine surface in $\mathbb{C}^3$ defined by a holomorphic map-germ $(\mathbb{C}^3, 0) \xrightarrow{f} (\mathbb{C}, 0)$ with an isolated critical point at $0 \in \mathbb{C}^3$, then one has the canonical 2-form ω of 3.2.1 above:

$$\omega = \frac{dz_1 \wedge dz_2}{\partial f/\partial z_3} = \frac{dz_2 \wedge dz_3}{\partial f/\partial z_1} = \frac{dz_3 \wedge dz_1}{\partial f/\partial z_2} .$$

This 2-form defines as above, multiplication by the quaternions i, j, k. The normal bundle of the link M in $V^* = V - \{0\}$ is trivial and canonically trivialized by choosing the unit outwards normal vector field ν (for the restriction to V^* of the ambient Hermitian metric). Multiplying ν by i, j, k at each point of $T(V^*)|_M$ we get a *canonical* trivialization of the bundle $T(M)$ tangent to M. This was called in [224] *the canonical framing* of the link. This also induces a canonical metric and a spin structure on M.

III.8   Remarks

There are several other lines of research, with very important results, that fall into the theme of this chapter, and which I am not including here for several reasons. However I would like to make a few comments in this respect and give some guidelines for further reading. Most of what I say here points towards articles of W.D. Neumann, either by himself or with some co-authors.

8.1. It is important, for many reasons, to know when the link of a surface singularity is a homology sphere. It was proved in [185, Th. 1] that every weighted homogeneous surface singularity whose link is a homology sphere, is equivalent to one given by a Brieskorn-Hamm complete intersection (which were described in II.7); and it is relatively easy to decide when the link of a Brieskorn-Hamm complete intersection is a homology sphere. This is explained in Milnor's book in the case of hypersurfaces in $\mathbb{C}^3$, and for the general case one can use, for instance, the description in [183] of the corresponding Seifert invariants. On the other hand, it is known (see Chapter IV below) that the link of every surface singularity *is a plumbed manifold*, and there is an operation one can make on plumbed manifolds, called *splicing* (see [75]), which allows us to construct every link of a singularity which is a homology sphere by iterated splicing of homology spheres which are Seifert manifolds.

8.2. In the previous sections we looked at 3-manifolds of the form $\Gamma\backslash G$ for G a Lie group and Γ a uniform subgroup, and we discussed the relation of these and surface singularities. There are however much deeper things one can say in this respect. On the one hand, one has the deformations of the complex structure on the corresponding surface singularities; on the other hand, by expressing M as $\Gamma\backslash G$ we are endowing this manifold with a geometric structure in the sense of Thurston, and this geometric structure can be also deformed. This corresponds to deforming the embedding of the group Γ in G. It is natural to ask which are all these moduli spaces of deformations, and how the deformations of the complex structure of a surface singularity relate to the deformations of the geometric structure on the link. These questions are beautifully answered in [186], where the author also gives references to several significant previous contributions by various authors, including himself.

8.3. Given a complex analytic surface singularity $(V,0)$ in $\mathbb{C}^n$, we know that the link M of V is the intersection $V \cap \mathbb{S}_\varepsilon$ of V with a sufficiently small sphere, and we know from Chapter I that the diffeomorphism type of M does not depend on the choice of sphere. In 8.2 above we explained that under certain hypotheses the link M has a geometric structure in the sense of Thurston. Still, it was observed in [220] that one *always has* another type of geometric structure: since M is a real hypersurface in a complex manifold, it is automatically a CR-manifold. Of course, the CR-structure on M depends on the choice of the sphere, and on the embedding of V in the ambient space. However J. Scherk proved in [220], remarkably, that if (V_1, x_1), and (V_2, x_2), are normal surface singularities whose links M_1, M_2 are CR-isomorphic (for appropriate links), then they are actually analytically equivalent. In fact Scherk proves more: this statement holds in all dimensions, not only in complex dimension 2, and if f is a CR-isomorphism between the links, then f itself extends to an analytic isomorphism between the analytic germs. This was used later in [73] to have a better understanding of the relation between the geometric structure of the link, regarded as a homogeneous space $\Gamma\backslash G$, and the analytic structure on the surface singularity. In [59] Dolgachev mentions that F. Ehlers and J. Scherk pointed out to him that the diffeomorphisms that he obtains (and that we explained above) between the links of normal Gorenstein surface singularities with $\mathbb{C}^*$-action and quotient spaces of the form $\Gamma\backslash G$, for $G = SU(2)$, $\widetilde{SL}(2, \mathbb{R})$ or N, also preserves the CR structures: on the link M it is determined by the analytic structure on V; on $\Gamma\backslash G$ it is induced by the unique G-invariant CR structure on G [51]. I presume the same statement holds for all quotients $\Gamma\backslash G$ for the remaining 3-dimensional Lie groups.

Chapter IV

Within the Realm of the General Index Theorem

The goal of this chapter is to show some of the ways in which the general index theorem of Atiyah-Singer has had impact in singularities theory. Mostly, I will restrict myself to some interesting applications of two "particular" cases of the index theorem: the Hirzebruch-Riemann-Roch formula and the Hirzebruch signature theorem. We consider also Rochlin's signature theorem.

The general philosophy is the following. The index theorem of Atiyah-Singer may be thought of as a beautiful and far-reaching generalization of the Hirzebruch-Riemann-Roch theorem, both in statement and in the spirit of the original proof (modified in [20, 21] to allow certain generalizations). Given a closed, oriented manifold M, vector bundles E and F over M and an elliptic operator D from the sections of E to those of F, one has that both the kernel and the cokernel of D are finite-dimensional, and the difference of these dimensions is by definition the analytic index of D. The index theorem gives a description of this integer in terms of topological data implicit in the elliptic operator, the so-called topological index. This establishes a very deep connection between analysis/geometry and topology.

Special cases are the signature theorem of Hirzebruch, the Hirzebruch-Riemann-Roch theorem, the Lefschetz fixed point formula, the relation of the Dirac operator with the $\widehat{A}$-genus for spin manifolds and several other fundamental theorems in mathematics. Rochlin's signature theorem can also be seen through index theory (see §3 below).

In the case of a compact manifold M with non-empty boundary ∂M, both sides of the index formula are still defined, the analytic index and its topological counterpart, but their difference need not vanish. This led Hirzebruch to introduce the *signature defect*, which has had significant applications to singularities. For elliptic operators in general, the signature defect was the starting point of a series of articles by Atiyah, Patodi and Singer, leading them to the definition of the

η-invariant, which is the boundary contribution that gives the correction term for the index formula.

This has been used in singularity theory in essentially two related general ways:

(i) If $(V, 0)$ is an isolated (real or complex) singularity germ in some affine space, then the diffeomorphism type of its link $M = V \cap \mathbb{S}_\varepsilon$ depends only on the analytic structure of V, and not on the choices of defining equations for V or the radius of the sphere (see Chapter I). Thus any invariant of closed manifolds gives automatically an invariant of singularities. This has led to a vast number of interesting results for complex singularities, especially in dimension 2. For example for the Rochlin invariant, the signature defect (we discuss both of these below), and in many more directions. Of course, the index theorem comes from the 1960s and since then there has been a tremendous evolution in mathematics, particularly in the low dimensions, where the theorems of Donaldson, Witten and others have given rise to powerful invariants of 3- and 4-dimensional manifolds, which are nowadays giving important information about surface singularities (see the last section of this chapter).

It is worth noticing that (to my knowledge), essentially nothing has been done in this direction for real analytic singularities, and I believe this can be interesting.

(ii) Consider now an isolated (real or complex) singularity germ $(V, 0)$ and a resolution of this singularity $\pi : \widetilde{V} \to V$ (see Section 5 below); these can be considered as manifolds with boundary the link M, and we can use them to compute the invariants associated to the link as in i) above. Furthermore, if $(V, 0)$ is a complex analytic complete intersection germ, or more generally a smoothable singularity (see below), one also has a Milnor fibre F (or a smoothing) of V. These can also be considered as manifolds with boundary the link, and the same invariants that we evaluated using a resolution can be computed via the smoothing. This leads to interesting relations among the invariants of $(V, 0)$ coming from a resolution and those coming from the smoothings. For instance the formula of Laufer (and its generalization to higher dimensions by Looigenga) for the Milnor number, and that of Durfee for the signature, are both obtained in this way.

IV.1   A review of characteristic classes

The material in this section is all standard, see for instance [170, 166, 242]. Let $v = (v_1, \ldots, v_n)$ be a vector field in an open set $U \subset \mathbb{R}^n$; the vector field is said to be continuous, smooth, analytic, etc., according as its components $\{v_1, \ldots, v_n\}$ are continuous, smooth, analytic, etc., respectively. A *singularity* P of v is a point where all of its components vanish, i.e., $v_i(P) = 0$ for all $i = 1, \ldots, n$, and a singularity of v is *isolated* if at every point x near P there is at least one component of v which is not zero.

Let v be a vector field as above and let P be an isolated singularity of v. Let $\mathbb{S}_\varepsilon$ be a small sphere around P. Then *the Poincaré-Hopf index* of v at P is the degree of the map $\frac{v}{||v||}$ from $\mathbb{S}_\varepsilon$ into the unit sphere in $\mathbb{R}^n$. It is clear that all these definitions extend to vector fields on smooth manifolds, and a fundamental property of this index is the following theorem:

1.1 Theorem. (*Poincaré-Hopf*) *Let M be a closed, oriented n-manifold and let v be a continuous vector field on M with isolated singularities. Define the total index of v, denoted $PH(v, M)$, to be the sum of all its local indices at the singular points. Then one has*

$$PH(v, M) = \chi(M),$$

independently of v, where $\chi(M)$ is the Euler characteristic.

The proof of this theorem can be found in many text books. If M is now an oriented manifold with boundary, one has a similar theorem:

1.2 Theorem. *Let M be a compact, oriented n-manifold with boundary ∂M, and let v be a non-singular vector field on a neighborhood U of ∂M. Then:*

 (i) *v can be extended to the interior of M with isolated singularities.*

 (ii) *the total index of v in M is independent of the way we extend it to the interior of M. In other words, the total index of v is fully determined by its behavior near the boundary.*

(iii) *if v is everywhere transversal to the boundary and pointing outwards from M, then one has $PH(v, M) = \chi(M)$. If v is everywhere transversal to ∂M but it points inwards M, then $PH(v, M) = \chi(M) - \chi(\partial M)$.*

There is another way of defining the index, which brings us closer to the topic that we discuss in a moment. Suppose first that M has no boundary and we want to construct a non-singular vector field on M, i.e., a non-zero section v of its tangent bundle TM. We do it step by step: let M^j be the j-skeleton of M for some triangulation (or cell decomposition). It is clear that we can always construct a non-zero vector field on M^0, which consists of isolated points. Now observe that if we have the vector field defined and non-zero on the boundary of a p-simplex σ, then the vector field can be regarded locally as a map $\partial \sigma \cong \mathbb{S}^{p-1} \to \mathbb{S}^{n-1}$; hence it extends to the interior of σ if $p < n$. In this way we see that a never-zero vector field can be constructed up to the $(n-1)$-skeleton of M. When we try to extend it to the n-skeleton, for each n-simplex σ we have the vector field on its boundary, a topological $(n-1)$-sphere, defining an element in the homotopy group $\pi_{n-1}(\mathbb{S}^{n-1}) \cong \mathbb{Z}$. The vector field can be extended to this cell iff the corresponding map is nulhomotopic. Thus we have for each n-simplex a local obstruction in $\mathbb{Z}$. In fact the vector field always extends to all of σ minus its barycentre $\hat{\sigma}$ and the corresponding local obstruction in $\mathbb{Z}$ is the Poincaré-Hopf index at $\hat{\sigma}$. This gives rise to a cochain of dimension n, which is in fact a cocycle, hence representing a cohomology class with integer (local) coefficients. The cohomology class that we obtain in this way is *the Euler class* of M, $e(M) \in H^n(M; \mathbb{Z})$, which evaluated on the orientation cycle $[M]$ gives a number: the Euler characteristic $\chi(M)$.

Suppose now that M has boundary ∂M and the vector field v is defined and non-singular around the boundary. The previous stepwise process tells us that v can be extended to all of $\partial M \cup M^{(n-1)}$. For each n-cell of M we have a local obstruction in $\mathbb{Z}$. Thus we get a cocycle $e(M, v)$ in the relative cohomology group $H^n(M, \partial M; \mathbb{Z})$. As a relative class $e(M, v)$ does depend on the choice of the vector field, generally speaking, though its image in $H^n(M; \mathbb{Z})$ depends only on M. This is the *Euler class of M relative to v*. Evaluating $e(M, v)$ on the orientation cycle of $(M, \partial M)$ we get a number, *the total index of v in M*.

Let us now extend these constructions to define the Chern classes via obstruction theory. This can be done with no extra work for a continuous complex vector bundle ξ of complex (fibre) dimension n over a finite simplicial complex K of real dimension $2m$; we assume for simplicity $n \leq 2m$, but this is not necessary. We denote by K^j the jth-skeleton.

1.3 Definition. Consider a set $F^{(r)} = \{v_1, \ldots, v_r\}$ of r continuous sections of ξ defined on a subcomplex A of K. If the vectors v_i are linearly independent over $\mathbb{C}$ at every point of A, we say that $F^{(r)}$ defines an *r-frame*, or a *non-singular r-field* on A. If at a point $a \in A$ one (at least) of the vectors v_i is zero, or if the r vectors are not linearly independent over $\mathbb{C}$, then we say that a is a *singular point* of the r-field $F^{(r)}$.

Let $W_r(n)$ be the Stiefel manifold of complex r-frames in $\mathbb{C}^n$. We know (see [242]) that this manifold is $(2n-2r)$-connected and its first non-zero homotopy group is $\pi_{2n-2r+1}(W_r(n)) \cong \mathbb{Z}$. The bundle of r-frames of ξ, denoted by $W_r(\xi)$, is the bundle associated with ξ whose fibre over $x \in K$ is the set of r-frames in ξ_x.

The Chern class $c_q(\xi) \in H^{2q}(K)$ is the first possibly non-zero obstruction to construct a section of $W_r(\xi)$. Let us construct this class using the same stepwise process we used to construct the Euler class of a manifold. Let σ be a k-simplex of K (or a k-cell of a given cell decomposition of a smooth manifold). If the section $F^{(r)}$ of $W_r(\xi)$ is already defined over the boundary of σ, it gives a map:

$$\partial \sigma \cong \mathbb{S}^{k-1} \xrightarrow{\ F^{(r)}\ } W_r(\xi)|_\sigma \cong \sigma \times W_r(n) \xrightarrow{\ pr_2\ } W_r(n),$$

thus an element of $\pi_{k-1}(W_r(n))$. If $k \leq 2n - 2r + 1$, this homotopy group is zero, so the section $F^{(r)}$ can be extended to σ without singularities. If $k = 2n - 2r + 2 = 2q$, we meet an obstruction. So we can always construct a section $F^{(r)}$ of $W_r(\xi)$ over the $(2q - 1)$-skeleton. When we try to extend $F^{(r)}$ to the $2q$-skeleton, we have for each simplex an r-frame on its boundary, which defines an element $I(F^{(r)}, \sigma) \in \pi_{2q-1}(W_r(n)) \cong \mathbb{Z}$. Let us define a cochain $\gamma \in C^{2q}(K; \pi_{2q-1}(W_r(n)))$ by $\gamma(\sigma) = I(F^{(r)}, \sigma)$ for each simplex σ, and then extend it by linearity. *This cochain is actually a cocycle and the cohomology class that it represents is the qth Chern class $c_q(\xi)$ of ξ in $H^{2q}(K; \mathbb{Z})$*. It is independent of the various choices involved in its definition. We have one such class for each $q = 1, \ldots, n$.

For an almost complex manifold M^{2m}, its Chern classes are those of its tangent bundle. Note that $c_m(M)$ coincides with the Euler class of the underlying real tangent bundle $T_{\mathbb{R}} M$.

Suppose now that L is a sub-complex of K such that the bundle ξ is trivial over L, and we are given a specific trivialization τ of this bundle. Then we can form the quotient space $\widehat{K} = K/L$; this is a finite CW-complex. The trivialization $\tau : \xi|_L \to L \times \mathbb{C}^n$ gives us a way to identify all the fibres of ξ over L, and so we get a complex vector bundle $\widehat{\xi}$ over $\widehat{K}$. The Chern classes

$$c_i(\widehat{\xi}) \in H^{2i}(\widehat{K}; \mathbb{Z}) \cong H^{2i}(K, L; \mathbb{Z}) \ , i = 1, \ldots, n \, ,$$

of $\widehat{\xi}$ are by definition *the Chern classes of ξ relative to the trivialization τ over L*, denoted $c_i(\xi; \tau)$. These classes do depend on the choice of τ, generally speaking, but their images in $H^*(K; \mathbb{Z})$ are the usual Chern classes, which depend only on ξ and K.

Let us now repeat the same constructions as before, but we consider now an oriented real vector bundle ζ of fibre dimension n over a finite simplicial complex K of dimension m; for simplicity we assume $n \leq m$. The previous arguments tell us that the obstruction for constructing a non-zero section of ζ is a cohomology class $e(\zeta) \in H^n(K; \mathbb{Z})$, which is (by definition) *the Euler class* of ζ. The image of $e(\zeta)$ in $H^n(K; \mathbb{Z}_2)$, denoted $w_n(\zeta)$, is by definition *the top-dimensional Stiefel-Whitney class* of ζ. Notice that the orientation of ζ is only necessary to define the Euler class, which has integral coefficients. The Stiefel-Whitney class $w_n(\zeta)$ is well defined even for non-orientable bundles.

Now let us mimic the previous construction of Chern classes: we want to construct at each point of K two linearly independent sections of ζ. This means that (up to homotopy) we want a section of the bundle $V_2(\zeta)$, whose fibre at each point is the Stiefel manifold $V_2(n)$ of oriented 2-frames in $\mathbb{R}^n$. The homotopy groups of $V_2(n)$ vanish up to dimension $i = n - 3$ and one has $\pi_{n-2}(V_2(n)) \cong \mathbb{Z}$ for n even, and $\pi_{n-2}(V_2(n)) \cong \mathbb{Z}_2$ for n odd, see [244, 25.6]. Thus, for n even, one gets that the primary obstruction $W_{n-1}(\zeta)$ for constructing a 2-frame of ζ lives in $H^{n-1}(K; \pi_{n-2}(V_2(n))) \cong H^{n-1}(K; \mathbb{Z})$, while for n odd the obstruction is a class $w_{n-1}(\zeta)$ in $H^{n-1}(K; \mathbb{Z}_2)$. For n even, let us denote also by $w_{n-1}(\zeta) \in H^{n-1}(K; \mathbb{Z}_2)$ the reduction modulo 2 of $W_{n-1}(\zeta)$. By definition, the class $w_{n-1}(\zeta)$ is the $(n-1)$th *Stiefel-Whitney class* of ζ, for all n.

More generally, if we try to construct a k-frame of ζ by the stepwise process, the first possibly non-zero obstruction will be a cohomology class of K in dimension $(n - k + 1)$ with coefficients in $\pi_{n-k}(V_k(n))$, which is isomorphic to $\mathbb{Z}$ when $n - k$ is even, and it is isomorphic to $\mathbb{Z}_2$ when $n - k$ is odd and $k > 1$, see [244, 25.6]. In all cases we can consider the corresponding class $w_i(\zeta) \in H^{n-k+1}(K; \mathbb{Z}_2)$; this is by definition the (i)th *Stiefel-Whitney class* of ζ, $i = n - k + 1 = 1, \ldots, n$.

We notice that by construction, when $n - k$ is even, or when $k = 1$, the corresponding Stiefel-Whitney class automatically has a lifting to the integral homology. However, for $n - k$ odd, such a lifting may not exist.

For a smooth manifold, its Stiefel-Whitney classes are those of its tangent bundle.

Of course just as for Chern classes, when one has a trivialization of the bundle ζ over a subcomplex $L \subset K$, one also has a theory of *relative Stiefel-Whitney* *classes*, which live in $H^*(K, L; \mathbb{Z}_2)$. Their images in the absolute cohomology are the usual Stiefel-Whitney classes, which depend only on the space K and the vector bundle ζ, while the relative classes do depend in general on the choice of trivialization of ζ over L.

There is finally a third type of characteristic classes associated to vector bundles, which is very useful and important in order to obtain further information of the bundles. These are the Pontryagin classes. To define these, we start with a real vector bundle ζ over K as before. For each fibre V_x of ζ consider the complexification $V_x \otimes \mathbb{C} = V_x \otimes_{\mathbb{R}} \mathbb{C}$. The union of all these vectors gives a new vector bundle over K, which is now a complex vector bundle that we denote by $\zeta \otimes \mathbb{C}$ and call it the complexification of ζ.

Notice that the underlying real vector bundle $(\zeta \otimes \mathbb{C})_{\mathbb{R}}$ is canonically isomorphic to the Whitney sum $\zeta \oplus \zeta$, with each fibre being canonically of the form $V_x \oplus i V_x$.

We now consider the Chern classes of the complex bundle $\zeta \otimes \mathbb{C}$. It is an exercise (see [170, §15]) to show that the odd-dimensional classes $c_{2i+1}(\zeta \otimes \mathbb{C})$ are all elements of order 2, i.e., twice the element is zero in cohomology. Ignoring these elements of order 2, the *i*th *Pontryagin class* of ζ, $p_i(\zeta) \in H^{4i}(K; \mathbb{Z})$, is defined to be the integral class $(-1)^i c_{2i}(\zeta \otimes \mathbb{C})$, the sign being just a useful convention.

There is a situation which is especially relevant for what follows in this chapter, that is when the bundle we start with is already a complex bundle, particularly the tangent bundle of a complex manifold. So we consider now a complex vector bundle ξ over K; we may forget the complex structure on ξ and regard the underlying real vector bundle $\xi_{\mathbb{R}}$. One has (see [170, 15.4]) that the complexification of this bundle is canonically isomorphic to the Whitney sum $\xi \oplus \overline{\xi}$, where $\overline{\xi}$ is the dual vector bundle. We also know that the Chern classes satisfy the Whitney product formula with respect to Whitney sums. This means that one has:

$$1 - p_1(\xi) + - \cdots \pm p_n(\xi) = (1 + c_1(\overline{\xi}) + \cdots + c_n(\overline{\xi})) \cdot (1 + c_1(\xi) + \cdots + c_n(\xi))$$

and the Chern classes of $\overline{\xi}$ satisfy: $c_i(\overline{\xi}) = (-1)^i c_i(\xi)$, see [170, 14.9]. Thus one has ([170, 15.5]):

1.4 Proposition. *For any complex n-plane bundle ξ, the Chern classes of ξ determine the Pontryagin classes of $\xi_{\mathbb{R}}$ by the formula (dropping ξ from the notation for simplicity):*

$$1 - p_1 + p_2 - + \cdots \pm p_n = (1 - c_1 + c_2 - + \cdots \pm c_n) \cdot (1 + c_1 + c_2 + \cdots + c_n)$$

and therefore $p_i(\xi_{\mathbb{R}})$ is: $\quad p_i = c_i^2 - 2c_{i-1}c_{i+1} + - \cdots \pm 2c_1 c_{2i-1} \mp 2c_{2i}.$

In particular, for a complex manifold M of dimension 2 one has:

$$p_1(M) = c_1(M)^2 - 2c_2(M).$$

1.5 Remark. Notice that we have only considered here characteristic classes of vector bundles and manifolds. These will be used later to study invariants of singular spaces, usually by looking at the characteristic classes of either a resolution of the singular variety or a smoothing of it (see the definitions below). These are useful invariants, however they are not intrinsic to the singular variety and depend on the choice of resolution or smoothing. There is a vast literature about "Chern classes for singular varities", which are intrinsic to varieties. This began with the work of M.H. Schwartz [222] and R. MacPherson [152]; we refer to [34] for an overview of the topic. But we shall not use these in the sequel.

IV.2   On Hirzebruch's theorems about the signature and Riemann-Roch

Let M be a compact, smooth and oriented manifold (possibly with non-empty boundary ∂M) of dimension $2n \geq 2$ and consider its middle cohomology with real coefficients, $H^n(M, \partial M; \mathbb{R})$. This is a finite-dimensional vector space over which one has a non-degenerate symmetric bilinear form:

$$b_M \; : \; H^n(M, \partial M; \mathbb{R}) \times H^n(M, \partial M; \mathbb{R}) \longrightarrow H^{2n}(M, \partial M; \mathbb{R}) \cong \mathbb{R}$$

given by the cup product. The Poincaré-Lefschetz duality isomorphism implies that this pairing is non-degenerate. This duality isomorphism also allows us to look at this bilinear pairing in homology,

$$b_M^* \; : \; H_n(M; \mathbb{R}) \times H_n(M; \mathbb{R}) \longrightarrow H_0(M; \mathbb{R}) \cong \mathbb{R} \, ,$$

where it is given by the intersection product of cycles. One usually refers to the corresponding quadratic form in $H^n(M, \partial M; \mathbb{R})$ as *the quadratic form of M*.

2.1 Definition. Let $n = 2k$. The *signature of M^{4k}*, $\sigma(M)$, is the signature of the bilinear form b_M. That is, $\sigma(M) = s^+ - s^-$, the number s^+ of positive eigenvalues, minus the number s^- of negative eigenvalues of b_M.

For a manifold of dimension 4k+2 the signature is defined to be 0. It is clear that one has $\sigma(M) = -\sigma(-M)$ where $-M$ is M with reversed orientation.

One has:

2.2 Theorem. (R. Thom [250].) *The signature satisfies:*

(i) *it is an invariant of oriented cobordism, in particular it vanishes when the manifold is an oriented boundary;*

(ii) *it is multiplicative, i.e.,*

$$\sigma(M^{4k} \times N^{4r}) \; = \; \sigma(M^{4k}) \cdot \sigma(N^{4r});$$

(iii) *it is additive under disjoint unions:*

$$\sigma(M^{4k} \sqcup N^{4r}) \; = \; \sigma(M^{4k}) + \sigma(N^{4r}).$$

Thom also proved in [250, IV.14, p. 81] that for $k = 1$ one has

$$\sigma(M) = \frac{1}{3} p_1(M)[M],$$

a third of the Pontryagin number. This gives a clue to what is coming next.

To give a precise statement of the *signature theorem of Hirzebruch* we need to introduce the *multiplicative sequences of polynomials* (we follow [101]). These are sequences $\{K_j\}$ of polynomials in the indeterminates $p_1, p_2, \ldots$. One sets $p_0 = 1$ and considers the ring $\mathbb{Q}[p_1, p_2, \ldots]$. This ring is graded by assigning weight i to each variable p_i and weight $i_1 + i_2 + \cdots + i_n$ to the product $p_{i_1} p_{i_2} \cdots p_{i_n}$. Each polynomial K_j is homogeneous of weight j. The sequence $\{K_j\}$ is called multiplicative if every identity of the form:

$$1 + p_1 z + p_2 z^2 + \cdots = (1 + p_1' z + p_2' z^2 + \cdots) \cdot (1 + p_1'' z + p_2'' z^2 + \cdots)$$

with z, p_i, p_i', p_i'' indeterminate, implies an identity

$$\sum_{i=0}^{\infty} K_i(p_1, \ldots, p_i) z^i = \left(\sum_{i=0}^{\infty} K_i(p_1', \ldots, p_i') z^i \right) \left(\sum_{j=0}^{\infty} K_j(p_1, \ldots, p_j) z^j \right).$$

Given one such sequence, the power series,

$$Q(z) = K(1 + z) = \sum_{i=0}^{\infty} b_i z^i, \qquad b_0 = 1, b_i = K_i(1, 0, \ldots, 0) \in \mathbb{Q},$$

is called the *characteristic power series* of the multiplicative sequence. It turns out [101, 1.2.1] that each multiplicative sequence of polynomials as above is completely determined by its characteristic power series $Q(z)$. Furthermore ([101, 1.2.2]), every formal power series $Q(z) = \sum_{i=0}^{\infty} b_i z^i, b_0 = 1, b_i \in \mathbb{Q}$, corresponds to a multiplicative sequence of polynomials. Hence, to describe such a sequence we only need to specify the corresponding power series $Q(z)$. So now consider the power series

$$Q(z) = \frac{\sqrt{z}}{\tanh \sqrt{z}} = 1 + \sum_{r=0}^{\infty} (-1)^{r-1} \frac{2^{2r}}{(2r)!} B_r z^r.$$

The coefficients B_r are the Bernoulli numbers. Hirzebruch in [101] gives the first of these coefficients, some of them are:

$$B_1 = \frac{1}{6}, \quad B_2 = \frac{1}{30}, \quad B_3 = \frac{1}{42}, \ldots, \quad B_8 = \frac{3617}{510}.$$

2.3 Definition. The multiplicative sequence corresponding to this power series is *the L-sequence of Hirzebruch*, denoted $\{L_j(p_1, \ldots, p_j)\}$.

The first of these polynomials are listed by Hirzebruch. These are:

$$L_1 = \tfrac{1}{3}\,p_1\,,$$
$$L_2 = \tfrac{1}{45}\,(7p_2 - p_1^2)\,,$$
$$L_3 = \tfrac{1}{3^3 \cdot 5 \cdot 7}(62p_3 - 13p_2p_1 + 2p_1^3).$$

We can now state the *signature theorem of Hirzebruch* [101, Thm. 8.2.2]:

2.4 Theorem. *Assume M is closed (i.e., $\partial M = \emptyset$) of dimension $4k > 0$. Then its signature $\sigma(M)$ is:*

$$\sigma(M) = L_k\big(p_1(M), \ldots, p_k(M)\big)[M]\,,$$

where the $p_i(M) \in H^{4i}(M;\mathbb{Z})$ are the Pontryagin classes of M.

The polynomial on the right-hand side is called *the L-genus* of the manifold M, denoted $L[M]$. Notice that just the mere fact this genus is an integer is already a miracle.

The proof of this theorem was given by Hirzebruch in [101], and as mentioned before, this can also be deduced directly from the general Index Theorem of Atiyah-Singer. The idea of the proof is the following (see [101, Chapter II] or [170, p. 225]). First one notices that by construction, the L-genus satisfies a multiplicative property similar to that satisfied by the signature (see Definition 2.3 above), and they are both additive under disjoint unions, which is the operation in cobordism. From this, together with the fact that the Pontryagin numbers are invariants of oriented cobordism, one deduces that the correspondences $M \mapsto \sigma(M)$ and $M \mapsto L[M]$ give rise to algebra homomorphisms $\Omega_* \otimes \mathbb{Q} \longrightarrow \mathbb{Q}$, where Ω_* is the oriented cobordism ring. Hence it is enough to prove Theorem 2.4 for a set of generators of the algebra $\Omega_* \otimes \mathbb{Q}$, which vanishes in dimensions $\not\equiv 0 \bmod 4$ (by [250]). Such a set is provided by the complex projective spaces $\mathbb{CP}^{2k}$. It is not hard to show that the signature of $\mathbb{CP}^{2k}$ is always $+1$. So one needs to show only that the L-genus of every complex projective space is also $+1$, and this is proved in [101, p. 225] or [170].

Let us restrict now to dimension 4. As mentioned above, in this case the theorem was given by R. Thom [250, IV.14, p. 81] and it says:

$$\sigma(M) = \frac{p_1(M)[M]}{3}. \tag{2.5}$$

If we now take M to be a complex manifold (even almost complex is enough), then we know that $p_1 = c_1^2 - 2c_2$, and so the previous formula takes the very nice shape:

$$\sigma(M) = \frac{K^2 - 2\,\chi(M)}{3}\,, \tag{2.6}$$

where K is the canonical divisor, representing a homology class which is the Poincaré dual of the first Chern class of the cotangent bundle T^*M, and $\chi(M)$ is the topological Euler characteristic.

Let us move now towards the Hirzebruch-Riemann-Roch theorem. This begins with Riemann's calculation that the holomorphic differentials on a compact Riemann surface S of genus $g \geq 0$ is a complex vector space of dimension g, i.e.,

$$\dim H^0(S; \mathcal{K}_S) = g,$$

where $\mathcal{K}_S$ is the cotangent (or canonical) bundle on S. The next step is to consider an arbitrary holomorphic line bundle $\mathcal{L}$ over S, then we would like to be able to determine the dimension of the space of global sections of it $H^0(S; \mathcal{L})$, which we denote $h^0(\mathcal{L})$ for simplicity; in particular one may ask whether this number is still determined by the topology. Before stating the theorem, notice that the above formula can be written as:

$$h^0(\mathcal{O}_S) - h^0(\mathcal{K}_S) = \frac{1}{2}(2 - 2g) = \frac{1}{2} c_1(S)[S]$$

where $h^0(\mathcal{O}_S) = 1$ are the holomorphic functions from S to $\mathbb{C}$ and c_1 is the Chern class of TS. The Riemann-Roch formula for line bundles over a Riemann surface in general says:

$$h^0(\mathcal{L}) - h^0(\mathcal{K}_S \otimes \mathcal{L}^{-1}) = \left(c_1(\mathcal{L}) + \frac{1}{2} c_1(S) \right)[S].$$

The Serre duality theorem for line bundles over Riemann surfaces gives the duality:

$$H^0(\mathcal{K}_S \otimes \mathcal{L}^{-1}) \cong H^1(\mathcal{L}),$$

and therefore the Riemann-Roch formula can be expressed as:

$$h^0(\mathcal{L}) - h^1(\mathcal{L}) = \left(c_1(\mathcal{L}) + \frac{1}{2} c_1(S) \right)[S].$$

That is, the *analytic Euler characteristic*

$$\chi(\mathcal{L}) = \sum_{i=0}^{1} (-1)^i h^i(\mathcal{L}),$$

is determined by the topological invariants on the right-hand side. We write this as:

$$\chi(S; \mathcal{L}) = \left(Td_1(S) + c_1(\mathcal{L}) \right)[S], \tag{2.7}$$

where $Td_1(S) = \frac{1}{2} c_1(S)$. This is the expression that generalizes to higher dimensions. In order to explain this we need to introduce another multiplicative sequence of polynomials, the *Todd sequence* $\{Td_k\}$. This is determined by the power series:

$$Q(x) = \frac{x}{1 - e^{-x}} = 1 + \frac{1}{2} x + \sum_{k=0}^{\infty} (-1)^{k-1} \frac{B_k}{(2k)!} x^{2k}.$$

The first polynomials in this sequence are [101]:

$$Td_1 = \frac{1}{2}\, c_1\,, \qquad Td_2 = \frac{1}{12}(c_2 + c_1^2)\,, \qquad Td_3 = \frac{1}{24}\, c_2 c_1\,,$$

$$Td_4 = \tfrac{1}{720}(-c_4 + c_3 c_1 + 3c_2^2 + 4c_2 c_1^2 - c_1^4)\,.$$

2.8 Definition. Let M be an almost complex manifold of real dimension $2k$. Then its *Todd genus* $Td(M)$ is:

$$Td(M) \;=\; Td_k(c_1(M),\dots,c_k(M))[M]\,,$$

where $c_i(M) \in H^{2i}(M;\mathbb{Z})$, $i = 1,\dots,k$, are the Chern classes of M.

2.9 Definition. Let M be a complex manifold of dimension k. Its *analytic Euler characteristic* (also called the *arithmetic genus*) is:

$$\chi(M;\mathcal{O}) = \sum_{i=0}^{k}(-1)^i\, h^i(M;\mathcal{O})\,;$$

and more generally, if $\mathcal{L}$ is a holomorphic bundle over M, then the *analytic Euler characteristic* of $\mathcal{L}$ is:

$$\chi(M;\mathcal{L}) = \sum_{i=0}^{k}(-1)^i\, h^i(M;\mathcal{L})\,.$$

Now we can state the simplest version of the Hirzebruch-Riemann-Roch theorem:

2.10 Theorem. *Let M be a complex manifold. Then*

$$\chi(M;\mathcal{O}) = Td(M)\,.$$

For algebraic manifolds of dimension 2, this is Noether's formula. Stated in this way, Theorem 2.10 only makes sense for complex manifolds and Hirzebruch's proof was only for algebraic manifolds. However one can re-phrase it using the index theorem, replacing the left-hand side by the index of the $\bar\partial$-operator, and the theorem holds for almost complex manifolds. This implies in particular that $Td(M)$ is an integer, which is another miracle. In Section 4 below we give a theorem [76] which shows that for complex manifolds, the parity of $Td(M)$ is detected by a certain "mod (2) index", which for complex surfaces is the Arf invariant in Rochlin's theorem, which we discuss in the following section.

More generally, given a holomorphic bundle $\mathcal{L}$, the Riemann-Roch formula says that the analytic Euler characteristic of $\mathcal{L}$ is given by the Todd genus of M and a polynomial in the Chern class $c_1(\mathcal{L})$ and the Chern classes of M.

Making $c = c_1(\mathcal{L})$, in the low dimensions this takes the form:

$$\chi(M^1; \mathcal{L}) = \left(c + \frac{1}{2} c_1(M)\right)[M],$$

$$\chi(M^2; \mathcal{L}) = \left(\frac{1}{2}(c^2 + c \cdot c_1(M)) + \frac{1}{12}(c_2(M) + c_1^2(M))\right)[M],$$

$$\chi(M^3; \mathcal{L}) = \left(\frac{1}{6} c^3 + \frac{1}{4} c^2 c_1(M) + \frac{1}{12} c\left(c_2(M) + c_1^2(M)\right) + \frac{1}{24} c_1(M) c_2(M)\right)[M].$$

The general Riemann-Roch theorem (see [101, Ch. 4] and [21, §4]) expresses the analytic Euler characteristic $\chi(M; \mathcal{E})$ of every holomorphic vector bundle over M ($\mathcal{E}$ of any dimension) in terms of the total Todd class of M (i.e., the collection of all cohomology classes in all dimensions, obtained by evaluating the Todd sequence of polynomials in the Chern classes of M), and the Chern character of the bundle (see [101, §10]), which is a class in the cohomology ring $H^*(M; \mathbb{Q})$.

IV.3   Spin and Spin^c structures on 4-manifolds. Rochlin's theorem

We refer to [124] for a clear account of spin and spin^c structures on manifolds. The group $\mathrm{Spin}(n)$, $n > 1$, is usually defined as a subgroup of the group of units of an appropriate (Clifford) algebra, but it can be equivalently defined as the non-trivial double cover of $SO(n)$; since for $n > 2$ the fundamental group of $SO(n)$ is $\mathbb{Z}/2$, it follows that in these dimensions $\mathrm{Spin}(n)$ is also the universal covering group of $SO(n)$. In low dimensions one has that $\mathrm{Spin}(2)$ is the circle $\mathbb{S}^1$ regarded as the 2-fold cover $\mathrm{Spin}(2) \to SO(2)$; $\mathrm{Spin}(3)$ is the 3-sphere, isomorphic to $SU(2) \cong Sp(1)$; $\mathrm{Spin}(4)$ is $SU(2) \times SU(2)$, since $SO(4) \cong SO(3) \times SU(2)$.

Now, given a closed, connected smooth manifold M of dimension n, the structure group of its tangent bundle TM is $GL(n, \mathbb{R})$. To have a *reduction* of the structure group of this bundle to a certain subgroup $G \subset GL(n, \mathbb{R})$ means that we have a smooth atlas $\{(U_i, \phi_i)\}$ for M, so that $TM|_{U_i}$ is trivial on each chart, and the gluing functions for TM on each chart,

$$\tilde{\phi}_{ij} : U_i \cap U_j \to GL(n, \mathbb{R})$$

can be all taken in G (c.f. Ch. III, §6). Since each basis of $\mathbb{R}^n$ is homotopic to an orthonormal one, the group $GL(n, \mathbb{R})$ has $O(n)$ as a deformation retract. This means that endowing M with a Riemannian metric, we can reduce the structure group of TM from $GL(n, \mathbb{R})$ to $O(n)$. Equivalently, the bundle TM is classified, up to isomorphism, by a continuous map (see [164])

$$\tau_M : TM \to BGL(n, \mathbb{R})$$

into the classifying space of $GL(n, \mathbb{R})$. There is a map induced by the inclusion $BO(n) \overset{\iota}{\hookrightarrow} BGL(n, \mathbb{R})$, and a reduction of the structure group of TM to $O(n)$

simply means a lifting of the classifying map τ_M from $BGL(n,\mathbb{R})$ to $BO(n)$. If M is an orientable manifold, we may reduce the structure group of the bundle TM further to $SO(n)$. This means choosing our atlas so that all the gluing maps have determinant 1, i.e., we are fixing an orientation on M, or equivalently, lifting the classifying map τ_M further to $BSO(n)$. We remark that such lifting of τ_M exists only because we are assuming M is orientable, and there are exactly two homotopy classes of such liftings, corresponding to the possible orientations on M. It is an exercise (see [124]) to show that a smooth compact manifold M is orientable iff its first Stiefel-Whitney class $w_1(M) \in H_1(M;\mathbb{Z}_2)$ vanishes.

The covering map $\mathrm{Spin}(n) \to SO(n)$ induces a map of classifying spaces $B\mathrm{Spin}(n) \to BSO(n)$, and we may ask to further lift τ_M from $BSO(n)$ to $B\mathrm{Spin}(n)$.

3.1 Definition. A compact, oriented manifold M *admits a spin structure* if the structure group of its tangent bundle can be lifted from $SO(n)$ to $\mathrm{Spin}(n)$; the various homotopy classes of such liftings are the different spin structures on M.

In this case one says, shortly, that M is a spin manifold. Alternatively, an orientation on M determines a principal $SO(n)$-bundle P over M, whose fibre at each point $x \in M$ is the set of all orthonormal, positively oriented basis of the corresponding tangent space T_xM; and a spin structure on M means a principal spin bundle $\tilde{P}$ over M which double covers P and at each $x \in M$ restricts to the double cover $\mathrm{Spin}(n) \to SO(n)$. One has (see [124, 1.7]):

3.2 Proposition. *A compact smooth manifold M admits a spin structure iff its Stiefel-Whitney classes $w_1(M)$ and $w_2(M)$ vanish; and if this happens, then the different spin structures on M are classified by $H^1(M;\mathbb{Z}_2)$.*

Thus, for instance, every parallelizable manifold is spin, since TM trivial implies we can reduce its structure group to 1, and all its Stiefel-Whitney classes vanish. Also notice that every homology sphere (of dimension > 2) admits a unique spin structure.

Let us assume now that M is spin. Then one can easily see that its quadratic form is automatically an even form, and therefore its signature must be divisible by 4. A general result in algebra tells us that it is actually divisible by 8, which is already a non-trivial fact. Rochlin's remarkable theorem in 1952 says more:

$$\sigma(M) \equiv 0 \quad \mathrm{mod}\,(16)\,. \tag{3.3}$$

This was a key result in 4-manifold theory, being closely related to other peculiarities of low-dimensional topology and triangulation theory. This was generalized by M. Kervaire and J. Milnor for oriented 4-manifolds which may not be spin provided there is an embedded 2-sphere S in M representing an integral homology class whose reduction modulo 2 is the Stiefel-Whitney class. In this situation they proved:

$$\sigma(M) \equiv S^2 \quad \mathrm{mod}\,(16) \tag{3.4}$$

where S^2 is the self-intersection of S in M. In order to say more about this result of Kervaire-Milnor, and its later generalization by Rochlin himself, it is convenient to speak about spin^c-structures on manifolds.

For this we recall that the oriented, differentiable 2-plane vector bundles over an oriented manifold are in 1-to-1 correspondence with the cohomology group $H^2(M;\mathbb{Z})$, the correspondence being given by associating to each such bundle its Euler class. Notice that $SO(2) \cong U(1)$, so we can identify oriented 2-plane bundles with complex line bundles. We also recall that $\mathrm{Spin}(n)^c$ is defined to be the subgroup of $\mathrm{Spin}(n+2)$ that double covers the group $SO(n) \times SO(2) \subset SO(n+2)$. One has an isomorphism

$$\mathrm{Spin}(n)^c \cong \mathrm{Spin}(n) \times_{\mathbb{Z}_2} SO(2)\,,$$

and a short exact sequence

$$0 \longrightarrow \mathbb{Z}_2 \longrightarrow \mathrm{Spin}(n)^c \longrightarrow SO(n) \times SO(2) \longrightarrow 1\,,$$

where $\mathbb{Z}_2 \subset \mathrm{Spin}(n)^c$ is generated by the element $[(-1,1)] = [(1,-1)]$.

3.5 Definition. An oriented n-manifold M admits a spin^c structure if there exists a complex line bundle $\mathcal{L}$ over M, called *the determinant bundle* for the spin^c structure, such that the structure group of $TM \oplus \mathcal{L}$ lifts to $\mathrm{Spin}(n)^c$.

One has that M admits a spin^c structure iff there exists an integral homology class $W \in H_{n-2}(M;\mathbb{Z})$ whose reduction modulo 2 is the Poincaré dual of $w_2(M)$, the 2nd Stiefel-Whitney class of M (see for instance [124, Theorem D.2, p. 391]). This class W is precisely the Chern class of the determinant bundle $\mathcal{L}$. It follows that if M admits a spin^c structure, then all other spin^c structures correspond bijectively with the elements in $H^2(M;\mathbb{Z})$, which can be regarded as *the group of $\mathbb{C}^\infty$ complex line bundles* over M.

Of course if M is a complex manifold, then it has a canonical spin^c structure given by $-K$, the anti-canonical class, which is the Chern class of its tangent bundle, which equals that of the complex line bundle $\bigwedge^2 TM$. The canonical bundle $\mathcal{K}_M = \bigwedge^2 T^*M$ gives another spin^c structure. Also every spin manifold is canonically spin^c with the trivial bundle as determinant (and $W = \emptyset$).

3.6 Definition. An oriented submanifold $W \subset M$ of codimension 2 *is characteristic* if it is the Chern class of a line bundle over M which is the determinant of some spin^c structure on M.

Of course this is equivalent to saying that W represents an integral homology class $W \in H_{n-2}(M;\mathbb{Z})$ whose reduction modulo 2 is the Poincaré dual of $w_2(M)$, the 2nd Stiefel-Whitney class of M, which is the usual definition of a characteristic submanifold.

The following generalization of the theorems of Rochlin (3.3) and Kervaire-Milnor (3.4) was given by Rochlin himself in [209]. An essentially complete proof can be found in [81]; a generalization to higher dimensions was given in [192].

3.9 Theorem. *Let M be an oriented closed 4-manifold and $W \subset M$ a characteristic submanifold, then one has:*

$$\sigma(M) - W^2 \equiv 8\operatorname{Arf}(W) \mod (16) \,,$$

where W^2 is the self-intersection number of W and $\operatorname{Arf}(W) \in \{0,1\}$ is the Arf invariant of a certain quadratic form on $H_1(W; \mathbb{Z}_2)$.

We leave it as an exercise for the reader to prove that every closed, oriented 4-manifold admits a spin^c structure. We recall that the Arf invariant of a quadratic form on a finite-dimensional vector space over $\mathbb{Z}_2$ takes values in $\{0, 1\}$ and can be defined to be 0 iff it takes the value 0 more times than it takes the value 1; two such quadratic forms are equivalent iff they have the same Arf invariant (see the appendix in [211] for details).

Of course if M is spin we can take $W = \emptyset$ and we recover (3.3). And if M is not spin but we can take W to be a 2-sphere, then $H_1(W; \mathbb{Z}_2) = 0$, so $\operatorname{Arf}(W) = 0$ and we recover (3.4).

IV.4   Spin **and** Spin^c **structures on complex surfaces. Rochlin's theorem**

We now discuss Spin and Spin^c structures on complex surfaces and give a variant of Rochlin's Theorem 3.1 when both M and W are complex analytic. Our approach is via algebraic geometry, and in doing so one is naturally led to considering divisors on M, rather than smooth submanifolds. But the definition of the Arf invariant in Rochlin's theorem only makes sense when W is smooth. Thus we were led in [76] to considering another mod (2) invariant, instead of $\operatorname{Arf}(W)$, that comes naturally from algebraic geometry. This invariant is an extension for divisors of the mod (2) index introduced in [16, 19], which is the mod (2) index of a Dirac operator. Essentially everything we say here is taken from [76] and holds for all complex dimensions of the form $4k + 2$; here we restrict to complex dimension 2 for simplicity.

We recall (c.f. III.6) that if M is a compact complex manifold of dimension 2, then its *canonical bundle* $\mathcal{K} = \mathcal{K}_M$, is the bundle of holomorphic 2-forms on M, whose Chern class is the negative of $c_1(TM)$. The *anti-canonical bundle* $\mathcal{K}_M^{-1}$ is the exterior product $\bigwedge^2 TM$. This bundle determines a canonical spin^c structure on M, for which $\mathcal{K}_M^{-1}$ is the *determinant bundle* (c.f. [192]). We denote by K the divisor of $\mathcal{K}$, *the canonical divisor* of M; $-K$ is then the *anti-canonical divisor*. Notice that $-K$ represents the Poincaré dual of the Chern class $c_1(\mathcal{K}) \in H^2(M; \mathbb{Z})$. Since for complex manifolds the Chern classes reduced modulo 2 give the corresponding Stiefel-Whitney classes (see Section 1), it follows that both $\pm K$ represent integral homology classes whose reduction modulo 2 is the second Stiefel-Whitney class of M. However in most cases they are singular, non-reduced and reducible, so they cannot be used as characteristic submanifolds for M in the sense of Section 3. Yet we note that a submanifold $W \subset M$ is characteristic iff it represents a homology class of the form $2D - K$ for some integral class $D \in H^2(M; \mathbb{Z})$. This group classifies the different spin^c structures on M, and it can be regarded also as the group of C^∞ complex line bundles over M. Given a divisor D, we can form a divisor $W = 2D - K$, which corresponds to the line bundle $\mathcal{D}^2 \otimes \mathcal{K}^{-1}$; the reduction modulo 2 of the homology class represented by W is the second Stiefel-Whitney class of M. Summarizing, one has:

4.1 Proposition. *Every complex manifold M has a canonical* spin^c *structure whose determinant bundle is the anti-canonical divisor* $-K$. *Furthermore, every line bundle* $\mathcal{L}_W$ *on M of the form* $\mathcal{D}^2 \otimes \mathcal{K}^{-1}$ *determines canonically a* spin^c *structure on M whose determinant bundle is* $\mathcal{L}_W$.

Notice that if W were a smooth submanifold of M, then W would be characteristic in Rochlin's sense. This justifies the following definition:

4.2 Definition. A *characteristic divisor* of M is a non-negative divisor $W \geq 0$ of M which can be expressed as

$$W = 2D - K,$$

where D is some (any) divisor of M and K is the canonical divisor.

We now want to introduce the aforementioned mod (2) invariant of characteristic divisors that will replace Rochlin's Arf invariant in our discussion. For this we recall (see [16, 3.2]) that on a Riemann surface X the spin structures correspond bijectively with the holomorphic square roots of the canonical bundle $\mathcal{K}_X$, that is with the (isomorphism classes of) holomorphic line bundles $\mathcal{L}$ over X with $\mathcal{L}^2 \cong \mathcal{K}$, where $\mathcal{L}^2 = \mathcal{L} \otimes \mathcal{L}$. These bundles are called *theta characteristics* on the manifold; they were first studied by Krazer (1903). In [16, 177] Atiyah and Mumford showed (independently) that for Riemann surfaces, the dimension of the space of sections of these bundles, $\dim H^0(X, \mathcal{L})$, reduced modulo 2, is stable under holomorphic deformations. To explain this, Atiyah showed that the parity of $\dim H^0(X, \mathcal{L})$ can be regarded as being the mod (2) index of the Dirac operator for the corresponding spin structure.

Similarly, given a compact complex surface M as above, and an effective divisor W on M, one has the canonical (or dualizing) sheaf ω_W. This exists for all effective divisors (possibly non-reduced, reducible) on complex manifolds, see for instance [24]. One may define:

4.3 Definition. A *theta characteristic* on W is the restriction to W of a holomorphic line bundle $\mathcal{D}$ on M, such that $\mathcal{D}^2|_W \cong \omega_W$.

One has:

4.4 Proposition. *Let* $W = 2D - K$ *be a characteristic divisor of M, and let* $\mathcal{L} = \mathcal{L}_W$, $\mathcal{D}$ *be the corresponding line bundles. Then* $\mathcal{D}|_W$ *is a theta characteristic on W.*

Proof. We recall that for a Riemann surface E in M the classical *adjunction formula* says

$$\mathcal{K}_E \cong \mathcal{K}_M|_E \otimes \nu E,$$

where νE is the normal sheaf. Similarly, given an effective divisor W in M defined by a section of a holomorphic line bundle $\mathcal{L} \cong \mathcal{L}_W$, the restriction of $\mathcal{L}$ to W can be identified with the normal sheaf $\mathcal{O}_W(W)$ of W, and one also has the *adjunction formula* (see [24]):

$$\omega_W \cong \mathcal{K}_M|_W \otimes \mathcal{L}|_W.$$

By hypothesis one has $\mathcal{L} \cong \mathcal{D}^2 \otimes \mathcal{K}_M^{-1}$. Therefore:

$$\omega_W \cong \mathcal{K}_M|_W \otimes \mathcal{L}|_W \cong \mathcal{K}_M|_W \otimes \mathcal{D}^2|_W \otimes \mathcal{K}_M^{-1}|_W \cong \mathcal{D}^2|_W. \qquad \square$$

Now define (following [16, 19, 177, 76]):

4.5 Definition. The mod 2 index of the characteristic divisor $W = 2D - K$ is:

$$\hbar(W) = \begin{cases} \dim H^0(W, \mathcal{D}) \mod 2, & \text{if } W \neq 0, \\ 0, & \text{if } W = 0; \end{cases}$$

i.e., for $W \neq 0$ it is the reduction modulo 2 of the space of sections of its associated theta characteristic $\mathcal{D}|_W$.

4.6 Remarks.

(i) Notice that given the divisor W, the corresponding bundle $\mathcal{D}$ is well defined modulo the 2-torsion in the Picard group Pic M of holomorphic line bundles on M. However the Hirzebruch-Riemann-Roch formula [101] of §2 above, together with Theorem 4.7 below, grants that this invariant depends only on the numerical equivalence class of $\mathcal{D}$ and is independent of the 2-torsion in Pic M.

(ii) We also remark that everything we have said so far holds for compact manifolds of complex dimension $4k + 2$, with obvious adaptations. In this case the characteristic divisor has dimension $4k + 1$ and the corresponding mod 2 index is the reduction modulo 2 of $\sum_{i=0}^{2k} \dim H^i(W, \mathcal{D})$, see [76, 1.4]. The following theorem also holds in this general setting ([76, 2.1]), but for simplicity we do it here only for $k = 0$.

4.7 Theorem. *Let M be a compact complex surface, let $W = 2D - K$ be a characteristic divisor of M, $\hbar(W)$ its mod 2 index, and let $\chi(M, \mathcal{D}) = \sum_{i=0}^{2}(-1)^i h^i(M, \mathcal{D})$ be the analytic Euler characteristic of M with coefficients in $\mathcal{D}$ (see [101]). Then one has:*

$$\hbar(W) \equiv \chi(M, \mathcal{D}) \mod 2.$$

Proof. Assume first that $W \neq 0$, and consider the exact sequence of sheaves over M:

$$0 \longrightarrow \mathcal{K}_M \otimes \mathcal{D}^{-1} \xrightarrow{s^*} \mathcal{D} \xrightarrow{r} \mathcal{D}|W \longrightarrow 0,$$

where s^* is multiplication by the section s of $\mathcal{L}$ that defines W and r is the restriction to W. One has the associated long exact sequence

$$0 \to H^0(\mathcal{K}_M \otimes \mathcal{D}^{-1}) \to H^0(\mathcal{D}) \to H^0(\mathcal{D}|_W) \xrightarrow{\alpha} H^1(\mathcal{K}_M \otimes \mathcal{D}^{-1}) \xrightarrow{\beta} H^1(\mathcal{D}) \to \cdots.$$

If we denote by $h^i(\cdot)$ the dimension of the corresponding cohomology group, then we want to prove:

$$h^0(W, \mathcal{D}|_W) \equiv \sum_{1=0}^{2} h^i(M, \mathcal{D}) \qquad \mathrm{mod}\ 2$$

and by exactness of the sequence above we know:

$$h^0(W, \mathcal{D}|_W) \equiv h^0(M, \mathcal{K}_M \otimes \mathcal{D}^{-1}) + h^0(M, \mathcal{D}) + \dim \mathrm{Im}\,(\alpha) \qquad \mathrm{mod}\ 2.$$

Serre's duality on M gives:

$$H^0(M, \mathcal{K}_M \otimes \mathcal{D}^{-1}) \cong H^2(M, \mathcal{D}),$$

so, again by exactness, the theorem will be proved if we show that

$$\dim Ker\,(\beta) \equiv h^1(M, \mathcal{D}) \qquad \mathrm{mod}\ 2,$$

but Serre's duality tells us that β is essentially defined by the cup-product in $H^1(M, \mathcal{K}_M \otimes \mathcal{D}^{-1})$, and in dimension 2 the cup-product is a skew-form, so it can be expressed as a direct sum of blocks of the form $\begin{pmatrix} 0 & b \\ -b & 0 \end{pmatrix}$, hence the theorem. This proves Theorem 4.7 for $W \neq 0$ characteristic. It remains to prove the theorem for $W = 0$; this is left as an exercise (see [76] for details). $\square$

Let us now recall that the Riemann-Roch formula says that the analytic Euler characteristic of M with coefficients in $\mathcal{D}$ is given by:

$$\chi(M, \mathcal{D}) = Td(M) + \frac{1}{2}(D^2 - D \cdot K)$$
$$= \frac{1}{12}(K^2 + \chi(M)) + \frac{1}{8}(W^2 - K^2),$$

since $W = 2D - K$, where $\chi(M)$ is the usual (topological) Euler-Poincaré characteristic of M. Thus Theorem 4.7 can be reformulated as:

$$\hbar(W) \equiv \frac{1}{24}(2\chi(M) - K^2) + \frac{1}{8}(W^2) \qquad \mathrm{mod}\ 2.$$

Since for a complex surface the Pontryagin class is:

$$p_1(M) = c_1^2 - 2c_2,$$

and the Hirzebruch signature theorem says $\sigma(M) = -\frac{1}{3}p_1(M)[M]$, it follows that Theorem 4.7 is equivalent to:

4.8 Theorem. *Let M be as above, a compact complex manifold of dimension 2, and let $W = 2D - K$ be a characteristic divisor of M. Then:*

$$\sigma(M) - W^2 \equiv 8 \dim H^0(W; \mathcal{D}|_W) \qquad \mathrm{mod} \ (16) .$$

This theorem is essentially an algebro-geometric variation of Rochlin's signature theorem, the difference being that in Theorem 3.7, W has to be a smooth characteristic submanifold and the term on the right-hand side is the Arf invariant of a quadratic form on $H_1(W; \mathbb{Z}_2)$ defined topologically, while here W is a (possibly singular, non-reduced and reducible) characteristic divisor.

Notice that W is defined by a holomorphic section of a line bundle $\mathcal{L}$. Every such section can be approximated by a C^∞ section s_t of $\mathcal{L}$ which is transversal to the zero-section. The zero set $\widehat{W}$ of s_t is then a smooth, real submanifold of M representing the same homology class as W. Hence $\widehat{W}$ is a characteristic submanifold in the sense of Section 1, and 4.8 together with Rochlin's theorem yield:

$$\mathrm{Arf}(\widehat{W}) \equiv \dim H^0(W; \mathcal{D}|_W) \qquad \mathrm{mod} \ (2) . \tag{4.9}$$

Is this a special case of a more general theorem? For instance, following [192] we see that the manifold $\widehat{W}$ is canonically spin:

4.10 Question. Is the mod (2) index $\hbar(W) = \dim H^0(W; \mathcal{D}|_W)$ equal to the mod (2) index of the Dirac operator on $\widehat{W}$? And in higher dimensions?

For instance, if a real manifold M of dimension $8k + 4$ has a spin^c structure and $W \subset M$ is a codimension 2 characteristic submanifold, then W is automatically spin. Each of these manifolds has a Dirac operator D_M, D_W; the C^∞ version of Theorem 4.7 in higher dimensions (see [76]) would be to ask whether the index of D_M reduced modulo 2 equals the mod (2) index of D_W. I do not know the answer to these questions.

4.11 Remark. It is worth pointing out some special cases of Theorem 4.7:

(i) If $K < 0$, then $W = -K$ is characteristic with $D = 0$. We obtain,

$$\dim H^0(-K, \mathcal{O}) \equiv \chi(M, \mathcal{O}) \quad \mathrm{mod} \ 2 .$$

(ii) If $K > 0$, then $W = K = D$ is characteristic and one has:

$$\dim H^0(K, \mathcal{K}_M|_K) \equiv \chi(M, \mathcal{O}) \quad \mathrm{mod} \ 2 ,$$

because $\chi(M, \mathcal{K}_M) = \chi(M, \mathcal{O})$ (by [101]).

(iii) If K is even, so that M is spin, then K^2 is divisible by 8 and:

$$Td(M) \equiv \frac{1}{8} K^2 \quad \mathrm{mod} \ 2 ,$$

that is, $Td(M)$ is always an integer (by [101]), but it may or may not be an even integer, and this is given by the parity of $\frac{1}{8} K^2$. Hence the Todd genus is even when $K = 0$ (c.f. [18, Cor. 2.ii]).

IV.5 A review of surface singularities

We want to use the previous results for compact manifolds to study germs of surface singularities. For this we need some preliminary background about resolutions of singularities. Since this important material is now standard in the literature, we only make a brief review, for completeness and to set up our notation and conventions.

Consider first a singular (reduced) complex curve C in some smooth complex surface X. An important result for plane curves, due to Max Noether (1883), is that by a finite sequence of blowing ups we can always resolve the singularities of C. Let us explain this with a little more care (see [44] for a clear account of the subject). Given a smooth point x in a complex surface X, take local coordinates so that we identify the germ of X at x with that of $\mathbb{C}^2$ at 0. Let us take a small disc U around x and consider the map: $\gamma : U - \{x\} \to \mathbb{C}P^1$ which associates to each $y \in U - \{x\}$ the point in $\mathbb{C}P^1$ represented by the line determined by x and y. The graph of γ is an analytic subset of $(U - \{x\}) \times \mathbb{C}P^1$, whose closure

$$\widetilde{X} = \overline{\mathrm{graph}(\gamma)} \subset (U - \{x\}) \times \mathbb{C}P^1$$

turns out to be a smooth complex surface. Notice that $\widetilde{X}$ is obtained by removing x from X and replacing it by the limits of lines converging to x. Thus we have replaced x by a copy of $\mathbb{C}P^1$. There is a projection map $\pi : \widetilde{X} \to X$ which is biholomorphic away from $E \cong \mathbb{C}P^1 = \pi^{-1}(x)$. This transformation is called *the blow-up* of X at x (in the literature this is called sometimes a *σ-process* or a *monoidal transformation*).

Now, given the reduced (maybe reducible) singular curve $C \subset X$, with X a smooth complex surface, let x be a point in C_{sing}, the singular set of C, and look at the blow-up of X at x, $\pi : \widetilde{X} \to X$. The closure $\overline{\pi^{-1}(C - x)}$ in $\widetilde{X}$ is called *the proper (or strict) transform* of C under the blow-up, and denoted $\widetilde{C}$. Notice that $\widetilde{C}$ is obtained by removing x from C and replacing it by the limits of lines which are tangent to $C - \{x\}$. This curve $\widetilde{C}$ is analytic in $\widetilde{X}$ and projects to C under π; this curve may still be singular, but its singularities are simpler. We may now repeat the process, choosing a singular point in $\widetilde{C}$, blowing up $\widetilde{X}$ at this point to get $\pi_2 : \widetilde{X}_2 \to \widetilde{X}$ and then consider the proper transform of C in $\widetilde{X}_2$, which is the closure of $(\pi_2 \circ \pi)^{-1}(C - x)$, and so on. The theorem is (see [24, II.7.1] for a short proof):

5.1 Theorem. *Let X be a smooth complex surface and $C \subset X$ an embedded reduced curve. Then there is a smooth complex surface Y and a proper map $\tau : Y \to X$ obtained by a finite sequence of blow-ups, such that the proper transform $\widetilde{C}$ of C in Y is smooth.*

The curve $E = \tau^{-1}(C)$, is called *the total transform* of C in Y. It consists of the proper transform $\widetilde{C}$ and the divisor $\tau^{-1}(C_{\mathrm{sing}})$. The theorem above can be refined to the following theorem, which will be used later in the text.

5.2 Theorem. *Let X and C be as above. Then by performing finitely many blow-ups more, if necessary, we can assume that the whole total transform $E = \tau^{-1}(C)$ of C in Y has only ordinary normal double points as singularities, and these are all away from the proper transform $\widetilde{C}$ of C.*

We recall that an ordinary normal double point is locally defined by the equation $\{xy = 0\}$.

Now consider the germ $(V, 0)$ of a normal complex surface singularity. The following important theorem has a long history. This was first stated by Jung but his proof was not complete, and was completed later by Hirzebruch. The first complete proof of Theorem 5.3 is due to O. Zariski; this was done using blowing ups and it was based on a previous proof by R. Walker (1935) which was not correct either. We refer to [24, Ch. III] for a proof. This result was later generalized by Hironaka to all dimensions and for (real or complex) analytic spaces with arbitrary singularities.

5.3 Theorem. *Let $(V, 0)$ be a normal complex surface singularity. For simplicity assume it is defined in a sufficiently small ball around the origin in some $\mathbb{C}^n$, so that $V^* = V - 0$ is non-singular. Then there exists a non-singular complex surface $\widetilde{V}$ and a proper analytic map $\pi : \widetilde{V} \to V$, such that:*

(i) *the inverse image of 0, $E = \pi^{-1}(0)$, is a (connected, reduced) divisor in $\widetilde{V}$, i.e., a union of 1-dimensional compact curves in $\widetilde{V}$; and*

(ii) *the restriction of π to $\pi^{-1}(V^*)$ is a biholomorphic map between $\widetilde{V} - E$ and V^*.*

The surface $\widetilde{V}$ is called a *resolution* of the singularity of V, and $\pi : \widetilde{V} \to V$ is the *resolution map*. Sometimes these are called *desingularizations* of the singularities instead of resolutions. The divisor E is called *the exceptional divisor.*

Notice that "the" resolution of $(V, 0)$ is not unique: given a resolution $\widetilde{V}$ we can obtain new resolutions by performing blow-ups at points in E. By Theorem 5.2 above, given a resolution, we can make blow-ups on it, if necessary, so that the divisor E in Theorem 5.3 is *good*, i.e.:

(iii) each irreducible component E_i of E is non-singular; and

(iv) E has normal crossings, i.e., E_i intersects E_j, $i \neq j$, in at most one point, where they meet transversally, and no three of them intersect.

5.4 Definition. A resolution $\pi : \widetilde{V} \to V$ is *good* if its exceptional divisor is good, i.e., if it satisfies conditions (iii) and (iv) above.

Some authors allow good resolutions to have irreducible components of E intersecting transversally in more than one point, and they reserve the name *very good* for resolutions as in Definition 5.4. This makes no big difference and we prefer to keep the notation of Definition 5.4.

We recall that given a non-singular complex surface $\widetilde{V}$ and a Riemann surface S in it, the *self-intersection* of S, usually denoted by $S \cdot S$ or simply by S^2, is the Euler class of its normal bundle $\nu(S)$ in $\widetilde{V}$ (which coincides with its Chern class) evaluated in the fundamental cycle $[S]$. Equivalently, $S \cdot S$ is the number of zeroes, counted with signs, of a generic section of the normal bundle $\nu(S)$. It is an exercise to see that every time we make a blow-up on a smooth complex 2-manifold, we get a copy of $\mathbb{C}P^1$ with self-intersection -1. It is remarkable that the converse is true: recall that a non-singular curve in a smooth complex surface is said to be *exceptional of the first kind* if it is a copy of $\mathbb{C}P^1$ embedded with self-intersection -1, one has:

5.5 Theorem. (*Castelnuovo's criterium*) *Let* $\tilde{X}$ *be a non-singular complex surface and* C *an exceptional curve of the first kind. Then* S *can be blown down analytically and we still get a non-singular surface* X.

This result is in fact a special case of a more general theorem of Castelnuovo for *exceptional divisors of the first kind*. We refer to [82, Ch. 3] for a proof. Notice that Definition 5.4 has the very important consequence of giving us a *minimal model*:

5.6 Definition. A resolution $X \xrightarrow{\pi} V$ is *minimal* if given any other resolution $X' \xrightarrow{\pi'} V$, there is a proper analytic map $X' \xrightarrow{p} X$ such that $\pi' = \pi \circ p$.

One has (see [24, III.6.2]):

5.7 Theorem. *Up to isomorphism, there exists a unique minimal resolution of* V, *and this is characterized by not containing non-singular rational curves with self-intersection* -1.

We remark that these statements are false in dimensions more than 2: there are no minimal resolutions in general.

Notice that the minimal resolution may not be good. For instance (c.f. [57]), for the Brieskorn singularity

$$z_1^2 + z_2^3 + z_3^7 = 0,$$

the minimal resolution has an exceptional divisor consisting of three non-singular rational curves meeting at one point, so it is not good; making one blow-up at that point we obtain a good resolution, which has now a central curve which is a 2-sphere with self-intersection -1, and three other spheres, each meeting the central curve in one point and with self intersections $-2, -3, -7$.

Something similar happens in general: we can make the minimal resolution good by performing blow-ups, if necessary, and there is a unique (up to isomorphism) *minimal good resolution*.

Consider now a divisor $E = \bigcup_{i=1}^{r} E_i$ in a complex 2-manifold X, whose irreducible components E_i are non-singular, they all meet transversally and no three of them intersect. To such a divisor we can associate an $r \times r$ integral matrix $A = ((E_{ij}))$, called *the intersection matrix* of E, as follows: on the diagonal Δ of A we put the self-intersection numbers E_i^2; and if a curve E_i meets E_j at E_{ij} points, we put this number as the corresponding coefficient of A. So this is necessarily a symmetric matrix, whose coefficients away from the diagonal Δ are integers ≥ 0 and in Δ we have the self-intersection numbers of the E_i, called *the weights* of these curves.

We have the following remarkable theorems of Mumford and Grauert (see [24, III.2]):

5.8 Theorem [176]. *If E is the exceptional divisor of a resolution $X \xrightarrow{\pi} V$, where V is a normal surface, then the intersection matrix A is negative definite (and the weights of the E_i are all negative numbers).*

5.9 Theorem [91]. *Conversely, if the divisor E in X is such that the intersection matrix A is negative definite, then we can blow down E analytically; we get a normal complex surface V, in general with a singularity at the image 0 of E, and the projection $\pi : X \to V$ is a good resolution of $(V, 0)$ with exceptional divisor E.*

A divisor E as above is usually called *an exceptional divisor*, meaning by this that it can be blown down. It is said to be of the first kind when the blow-down is smooth.

Notice that we can associate a weighted graph $\mathcal{G} = \mathcal{G}(E)$ to a good exceptional divisor E in a complex 2-manifold X as follows: to each irreducible component E_i of E we associate a vertex v_i, and if the curves E_i and E_j meet, then we join the vertices v_i and v_j by an edge. Each vertex has two integers attached to it: one is the genus $g_i \geq 0$ of the corresponding Riemann surface E_i; the other is the weight $w_i = E_i^2 \in \mathbb{Z}$, which is the self-intersection number of E_i in X. This weighted graph is called *the dual graph* of the exceptional divisor E, or *the dual graph of the resolution* when E is regarded as the exceptional set of a good resolution of a normal singularity.

We observe that every finite graph Σ has associated a matrix $\mathcal{I}(\Sigma)$ called *the matrix of adjacencies* of the graph: it has zeroes in the diagonal and if a vertex v_i is joined to v_j by δ_{ij} edges, then we put this number in the corresponding place of $\mathcal{I}(\Sigma)$. It follows that the intersection matrix of the exceptional divisor E is the result of taking the matrix of adjacencies of the dual graph and replacing its diagonal by the vector of weights $w_1, \dots, w_m$, $w_i = E_i \cdot E_i$.

A beautiful aspect of these constructions is that the dual graph of a resolution allows us to re-construct the topology of the resolution, and hence that of the link of the singularity. For this we need to introduce a construction known as *plumbing*. This was used already by Milnor to construct his first examples of exotic spheres and by Von Randow (1962) in relation with Seifert manifolds, though it was Hirzebruch who made this construction systematic. The plumbing construction is very nicely explained in [105] (see also [187, 75]), we just recall it here briefly.

Let E be a real 2-dimensional oriented vector bundle over a Riemann surface S, and denote by $D(E)$ *its unit disc bundle* for some metric. The total space of $D(E)$, that we denote by the same symbol, is a 4-dimensional smooth manifold with boundary the unit sphere bundle $S(E)$. Notice that, restricted to a small disc $\mathbb{D}_\varepsilon$ in S, the manifold $D(E)$ is a product of the form $\mathbb{D}^2 \times \mathbb{D}^2$, where the first disc is $\mathbb{D}_\varepsilon \subset S$ and the second disc is in the fibres of E. Now suppose we are given two such bundles E_i, E_j, over Riemann surfaces S_i, S_j. To *perform plumbing* on them we consider the total spaces of the corresponding unit disc bundles $D(E_i)$, $D(E_j)$, we choose small discs $\mathbb{D}_{i,\varepsilon}$, $\mathbb{D}_{j,\varepsilon}$ in S_i, S_j, and take the restriction of $D(E_i)$, $D(E_j)$ to these discs. Each of them is of the form $\mathbb{D}^2 \times \mathbb{D}^2$ as above. We now identify each point $(x, y) \in \mathbb{D}_{i,\varepsilon} \times \mathbb{D}^2 \subset D(E_i)$ with the corresponding point $(y, x) \in \mathbb{D}_{j,\varepsilon} \times \mathbb{D}^2 \subset D(E_j)$, i.e., interchanging base points in one of them with fibre points in the other. The result is a 4-dimensional, oriented manifold with boundary and with corners, which can be smoothed off in a unique way up to isotopy (see [60]). We denote this manifold by $P(E_i, E_j)$. One says that $P(E_i, E_j)$ is obtained by plumbing the bundles E_i and E_j over the Riemann surfaces S_i and S_j.

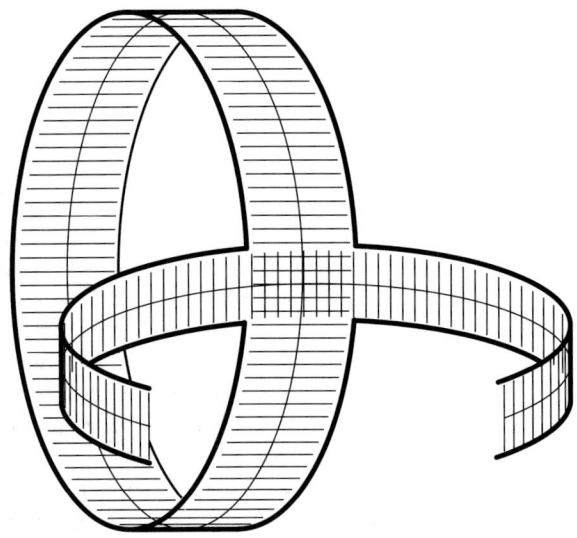

Figure 9: Plumbing line bundles over circles.

The boundary $S(E_i, E_j) = \partial P(E_i, E_j)$ of this 4-manifold is obtained by plumbing the corresponding sphere bundles $S_i(E)$ and $S_j(E)$: we remove from $S_i(E)$ and $S_j(E)$ the interior of the solid tori $\partial D_i \times \mathbb{D}^2$ and similarly for E_j. Thus we get two 3-manifolds with boundary a torus $\mathbb{S}^1 \times \mathbb{S}^1$ in each; we then identify these boundaries by gluing the meridians in one torus to the parallels in the other. The result is a 3-manifold with corners, which can be smoothed off in a unique way up to isotopy. The surfaces S_i, S_j are naturally embedded in $P(E_i, E_j)$ as the zero-sections of the corresponding bundles, and they meet transversally in one point.

Notice that the manifolds one gets in this way are entirely described, up to diffeomorphism, by the genera of the Riemann surfaces S_i, S_j, and by the Euler classes of the corresponding bundles, since these classes determine the isomorphism class of the bundles.

5.10 Definition A *plumbing graph* is a triple (Σ, w, g) consisting of a finite graph Σ with vertices $v_1, \ldots, v_r$, $r \geq 1$ and with no loops, a vector w of weights, $w = (w_1, \ldots, w_r)$, $w_i \in \mathbb{Z}$, and a vector $g = (g_1, \ldots, g_r)$ of genera, $g_i \in \mathbb{N}$.

So the dual graph of a good resolution of a normal singularity is a plumbing graph with negative definite intersection matrix. In this definition, by a loop we mean an arrow that begins and ends at the same vertex, and we do not allow this (geometrically this means a singular curve in the exceptional divisor that has a double crossing). There can be cycles, i.e., a chain of vertices and edges that returns to itself after a certain time. In [184] Neumann considers a slightly more general situation of plumbing graphs than the one envisaged here, which is important for the plumbing calculus developed there, but it does not really make a difference for the present work.

Now, given a plumbing graph we may *perform plumbing according to the graph*: for each vertex v_i take a Riemann surface S_i of genus g_i and an oriented 2-plane bundle E_i over S_i with Euler class w_i. If there is an edge between the vertices v_i and v_j, we plumb the corresponding bundles as above. If a vertex v_i is joined with other vertices, we choose pairwise disjoint small discs in each surface, as many as one has adjacent vertices, and perform plumbing by pairs as above. The result is a 4-dimensional manifold $\mathcal{P}(E)$ with boundary $\mathcal{S}(E)$. It follows from the construction that the manifold $\mathcal{P}(E)$ contains the union $E = \bigcup S_i$ as a deformation retract, and these surfaces are contained in $\mathcal{P}(E)$ with self-intersection w_i. Hence the homology of $\mathcal{P}(E)$ is that of E, and the intersection form on $\mathcal{P}(E)$ is given by the intersection matrix of its corresponding graph.

A manifold obtained in this way is known as a *plumbed manifold*, and this term may refer either to the 4-manifold $\mathcal{P}(E)$ with boundary, or to its boundary, which is a 3-manifold. Notice that if the plumbing graph (Σ, w, g) is the dual graph of a resolution $\pi : \widetilde{V} \to V$, then the manifold $\mathcal{P}(E)$ is diffeomorphic to a regular neighborhood of the exceptional set E in the resolution, which may be taken to be of the form $\pi^{-1}(V \cap \mathbb{D}_\varepsilon)$, where $\mathbb{D}_\varepsilon$ is a small closed disc in the ambient space $\mathbb{C}^n$ with centre at 0. Since the resolution map is a biholomorphism away from E, it follows that the boundary $\mathcal{S}(E)$ is diffeomorphic to the link of V, a fact that we state as a theorem:

5.11 Theorem. *Let* $\pi : \widetilde{V} \to V$ *be a good resolution of* $(V, 0)$, *a normal surface singularity. Then the irreducible components* $E_1, \ldots, E_r$ *of the exceptional divisor* E *determine a plumbing graph* $\mathcal{G}(E)$, *called the dual graph of the resolution, and performing plumbing according to this graph we obtain a 4-manifold homeomorphic to* $\pi^{-1}(V \cap \mathbb{D}_\varepsilon) \subset \widetilde{V}$, *whose boundary is the link* M, *where* $\mathbb{D}_\varepsilon$ *is a small closed disc in the ambient space* $\mathbb{C}^n$ *with centre at 0.*

Hence Theorems 5.8 and 5.9 say that a plumbing graph is the dual graph of a resolution if and only if its intersection matrix is negative definite. This gives a necessary and sufficient condition for an oriented 3-manifold to be the link of a surface singularity. A natural question is:

5.12 Question. Which plumbed 3-manifolds with negative-definite intersection matrix arise as links of isolated singularities of hypersurfaces in $\mathbb{C}^3$?

This is for me one of the most beautiful open questions in singularities. It is clear that if a 3-manifold M can be obtained in this way, then M can be embedded in $\mathbb{S}^5$ with trivial normal bundle. One may think that this could be a first step towards answering Question 5.12. However every orientable, closed 3-manifold embeds in $\mathbb{S}^5$ with trivial normal bundle, so this is not an obstruction; hence the problem is not only topological. This is actually a very subtle problem and a number of people have worked on it obtaining interesting partial results (see for instance [265]).

5.13 Remark. It is worth noticing that the condition on a plumbing graph to be the dual graph of a resolution depends only on the graph and its weights, but not on the genera assigned to the vertices. I do not know if there is any interesting relationship between singularities with the same weighted graph but different genera. We also remark that given any finite graph, for almost every set of negative weights that we assign to the vertices, the corresponding intersection matrix is negative definite (see [119]) and therefore is the dual graph of a surface singularity.

IV.6 Gorenstein and numerically Gorenstein singularities

The concept of numerically Gorenstein singularities was introduced by A. Durfee in [64]; these could as well be called "topologically Gorenstein" surface singularities, by Corollary 6.3 below.

6.1 Definition. A normal surface singularity germ $(V,0)$ is *numerically Gorenstein* if the complex tangent bundle TV^* of $V^* = V - 0$ is $\mathbb{C}^\infty$ trivial.

It is worth noting that this definition is less innocent than it looks. Obviously Definition 6.1 is equivalent to demanding that $TV^*|_M$ be trivial. As a real vector bundle $TV^*|_M$ splits as the direct sum $TM \oplus \nu(M)$, where M is the link of V, TM its tangent bundle and $\nu(M)$ its normal bundle for some metric. The bundle TM is trivial, because every oriented 3-manifold is parallelizable; and $\nu(M)$ is also trivial, because a hypersurface in an oriented manifold is orientable iff it has trivial normal bundle, and we know that M is orientable. Hence TV^* is always trivial as a real vector bundle. Thus Definition 6.1 does ask for TV^* to be trivial as a complex bundle; but it does not ask for TV^* to be holomorphically trivial: the Zariski-Lipman conjecture (still open) claims that this happens iff V is smooth at 0 (this conjecture is a theorem in many cases, as for instance for hypersurface germs).

The following lemma was noticed in [64, 1.1] and summarizes well-known properties of vector bundles.

6.2 Lemma. *Let ξ be a 2-dimensional complex bundle over a connected, open C^∞-manifold X. The following conditions are equivalent:*

(i) *ξ is trivial;*

(ii) *ξ is stably trivial;*

(iii) *the first Chern class of ξ vanishes;*

(iv) *the second exterior power $\bigwedge^2 \xi$ is a trivial bundle.*

The proof is easy: we recall that a bundle is stably trivial if its tangent bundle plus a trivial bundle is trivial. For instance the tangent bundle of the 2-sphere is not trivial, but when we add to it the normal bundle of $\mathbb{S}^2$ in $\mathbb{R}^3$ we get $T\mathbb{R}^3|_{\mathbb{S}^2}$, which is trivial. Hence $T\mathbb{S}^2$ is stably trivial. So (i) implies (ii) by definition. But also (ii) implies (i), because it is easy to see that, in general, if ζ is a complex n-plane bundle over a manifold N of dimension $2n$ and ζ is stably trivial, then there is a unique obstruction to have it actually trivial, and this is a cohomology class in top dimension $2n$ (see [113]). Since in our case we are assuming X is open and connected, then one has $H^4(M) = 0$ and therefore (ii) implies (i). That (i) implies (iii) is by definition of the Chern classes. Now, for every complex vector bundle one has that its first Chern class is, up to sign, the Chern class of its second exterior power $\bigwedge^2 \xi$. Hence (iii) implies $c_1(\bigwedge^2 \xi) = 0$ and this implies $\bigwedge^2 \xi$ is trivial, because a line bundle is trivial iff its Chern class vanishes; so (iii) and (iv) are equivalent. Finally, since $H^4(M) = 0$, the construction we gave above of the Chern classes via obstruction theory shows that one can always construct a never-zero section s_1 of ξ on all of X. Now, $c_1(\xi) = 0$ is equivalent to saying that we can construct a second section s_2 of ξ which is never-zero and linearly independent of s_1 on the 2-skeleton of X (for some triangulation). Then we can extend this section s_2 to all of X using the stepwise process of Section 1, since $\pi_i(\mathbb{S}^1)$ vanishes for $i > 1$.   □

6.3 Corollary. *The following conditions are equivalent:*

(i) *the germ $(V, 0)$ is numerically Gorenstein;*

(ii) *the bundle $\bigwedge^2 TV^*$ is topologically trivial;*

(iii) *the structure group of the bundle TV^* can be reduced to $SU(2)$;*

(iv) *if $\pi : \widetilde{V} \to V$ is any resolution of V, then the restriction of the tangent bundle $T\widetilde{V}$ to $\widetilde{M} = \pi^{-1}(M)$, where $M = V \cap \mathbb{S}_\varepsilon^{2n-1}$ is the link, is a trivial bundle;*

(v) *the Chern class $c_1(T\widetilde{V})$ can be represented by a relative cohomology class in $H^2(\widetilde{V}, \widetilde{M}; \mathbb{Z})$, whose Lefschetz dual $-K \in H_2(\widetilde{V}; \mathbb{Z})$ is uniquely characterized by the adjunction formula: for every compact, non-singular curve C in $\widetilde{V}$ one has:*

$$2g_C - 2 = C^2 + K \cdot C. \tag{6.4}$$

Proof. The equivalence (i) $\Leftrightarrow$ (ii) is a special case of the equivalence (i) $\Leftrightarrow$ (iv) in Lemma 6.2. The equivalence (ii) $\Leftrightarrow$ (iii) was proved in III.6. Now, it is clear that (i) implies (iv) since π is a biholomorphism away from $E = \pi^{-1}(0)$; conversely, it is clear that (iv) implies that $TV^*|_M$ is trivial, hence TV^* is trivial by the conical structure of V. Also, if $T\widetilde{V}|_{\widetilde{M}}$ is trivial, then a choice of a trivialization $\mathcal{T}$ of this bundle determines representatives of the Chern classes of $T\widetilde{V}$ that vanishes over M; the top relative class $c_2(\widetilde{V};\mathcal{T})$ lives in $H^4(\widetilde{V},\widetilde{M})$ and it is determined by the degree of $\mathcal{T}$, i.e., by the number of zeroes (counted with multiplicities) of an extension to $\widetilde{V}$ of one of the sections that determine $\mathcal{T}$. The class $c_2(\widetilde{V};\mathcal{T})$ lives in $H^2(\widetilde{V},\widetilde{M})$, as stated in (v) above, and it must satisfy the adjunction formula because for every C as above one has a splitting:

$$T\widetilde{V}|_{\widetilde{M}} \cong TC \oplus \nu(C)$$

as C^∞ bundles, where ν is the normal bundle. Thus one has

$$c_1(T\widetilde{V}|_{\widetilde{M}})[C] = c_1(TC)[C] + c_1(\nu(C))[C],$$

where $[C]$ is the fundamental cycle. Hence,

$$-K \cdot C = (2 - 2g_c) + C^2$$

as claimed. Notice that these equations characterize uniquely the class K, since the intersection matrix is negative definite. Thus (iv) $\Rightarrow$ (v). Finally, (v) implies that the first Chern class of the bundle $T\widetilde{V}|_{\widetilde{M}\times(-\varepsilon,\varepsilon)}$ vanishes, where $\widetilde{M}\times(-\varepsilon,\varepsilon)$ is a tubular neighbourhood of $\widetilde{M}$ in $\widetilde{V}$. Hence (v) and Lemma 6.2 imply (iv). $\quad\square$

Notice that the manifold $\widetilde{V}$ has the exceptional set $E = \pi^{-1}(0)$ as a deformation retract. Hence, if the resolution is good, then the irreducible components $E_1,\ldots,E_r$ of E are non-singular and they form a basis for the homology of $\widetilde{V}$. Thus, in order to characterize uniquely the canonical class K, it is enough to consider these curves in the adjunction formula (6.4).

6.5 Definition. The class K is the *canonical class* of the resolution.

6.6 Corollary. *Every Gorenstein singularity is numerically Gorenstein. More precisely, Gorenstein means that the bundle $\bigwedge^2 T^*(V-0)$ is holomorphically trivial, while numerically Gorenstein means that this bundle is topologically trivial.*

Notice that the canonical class K corresponds to the zero-set of a differentiable section of the bundle $\bigwedge^2 T^*(V-0)$. If V is assumed to be Gorenstein, then K can be taken to be a divisor, i.e., defined by a meromorphic 2-form. We also remark that regardless of whether or not $(V,0)$ is Gorenstein, the intersection matrix is negative definite and therefore the adjunction formula determines a unique canonical class, which may have rational coefficients (see [64, 119]). The geometric reason behind this is that the class determined by the adjunction formula is, by construction, concentrated in the exceptional divisor, it is always a rational linear combination of the form:

$$K = n_1 E_1 + \cdots + n_r E_r\,;$$

however, when the singularity is not numerically Gorenstein, one cannot possibly have a representative of the first Chern class of $T\widetilde{V}$ that is localized on E.

6.7 Examples.

(i) First consider a plumbing graph with only one vertex:

$$\overset{\bullet}{(w, g)}$$

where $w \in \mathbb{Z}$ is the weight and $g \in \mathbb{N}$ is the genus. The intersection matrix is now 1×1, so it is negative definite iff $w < 0$; hence all of these are graphs of normal surface singularities. By the adjunction formula, the canonical class is

$$K = \left(\frac{2g - 2}{w} - 1 \right) \cdot E \ ,$$

where E is the curve of genus g represented by the vertex. Hence the corresponding singularities are numerically Gorenstein iff $2g - 2 \equiv 0 \mod w$.

(ii) If the graph Σ is one of the classical Dynkin diagrams A_n, D_n, E_6, E_7 or E_8, with all weights equal to -2 and all genera 0, then the intersection matrix is negative definite, the corresponding singularities are the Klein singularities studied in Chapter II, i.e., those of the form $\Gamma \backslash \mathbb{C}^2$ for a finite subgroup of $SU(2)$, and a trivial computation shows that $K \equiv 0$, see for instance [65].

(iii) If the graph is a star, with a central curve E_0 of genus $g \geq 0$, weight $w = 2 - 2g - n$, and all other vertices represent spheres E_i with weights $\{-\alpha_1, \ldots, -\alpha_r\}$ satisfying $\alpha_i > 1$ and $w < \sum_{i=1}^{r} \frac{-1}{\alpha_i}$, then the intersection matrix is negative definite and the corresponding plumbing graph represents the minimal good resolution of a Dolgachev singularity [57], i.e., one of the singularities considered in III.3.3, which corresponds to taking a cocompact Fuchsian group $\Gamma \subset PSL(2, \mathbb{R})$, making it act on $T\mathbb{H} = \mathbb{H} \times \mathbb{C}$ and then blown down the zero-section. By the adjunction formula, the canonical class is now

$$K = -2E_0 - \sum_{i=1}^{r} E_i \ .$$

(iv) Now start with a discrete subgroup $\Gamma \subset SL(2, \mathbb{R})$ with compact quotient $\Gamma \backslash SL(2, \mathbb{R})$ which does not contain the centre $\pm Id$, and construct a surface singularity as in III.3.3, i.e., we make Γ act on $\mathbb{H} \times \mathbb{C}_2$, where $\mathbb{C}_2$ denotes the 2-fold cyclic cover of $\mathbb{C}$ branched at 0. The quotient contains $(\Gamma \backslash \mathbb{H} \times \{0\})$ as a divisor, which can be blown down to get a normal singularity. The resolution graph is now of the form (see [194] or [231, p. 351]):

where $\{g; \alpha_1, \ldots, \alpha_r\}$ is the signature of the Fuchsian group obtained by mapping Γ to $PSL(2, \mathbb{R})$. All the α_i are (automatically) odd numbers and all vertices represent spheres, except the centre E_0 which has genus $g \geq 0$. In this case one has that the canonical class is

$$K = -3E_0 - 2\sum_{i=1}^{r} E_{i,1} - \sum_{i=1}^{r} E_{i,2}.$$

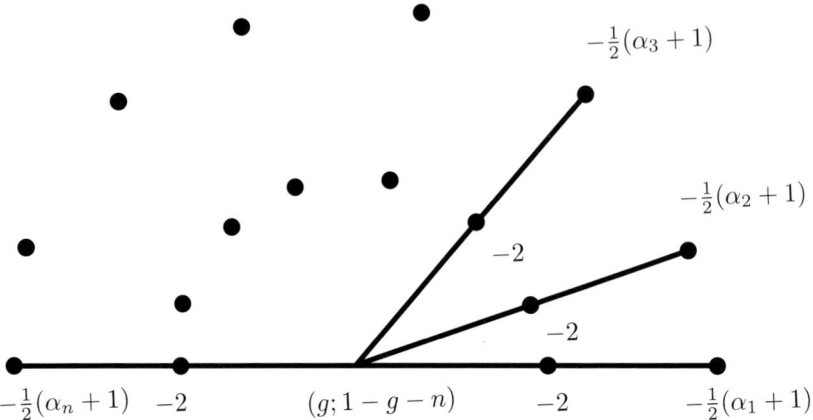

Figure 10: Resolution graph for $SL(2, \mathbb{R})$-singularities.

v) More generally, we recall from Chapter III that if $(V, 0)$ is a normal singularity with a good $\mathbb{C}^*$-action, then its link is a Seifert manifold of the form $\Gamma \backslash G$ where G is either $SU(2)$, the Heisenberg group N or $\widetilde{SL}(2, \mathbb{R})$, the universal cover of $PSL(2, \mathbb{R})$, and Γ is a uniform subgroup of G. Consider the case $\Gamma \subset \widetilde{SL}(2, \mathbb{R})$, which is the most interesting. The minimal good resolution of these singularities is given in [194]. It is always a star with a central curve E_0 of some genus $g \geq 0$, as many branches as exceptional fibres of the corresponding Seifert fibration and all other vertices represent spheres; the lengths of the branches and their weights are all determined by the corresponding Seifert invariants. The canonical class was essentially computed in [231]. The remarkable thing is ([231, 2.4], c.f. III.2):

6.8 Proposition. *Let* $-m_o$ *be the coefficient of the canonical class K of the minimal resolution of an* $\widetilde{SL}(2, \mathbb{R})$-*singularity. Let* $r = m_0 - 1$. *Then the group Γ has index* r *in* $p^{-1}(p(\Gamma))$. *Or equivalently, if we denote by Γ_r the projection of Γ to the r-fold cyclic cover* $PSL(2, \mathbb{R})_r$ *of* $PSL(2, \mathbb{R})$, *then:*

(i) $\Gamma_r \backslash PSL(2, \mathbb{R})_r$ *is diffeomorphic to* $\Gamma \backslash \widetilde{SL}(2, \mathbb{R})$; *and*

(ii) Γ_r *is isomorphic to its image in* $PSL(2, \mathbb{R})$.

6.9 Remark. We pointed out earlier that the property for a plumbing graph to have negative definite intersection matrix depends only on the weights and not on the genera assigned to the vertices. However the property of being the dual graph of a numerically Gorenstein singularity does depend on the genera too, since these do appear in the adjunction formula. One can prove (see [119, 3.10]) that given any graph Γ and weights $w = (w_1, \ldots, w_r)$ so that the intersection matrix $\mathbb{E}$ is negative definite, there are infinitely many different possible genera for the vertices that make it correspond to a numerically Gorenstein singularity. In fact one can say more in some cases. For example, we know that K represents the 1st Chern class of the resolution $\widetilde{V}$. Hence K is an even class iff $\widetilde{V}$ admits a spin structure.

One can prove that if the weights are even, then there are infinitely many different genera for which the class K is even. Furthermore, if the weights are even and the determinant $det(\mathbb{E})$ of $\mathbb{E}$ is odd, then K is integral iff it is even. As a consequence, if a complete intersection surface singularity has a resolution with even weights and $det(\mathbb{E})$ is odd, then K is even.

IV.7   An application of Riemann-Roch: Laufer's formula

There is an important invariant of surface singularities called the geometric genus, which extends to surface singularities the genus of Riemann surfaces (regarded as the number of linearly independent differential forms). M. Artin introduced it in [1] to define the concept of *rational singularities*, which means geometric genus 0. This has been much used and studied by several authors for various purposes, as for instance F. Hirzebruch, H.B. Laufer, K. Saito, S.T. Yau and many more.

To define the *geometric genus* $\rho_g = \rho_g(V, 0)$ of a normal surface singularity $(V, 0)$, let $\pi : \widetilde{V} \to V$ be a resolution of V. Then $\rho_g(V, 0)$ is:

$$\rho_g(V, 0) \; = \; \dim \; R^1\pi^*\mathcal{O}_{\widetilde{V}}$$

the dimension of the direct image sheaf $R^1\pi^*\mathcal{O}_{\widetilde{V}}$, which is a coherent sheaf concentrated at 0. It was noticed in [14] that this number is independent of the choice of resolution (since it is invariant under blowing ups) and it can also be computed as:

$$\rho_g \; = \; \dim H^1(X, \mathcal{O}_X) \, ,$$

where the term on the right must be understood as an inverse limit over a fundamental set of open neighborhoods of the exceptional divisor $E = \pi^{-1}(0)$.

The formula of Laufer in [123] (and later generalized by others, as we explain below) gives a very interesting relationship between the geometric genus and invariants of the singularity which are associated to deformations of it, namely the *Milnor number* in the case of complete intersection germs, or an extension of it for more general normal singularities which are Gorenstein and smoothable (see below).

Let us look first at the classical case of a hypersurface germ in $\mathbb{C}^3$, as considered by Laufer. We notice that every such germ is automatically Gorenstein (see for instance III.3.2 above), so given a resolution $\pi : \widetilde{V} \to V$, then a never-zero holomorphic 2-form on $V - 0$ lifts by π to a never-zero holomorphic 2-form on $\widetilde{V} - E$, which extends to a meromorphic form on $\widetilde{V}$, possibly with poles and zeroes in the divisor E. Thus we know that the canonical bundle $\mathcal{K}_{\widetilde{V}}$ is holomorphically trivial away from E and we can choose the canonical divisor $\widetilde{K}$ to be *vertical*, i.e., with support in E. If we assume that the resolution $\widetilde{V}$ is good, then the irreducible components $E_1, \cdots, E_r$ of E are non-singular curves and $\widetilde{K}$ will necessarily be an *integral linear combination* of these (see the previous section).

Let $f : (U \subset \mathbb{C}^3,) \to (\mathbb{C}, 0)$ be a function-germ that defines V. Then one has the Milnor number μ of f defined in I.7. We know (see for instance [120, 7.1]) that up to a local change of coordinates, we can assume f is a polynomial map. Compactify $\mathbb{C}^3$ to $\mathbb{C}P^3$, denote by $\bar{V}$ the closure of V in $\mathbb{C}P^3$ and let $\bar{V}_t$ be the compactification of $V_t = f^{-1}(t)$ for some $t \neq 0$, $|t|$ small. As noted by Laufer, we can assume that $\bar{V}$ is non-singular away from 0 and $\bar{V}_t$ is non-singular. One obviously has

$$\chi(\bar{V}_t) = \chi(\bar{V}) + \mu,$$

where $\mu = \chi(V_t) - 1$ is the Milnor number. Now, given a resolution $\pi : \tilde{V} \to V$ of V, let $\tilde{V}^*$ be the resolution of $\bar{V}$ that has $\tilde{V}$ as open set. Topologically this means that $\tilde{V}$, V and V_t have all been compactified by adding "the same divisor" K_∞ at ∞. Letting $E = \pi^{-1}(0)$, one has:

$$\chi(\tilde{V}^*) = \chi(\bar{V}) + \chi(E) - 1,$$

since $\tilde{V}^*$ is obtained by removing 0 from $\bar{V}$ and replacing it by E. Also, the Riemann-Roch formula of Section 2 says that the analytic Euler characteristic is:

$$\chi(\tilde{V}^*, \mathcal{O}) = \frac{1}{12}(K^2_{\tilde{V}^*} + \chi(\tilde{V}^*)) = \frac{1}{12}(K^2_\infty + K^2_{\tilde{V}} + \chi(\tilde{V}^*)),$$

and

$$\chi(\bar{V}_t, \mathcal{O}) = \frac{1}{12}(K^2_\infty + \chi(\bar{V}_t)),$$

because $K^2_{V_t} = 0$ since the canonical bundle of V_t is trivial.

Finally, using Mayer-Vietoris ([7, p. 236]) for the cohomology of $\tilde{V}^* = \tilde{V} \cup \tilde{V}^* - E$ and $\bar{V} = V \cup \bar{V} - 0$ one gets:

$$\chi(\tilde{V}^*, \mathcal{O}) = \chi(\bar{V}, \mathcal{O}) - h^1(\tilde{V}, \mathcal{O}).$$

Putting these equations together Laufer[123] arrives at his formula:

$$\mu + 1 = \chi(\tilde{V}) + K^2 + 12\rho_g , \qquad (7.1)$$

where $\chi(\tilde{V})$ is the topological Euler-Poincaré characteristic of the resolution and K^2 is the self-intersection number of the canonical divisor.

This result of Laufer shows also that the geometric genus cannot in general be defined by the topology of V alone (c.f. [181, 182]). For instance, the Brieskorn singularities with (p, q, r) being $(3, 5, 15)$ and $(2, 9, 18)$ are homeomorphic but their Milnor numbers are 112 and 136, respectively, so their geometric genera are different.

Laufer's proof extends with minor changes to complete intersection germs, and he conjectured that the same formula should hold for normal Gorenstein smoothable singularities, just replacing the left-hand side by the Euler-Poincaré characteristic of the smoothing. Let us explain this.

Given a normal surface singularity $(V, 0)$, a *smoothing* of V means a flat deformation of V where all nearby fibres are smooth. To be precise:

7.2 Definition. An isolated complex singularity germ (X, P) of dimension $n \geq 1$ is *smoothable* if there exists a complex analytic space $(\mathcal{W}, 0)$ of dimension $n + 1$ and a proper analytic map:

$$\mathcal{F} : \mathcal{W} \longrightarrow \mathbb{D} \subset \mathbb{C} \, ,$$

where $\mathbb{D}$ is an open disc with centre at 0, such that:

(i) it is flat, i.e., it is not a zero divisor in the local ring of $\mathcal{W}$ at 0;

(ii) $\mathcal{F}^{-1}(0)$ is isomorphic to V; and

(iii) $\mathcal{F}^{-1}(t)$ is non-singular for $t \neq 0$.

It follows that $\mathcal{W}$ has (at most) an isolated singularity at 0, which is normal when X is normal at P, and one has (by Theorem I.4.3) a fibre bundle

$$\mathcal{F}|_{\mathcal{F}^{-1}(\mathbb{D}-0)} \, : \, \mathcal{F}^{-1}(\mathbb{D} - 0) \longrightarrow \mathbb{D} - 0$$

with non-singular fibres $X_t = \mathcal{F}^{-1}(t)$. For simplicity we call *smoothings* of X to both, the map $\mathcal{F}$ and the fibres X_t. If $\mathcal{W}$ is smooth, then its germ at 0 is essentially that of $\mathbb{C}^{n+1}$ at 0. The picture one has in general is similar to the one of hypersurfaces and their corresponding Milnor fibration, but now the ambient space $\mathcal{W}$ can be itself singular at 0.

For hypersurfaces and complete intersection germs the smoothing is unique up to isomorphism (because the base of the versal deformation space is connected) and it is given by the Milnor fibration. However there are normal singularities which are not smoothable (e.g., some triangle singularities of Dolgachev, see for instance [199, 147]), and there are also surface singularities which are smoothable but they have different smoothings whose fibres have distinct Euler-Poincaré characteristics (see [199]).

Laufer's formula explained above was later extended by Steenbrink [241] to surface singularities which are Gorenstein and smoothable. His proof is via mixed Hodge structures, but it can be proved exactly with the same method of Laufer, using a result in [225] and a globalization theorem of Looijenga, conjectured by J. Wahl in [258].

7.3 Theorem. (E. Looijenga [148]) *Let $\mathcal{F} : (\mathcal{W}, x) \to (\mathbb{C}, 0)$ be a smoothing of an isolated singularity (X, x). Then there is, up to isomorphism, a global projective compactification of the smoothing. That is, there exists a flat projective morphism $F : \mathcal{Z} \to \mathbb{C}$, a point $z \in \mathcal{Z}_0 = F^{-1}(0)$ and an isomorphism $h : (\mathcal{W}, x) \to (\mathcal{Z}, z)$ such that $\mathcal{F} = F \circ h$ and F is smooth along $\mathcal{Z}_0 - z$.*

We refer to the appendix in [148] for the proof of this theorem. As pointed out by Wahl in [258, p. 220], this assertion allows us to use the Riemann-Roch theorem to compare the cohomology of the singular variety, the smoothing and the resolution, just as in Laufer's proof. Everything works in exactly the same way as for hypersurfaces, except for one more point that has to be proved: for hypersurfaces (or complete intersections) of any dimension, the canonical bundle of the smoothing (the Milnor fibre) is holomorphically trivial. To see this one may, for instance, take the form $dz_1 \wedge \cdots \wedge dz_{n+k}$ in the ambient space, and contract it using the gradient vector fields of the function germs that define the complete intersection singularity. The same statement holds for smoothings of isolated, normal singularities in all dimensions.

7.4 Proposition [225]. *Let $\mathcal{F} : (\mathcal{W}, 0) \to \mathbb{C}$ be a smoothing of a normal, Gorenstein singularity $(V, 0)$ of dimension $n \geq 1$, and let $T_{\mathcal{F}}$ denote the tangent bundle along the fibres of $\mathcal{F}$ on $\mathcal{W} - V$. Then the bundle $\bigwedge^n T_{\mathcal{F}}^*$ is holomorphically trivial.*

In the following section we will see that Proposition 7.4, which is obvious for the algebraic-geometers, has an important topological implication in the case of surfaces, and this proved a conjecture of Durfee in [64].

Proof. Consider the smoothing $\mathcal{F} : \mathcal{W} \longrightarrow \mathbb{D} \subset \mathbb{C}$, so each fibre $V_t = \mathcal{F}^{-1}(t)$, $t \neq 0$, is non-singular. Since V is Gorenstein at 0 and $\mathcal{F}$ is flat, it follows that $\mathcal{W}$ is also Gorenstein at 0 (see for instance [97, V.9.6]). Hence there exists a nowhere-vanishing holomorphic $(n+1)$-form $\tilde{\omega}$ on $\mathcal{W} - 0$. Since $\mathcal{F}$ is flat, we have an exact sequence of vector bundles over $\mathcal{W} - 0$:

$$0 \to T_{\mathcal{F}} \to T(\mathcal{W} - 0) \to \mathcal{F}^*(T\mathbb{D}) \to 0\,,$$

where $T_{\mathcal{F}}$ denotes the tangent bundle along the fibres of $\mathcal{F}$. Hence,

$$\bigwedge^{n+1} T^*(\mathcal{W} - 0) \cong \bigwedge^n T_{\mathcal{F}}^* \otimes \mathcal{F}^*(T^*\mathbb{D})$$

and the above holomorphic $(n+1)$-form on $\mathcal{W} - 0$ defines a nowhere-vanishing holomorphic n-form on each non-singular fibre of the smoothing and trivializes the whole bundle $\bigwedge^n T_{\mathcal{F}}^*$. $\qquad \square$

The fact that $\mathcal{W}$ is Gorenstein at 0 is fairly standard in algebraic geometry and can be proved as follows: the local ring of V at 0 is:

$$\mathcal{O}_{V,0} \cong \left(\mathcal{O}_{\mathcal{W},0} / (\mathcal{F}) \right)$$

by hypothesis. Since V is Gorenstein, it is Cohen-Macaulay, so there are elements $\bar{g}_1, \ldots, \bar{g}_n \in \mathcal{O}_{V,0}$ which form a regular sequence, i.e., $\bar{g}_1$ is not a zero-divisor in $\mathcal{O}_{V,0}$, $\bar{g}_2$ is not a zero-divisor in $\mathcal{O}_{V,0}/(\bar{g}_1)$ and so on, and such that

$$\mathrm{Hom}_{\mathcal{O}_{V,0}} \{ \mathbb{C}\,,\, \mathcal{O}_{V,0}/(\bar{g}_1, \bar{g}_2, \ldots, \bar{g}_n) \} \cong \mathbb{C}\,.$$

Hence if $g_1, \ldots, g_n \in \mathcal{O}_{\mathcal{W},p}$ are extensions of $\bar{g}_1, \ldots, \bar{g}_n$ (which exist by normality), then $(\mathcal{F}, g_1, \ldots, g_n)$ form a regular sequence in $\mathcal{O}_{\mathcal{W},p}$ and

$$\mathrm{Hom}_{\mathcal{O}_{\mathcal{W},0}} \{ \mathbb{C}\,,\, \mathcal{O}_{\mathcal{W},0}/(\mathcal{F}, g_1, \ldots, g_n) \} \cong \mathbb{C}\,. \qquad \square$$

With this, one obtains the Laufer-Steenbrink formula in general:

7.5 Theorem. *Let $(V,0)$ be a 2-dimensional normal singularity which is Gorenstein and smoothable. Let V_t be a smoothing of V and let $\widetilde{V} \to V$ be a resolution. Then one has:*

$$\chi(V_t) = \chi(\widetilde{V}) + K^2 + 12\rho_g(V,0),$$

where $\chi(\cdot)$ is the usual (topological) Euler-Poincaré characteristic, K is the canonical class of $\widetilde{V}$ and $\rho_g(V,0) = \dim H^1(\widetilde{V}, \mathcal{O}_{\widetilde{V}})$ is the geometric genus of $0 \in V$.

7.6 Remarks.

(i) It is obvious that $\chi(\widetilde{V})$ depends on the choice of resolution, as can be seen just by noticing that this increases by 1 when we make a blow-up. But it is an exercise to show that the sum $\chi(\widetilde{V}) + K^2$ does not depend on the choice of resolution, and we know that neither does $\rho_g(V,0)$. Hence Theorem 7.3 implies that for normal surface, smoothable singularities, being Gorenstein implies that at least $\chi(V_t)$ does not depend on the choice of the smoothing (which is not the case for surface singularities in general, by [199]).

Notice that the right-hand side of the formula in Theorem 7.5 makes sense even for non-smoothable singularities, and it depends only on the analytic structure of V, not on the choice of resolution. It is thus natural to ask what plays the role of $\chi(V_t)$ when there is not a smoothing? In other words, what should we put in the left-hand side of Theorem 7.5 when there is no smoothing? This question was asked of me by Dolgachev and I do not know the answer. Presumably this will be in terms of the Seiberg-Witten invariant of the link (see §11 below).

(ii) There is an interesting generalization in [148] of the above formula to higher dimensions, also based on the generalized Riemann-Roch theorem. To state this, the first observation that Looijenga makes is that using a deep theorem of O. Gabber (or also by work of Steenbrink) one has that given an isolated singularity $(V,0)$ of dimension $n \geq 1$, the subalgebra of $H^*(V - 0)$ generated by the Chern classes of the tangent bundle $T(V - 0)$ is trivial in dimensions $\geq n$. For complete intersection germs this is an obvious fact, since the bundle $T(V - 0)$ is itself trivial, but for singularities in general it is not so obvious. Now let X denote either a resolution of the singularity or a smoothing of it (if it is smoothable), and think of it as being a compact manifold with boundary the link $M = \partial X$. Given any polynomial $P(c_1, \ldots, c_n)$ in the Chern classes of X of top degree n, the previous observation tells us that we can lift the corresponding cohomology class to $\widetilde{P} \in H^*(X, \partial X)$. Elementary properties of the cup product of relative cohomology classes yield that if the polynomial P is decomposable (i.e., if it is the product of at least two Chern classes of degree less than n), then the lifting $\widetilde{P}$ does not depend on the choice of lifting, so it is canonically defined. In the case of the top class c_n this is not so, and Looijenga shows that one can always do it in such a way that the class one gets is the Lefschetz dual of $\chi(X)$ in $H^{2n}(X, \partial X)$. With this one may now define the *Todd genus* of X, just as in Section 2 above, by taking the nth Todd polynomial in these classes, and evaluating it on the fundamental cycle of X. Taking the representative of $(V, 0)$ small enough, we can assume that the smoothing V_t is Stein and therefore (by the theorem of Andreotti-Frankel) all its cohomology groups vanish in dimensions $> n$. Thus $Td(V_t)$ is determined only by $\chi(V_t)$ and, if $n = 2m$ is even, by c_m^2. Now denote by $\pi : \widetilde{V} \to V$ a resolution of V. Then Looijenga's theorem says:

$$Td_n(V_t) = Td_n(\widetilde{V}) + \sum_{i>0}(-1)^{i-1}h^i(\widetilde{V}; \mathcal{O}_{\widetilde{V}}) - \dim \pi_*\mathcal{O}_{\widetilde{V}}/\mathcal{O}_V .$$

IV.8 Geometric genus, spin^c structures and characteristic divisors

This section is based on [76], where Riemann-Roch is used to explore the relationship between the geometric genus of normal, Gorenstein surface singularities and the dimension of the space of sections of holomorphic line bundles associated with spin^c-structures on a resolution of the singularity. Essentially everything we say here holds in complex dimensions $4k + 2$, with some obvious modifications; we restrict to $k = 0$ for simplicity, and we refer to [76] for details.

We know that since V is Gorenstein, given any resolution $\pi : \widetilde{V} \to V$, we may consider the canonical divisor K to have support in the exceptional divisor $\pi^{-1}(0)$. And we know from §4 that on a complex manifold, the anti-canonical divisor determines a canonical spin^c-structure. In the sequel we shall consider divisors in $\widetilde{V}$ of the form $W = 2D - K_{\widetilde{V}}$ with D *vertical*, i.e., with $\pi(|D|) = 0$. The bundle $\mathcal{L}$ of such a divisor determines a spin^c structure on $\widetilde{V}$ for which $\mathcal{L}$ is the determinant bundle. We call these *characteristic divisors of the resolution*.

Given a characteristic divisor $W = 2D - K_{\widetilde{V}}$ and its bundle $\mathcal{L}$, we define the invariant:

$$\hbar(W) = \begin{cases} 0 & , & \text{if } W = 0 \\ \dim H^0(W, \mathcal{D}|_W) & , & \text{if } W \neq 0; \end{cases}$$

The reduction of $\hbar(W)$ modulo 2 is the invariant introduced in §4. One has ([76, 3.5 & 4.1]):

8.1 Theorem. *Let $(V, 0)$ be as above, a normal Gorenstein surface singularity, and let $\rho_g(V, 0)$ be its geometric genus. Let $\pi : \widetilde{V} \to V$ be a resolution of V and $K_{\widetilde{V}}$ its canonical divisor, chosen so that it is vertical. For every characteristic divisor $W = 2D - K$ with D vertical, one has:*

(i) *The integer $(W^2 - K_{\widetilde{V}}^2)$ is divisible by 8, and*

$$\hbar(W) + \frac{1}{8}(W^2 - K_{\widetilde{V}}^2) \equiv \rho_g(V, 0) \qquad \mod (2) .$$

(ii) *Moreover, if $D = 0$ or $D > 0$ and $-D$ is relatively ample for π, then:*

$$\rho_g(V, 0) = \hbar(W) + \frac{1}{8}(W^2 - K_{\widetilde{V}}^2) .$$

(iii) *If $K_{\widetilde{V}} \leq 0$, then for all $D \geq 0$, setting $W = 2D - K_{\widetilde{V}}$, one has:*

$$\rho_g(V, 0) = \hbar(W) + \frac{1}{8}(W^2 - K_{\widetilde{V}}^2).$$

The proof of this theorem in [76, §3,S4] is based on the Hirzebruch-Riemann-Roch theorem and on several vanishing theorems. The idea is to compactify the resolution of the singularity, then use Riemann-Roch and cancel down the terms that come from contributions away from the exceptional divisor. More precisely, given any resolution $\pi : \widetilde{V} \to V$, a smooth compactification $\widetilde{V}^*$ of $\widetilde{V}$ and a vertical divisor D, for every $a \in \mathbb{Z}$ we can define an invariant $g(a \cdot D) \in \mathbb{Z}$ as *the difference of the analytic Euler characteristic*

$$g(a \cdot D) = \chi\big(\mathcal{O}_{\widetilde{V}^*}(a \cdot D)\big) - \chi\big(\mathcal{O}_{\widetilde{V}^*}\big).$$

It is an exercise to verify that the additivity of Euler characteristics for short exact sequences and the Riemann-Roch theorem imply:

8.2 Properties of $g(a \cdot D)$:

   (i) it is independent of the compactification $\widetilde{V}^*$ chosen;
   (ii) if $K_{\widetilde{V}}$ is the canonical divisor of $\widetilde{V}$, then $g(a \cdot D) = \frac{1}{2}(a \cdot D) \cdot (a \cdot D - K_{\widetilde{V}})$;
   (iii) for $W = 2D - K_{\widetilde{V}}$ one has: $g(D) = \frac{1}{8}(W^2 - K_{\widetilde{V}}^2)$.

Having this one can prove that for every effective vertical divisor Δ:

$$\hbar(W) + g(D) \equiv \hbar(W + 2\Delta) + g(D + \Delta) \qquad \mathrm{mod}\,(2). \tag{8.3}$$

Thus statement (ii) in Theorem 8.1 implies statement (i), since we can always choose some effective vertical divisor Δ such that $D + \Delta > 0$ and $-(D + \Delta)$ is relatively ample for π. The proof of statements (ii) and (iii) in Theorem 8.1 does rely on several vanishing theorems given in [76] and we refer to that article.

Since, by [176], for the minimal resolution of a surface singularity it is always the case that the canonical divisor is ≤ 0, one has:

8.4 Corollary. *If the resolution $\widetilde{V}$ is minimal, then*

$$\rho_g(V, 0) = \dim H^0(-K_{\widetilde{V}}, \mathcal{O})$$

$$= \dim H^0(W, \mathcal{D}|_W) + \frac{1}{8}(W^2 - K_{\widetilde{V}}^2)$$

for all $W = 2D - K_{\widetilde{V}}$ with $D \geq 0$.

8.5 Remarks.

(i) If the canonical divisor $K_{\widetilde{V}}$ of some resolution is even, one has that the resolution must be minimal (by [119]) and admits a spin structure; in this case one has (essentially from Theorem 8.1) that $K_{\widetilde{V}}^2$ is divisible by 8 and

$$\frac{1}{8}K_{\widetilde{V}}^2 \equiv \rho_g(V, 0) \qquad \mathrm{mod}\,(2).$$

If $K_{\widetilde{V}}$ is not even, the most natural choice of a characteristic divisor is $W_0 = \sum E_i$, where the sum is taken over all components of the exceptional divisor $E = \pi^{-1}(0)$ which have odd multiplicity in $K_{\widetilde{V}}$. With this choice, assuming that the resolution is good and the Betti number $b^1(E) = \dim H^1(E; \mathbb{Z})$ is 0, one gets:

$$\frac{1}{8}(W_0^2 - K_{\widetilde{V}}^2) \equiv \rho_g(V, 0) \qquad \mod(2)$$

since $b^1(E) = 0$ implies that $H^0(E, \omega_E) = 0$ and hence $H^0(W_0, \omega_{W_0}) = 0$, where ω_* is the canonical sheaf (c.f. §4 above). Hence $H^0(W_0, \mathcal{O}_{W_0}(D)) = 0$. This remark, together with the combinatorial argument contained in [140], imply that Theorem 2 in [140] holds without the assumption that the singularity is smoothable, i.e., if the determinant of the intersection matrix of E is odd, then one always has:

$$K_{\widetilde{V}}^2 + 8\rho_g(V, 0) \equiv W_0^2 \qquad \mod(16) \ .$$

(ii) In [140] the authors present four examples of singularities with $K_{\widetilde{V}}$ even and for which the relation $\frac{1}{8} K_{\widetilde{V}}^2 \equiv \rho_g(V, 0) \mod(2)$ does not hold. Alas their computations were mistaken (c.f. [141]), as proven by Theorem 8.1.

IV.9   On the signature of smoothings of surface singularities

In Section 7 we explained Laufer's formula expressing the Euler-Poincaré characteristic of a smoothing of a normal Gorenstein smoothable singularity in terms of invariants associated to a resolution of it. Section 8 gives relations between the geometric genus and the spaces of sections of characteristic line bundles over a resolution, particularly with the canonical bundle. All these formulae are proved by using the Hirzebruch-Riemann-Roch theorem, applied to projective compactifications of $\widetilde{V}$ (and in Section 7 of a "Milnor fibre" $V_t = f^{-1}(t)$, where f is the smoothing), and then cancelling down terms that come from the contributions away from the origin. In this section we are concerned with a formula obtained by A. Durfee in [64], which is in the same spirit as Laufer's but uses the *signature defect*, introduced by Hirzebruch, instead of the Hirzebruch-Riemann-Roch formula. The result is an interesting and useful formula for the signature of smoothings of Gorenstein singularities. One may notice that, since for compact complex surfaces the Todd genus and the signature are both expressible in terms of the canonical class and the Euler-Poincaré characteristic, the formulas of Durfee and Laufer turn out to be essentially equivalent.

Let us define first the *signature defect* of a framed 3-manifold. This invariant of 3-manifolds was defined by Hirzebruch via his signature theorem, and it was later generalized by Morita and (notably) by Atiyah, Patodi and Singer, giving rise to the so-called η-invariants. There is also a generalization in [147] which is an improvement of Morita's work. We define it here only in the classical case, which is all we need in the sequel.

We recall that if X is a smooth, oriented closed 4-manifold, then the signature theorem of Thom-Hirzebruch says that its signature is given by the Pontryagin number:

$$\sigma(X) = \frac{1}{3} p_1(X)[X] .$$

Now let X be a compact manifold with non-empty boundary M; both sides of this equation still make sense, but $p_1(X)$ lives in $H^4(X) \cong 0$, so one may not expect an equality in general. We notice however that the bundle $TX|_M$ is trivial, and a choice of a trivialization τ leads to a representative of $p_1(X)$ that vanishes over M and so defines a relative Pontryagin class $p_1(X, \tau) \in H^4(X, M) \cong \mathbb{Z}$, see §1 above. This class depends on the choice of the trivialization τ of $TX|_M$, so it cannot give the signature of X for arbitrary τ. Notice that the trivialization τ determines an orientation of M, and we assume this is the same as the orientation M gets as being the boundary of X. One has that the difference:

$$d_\sigma(X, \tau) = \frac{1}{3} p_1(X)[X] - \sigma(X) ,$$

is actually independent of X: it depends only on its boundary M and the trivialization τ of the bundle

$$TX|_M = T(M) \oplus (1) ,$$

where (1) denotes the trivial 1-dimensional bundle, isomorphic to the normal bundle of M in X. In fact, suppose we are given another compact, oriented 4-manifold X' with the same boundary M (as oriented manifolds). Then one can form the union $Y = X \cup (-X')$, where $(-X')$ is X' with the opposite orientation; Y is a smooth, closed, oriented 4-manifold, and one has:

$$\sigma(Y) = \sigma(X) - \sigma(X')$$

and

$$\frac{1}{3} p_1(Y)[Y] = \frac{1}{3} \Big(p_1(X, \tau)[X] - p_1(X', \tau)[X'] \Big) ,$$

because $p_1(Y)$ vanishes over M. Hence one has:

$$d_\sigma(X, \tau) - d_\sigma(X', \tau) = \left(\frac{1}{3} p_1(X)[X] - \sigma(X) \right) - \left(\frac{1}{3} p_1(X')[X'] - \sigma(X') \right)$$

$$= \left(\frac{1}{3} \Big(p_1(X)[X] - p_1(X')[X'] \Big) \right) - \Big(\sigma(X) - \sigma(X') \Big) = 0 .$$

Therefore $d_\sigma(X, \tau)$ depends only on the framed boundary (M, τ) and not on the choice of X. Since the group of oriented cobordism is trivial in dimension 3 (see, e.g., [244]), it follows that every framed 3-manifold M as above is the boundary of a compact, oriented 4-manifold, with the orientation being compatible with the one defined by τ.

9.1 Definition. The invariant $d_\sigma(X, \tau) = \frac{1}{3}p_1(X)[X] - \sigma(X)$ is called *the signature defect* of the framed manifold (M, τ).

We have:

9.2 Theorem. (Durfee) *Let $(V, 0)$ be a normal, numerically Gorenstein surface singularity, which is smoothable. Assume further that V_t is a smoothing of V with (topologically) trivial tangent bundle. Then:*

$$\sigma(V_t) = -\frac{1}{3}\left(2(\chi(V_t) - 1) + K^2 + s + 2b_1\right),$$

where b_1 is the first Betti number of the resolution and s is the second Betti number.

Proof. Observe first that if $(V, 0)$ is numerically Gorenstein, then by Corollary 6.3.iii above we know that the structure group of the bundle $TV_t|_M$ can be reduced to $SU(2)$. Since $SU(2) \cong Sp(1)$, this means we have multiplication of tangent vectors in this bundle by the quaternions i, j, k. Now let ν denote the unit outwards normal vector field of M in V_t (with respect to some metric). Then multiplying ν by i, j, k at each point of M we obtain three linearly independent vector fields on M, which define a trivialization of TM which is compatible with the complex structure on $TV_t|_M$, in the sense that they determine with ν a trivialization of this bundle as a complex bundle.

Now, let M be the link of V and let ρ be a trivialization of TV_t obtained as above, i.e., multiplying by i, j, k a unit normal vector field on $M = \partial V_t$. Let $\pi : \widetilde{V} \to V$ be a resolution of V. By the previous discussion we have:

$$\frac{1}{3}p_1(V_t)[V_t] - \sigma(V_t) = \frac{1}{3}p_1(\widetilde{V})[\widetilde{V}] - \sigma(\widetilde{V}).$$

But both TV_t and $T\widetilde{V}$ are complex bundles and the trivialization ρ is compatible with their complex structure, so the Pontryagin number can be expressed in terms of the Chern numbers relative to ρ. One has:

$$p_1(V_t)[V_t] = (c_1(V_t)^2 - 2c_2(V_t))[V_t] = -2\chi(V_t)$$

because $c_1(V_t)^2 = 0$, since TV_t is trivial by hypothesis, and $c_2(V_t)[V_t] = \chi(V_t)$ because ρ is given by a parallelism on the boundary M. Similarly,

$$p_1(\widetilde{V})[\widetilde{V}] = K^2 - 2\chi(\widetilde{V})$$

and $\sigma(\widetilde{V}) = -b_2(\widetilde{V})$, the 2nd Betti number of $\widetilde{V}$, because the intersection matrix on $\widetilde{V}$ is negative definite by Mumford's theorem (see §4 above). Putting these equations together we arrive at Durfee's formula. $\square$

Notice that Gorenstein singularities are numerically Gorenstein. So Theorem 9.2 applies to these singularities provided the tangent bundle of the smoothing is topologically trivial, a fact conjectured by Durfee in [64, 1.6] and proved in [225]:

9.3 Proposition. *Let V_t be a smoothing of a normal, Gorenstein surface singularity. Then the complex tangent bundle of V_t is topologically trivial,.*

The proof of 9.3 is immediate from Lemma 6.2 and Proposition 7.4. Thus one has,

9.4 Corollary. *Let $(V, 0)$ be a normal, smoothable Gorenstein surface singularity. Then for every smoothing of V and every resolution one has:*

$$\sigma(V_t) = -\frac{1}{3}\left(2(\chi(V_t) - 1) + K^2 + s + 2b_1\right),$$

where these invariants are as in Theorem 9.2.

IV.10 On the Rochlin μ invariant for links of surface singularities

We know (Rochlin's theorem) from Section 3 that if X is a closed, spin 4-manifold, then its signature is divisible by 16. This allows us to define an invariant of spin 3-manifolds as follows: let $(M, \mathcal{S})$ be a spin 3-manifold. Since the group of spin cobordism is trivial in dimension 3 (see, e.g., [244]), it follows that there exists a compact, spin 4-manifold X whose spin-boundary is $(M, \mathcal{S})$; let $\mu(M, \mathcal{S})$ be defined by

$$\mu(M, \mathcal{S}) = \sigma(X) \qquad \mathrm{mod}\,(16).$$

An argument similar to the one we used in the previous section to define the signature defect shows that this invariant is well defined modulo 16. This is the Rochlin μ invariant of the spin manifold $(M, \mathcal{S})$, which played an important role in 3-manifolds theory several decades ago, and still does, being related to the new 3-manifolds invariants coming from gauge theory (see §11 below). This is of particular interest when M is a homology sphere, since in this case the spin structure on M is unique.

Notice that Rochlin's Theorem 3.7 allows us to compute this invariant by having $(M, \mathcal{S})$ as an oriented boundary, provided we are given a characteristic submanifold (c.f. [224, 231]):

10.1 Definition. Let X be a compact oriented manifold with spin-boundary $(M, \mathcal{S})$. A *characteristic submanifold of X relative to the spin structure on the boundary* means an oriented 2-submanifold W in the interior of X, representing a homology class in $H_2(X; \mathbb{Z})$ whose reduction modulo 2 is the Poincaré-Lefschetz dual of the second Stiefel-Whitney class $w_2(X, \mathcal{S}) \in H^2(X, M; \mathbb{Z}_2)$ of X relative to the spin structure on its boundary.

Notice that a submanifold W as above actually defines a spin^c structure on X which is compatible with the given spin structure on the boundary. Now, given the spin 3-manifold $(M, \mathcal{S})$, consider it as a spin^c-boundary of X as above, and let W be a characteristic submanifold of X for this spin^c structure.

Then one has:

10.2 Theorem. *The Rochlin invariant of* $(M, \mathcal{S})$ *is:*

$$\mu(M, \mathcal{S}) = \sigma(X) - (W^2 + 8\operatorname{Arf}(W)) \qquad \mod (16),$$

where $\operatorname{Arf} \in \{0, 1\}$ *is the Arf invariant of a certain quadratic form on* $H_1(W; \mathbb{Z}_2)$.

The proof is left as an exercise (using Theorem 3.7). Of course if X is spin, then we can take $W = \emptyset$ and we recover the usual definition of the Rochlin invariant.

Now consider a normal Gorenstein surface singularity $(V, 0)$ and let Ω be a never-zero holomorphic 2-form on $V^* = V - 0$. Just as before, we see that this 2-form defines a reduction of the structure group of TV^* to $SU(2) \cong Sp(1) \cong \operatorname{Spin}(3)$. Hence it defines a spin structure on V^*. But it actually gives more: a multiplication by the quaternions i, j, k at each fibre of the tangent bundle TV^*, varying smoothly, just as in III.6. If ν denotes a unit outwards, normal vector field of the link M in V^*, then multiplication by i, j, k *determines a trivialization* of the bundle $TV^*|_M$ compatible with its complex structure. In fact this is the same as multiplying ν at each point by the unitary matrix $\begin{pmatrix} 0 & 1 \\ -1 & 0 \end{pmatrix}$, which defines a section of TV^*, everywhere linearly independent of ν over $\mathbb{C}$. We denote this trivialization of TM by ρ. This gives a spin structure on M.

Now let $\pi : \widetilde{V} \to V$ be a good resolution of V. For convenience we think of $\widetilde{V}$ indistinctly as a complex manifold or as a compact almost-complex manifold with boundary the link M and a complex structure in its interior. The trivialization ρ defines Chern classes of $\widetilde{V}$ relative to the boundary; they live in $H^{2i}(\widetilde{V}, M; \mathbb{Z})$. The Poincaré-Lefschetz dual of $c_2(X; \rho)$ is $\chi(\widetilde{V})$ by construction; the Poincaré-Lefschetz dual of $c_1(X; \rho)$ is the anti-canonical class $-K$, and it is independent of the choice of complex trivialization of $T\widetilde{V}|_M$, because the intersection matrix on $\widetilde{V}$ is negative definite. We call such a parallelism on M (that defines a trivialization of $T\widetilde{V}|_M$ as a complex bundle) a *complex parallelism* on M. Thus, if K is non-singular, then K is a characteristic submanifold for X, relative to any spin structure on the boundary which is given by a complex parallelism on M. More generally, choose the canonical divisor K to be vertical, and let $W = 2D - K$, D vertical, be any characteristic divisor as in §9. Then W represents an integral homology class whose reduction modulo 2 is the dual of $w_2(X, \rho)$. If $\mathcal{L}$ is the bundle of W, then we can approximate W by a smooth C^∞ manifold $\widetilde{W}$ defined by a C^∞ section of $\mathcal{L}$ which is transversal to the zero-section. Then $\widetilde{W}$ inherits a spin structure from the spin^c structure on $\widetilde{V}$: this follows from [192] in the general case, and also from [16] if $\widetilde{W}$ is complex analytic, since the line bundle $\mathcal{D}$ of D is a theta characteristic for $\widetilde{W}$. As noted in [231], the Arf invariant in Theorem 10.2 turns out to be a spin cobordism invariant, so it is 0 iff W (with its induced spin structure) is zero in the group $\Omega^2_{\text{spin}} \cong \mathbb{Z}_2$ of spin cobordism.

One has:

10.3 Theorem. *Let M be the link of a normal Gorenstein surface singularity $(V,0)$, let $\widetilde{V}$ be a good resolution of V and let β be a parallelism on M compatible with the complex structure on V. Let $W = 2D - K$ be a characteristic divisor of $\widetilde{V}$ and let $\widetilde{W}$ be a C^∞ smoothing of it. Then:*

(i) *$\mathrm{Arf}(\widetilde{W}) \equiv h^0(W; \mathcal{D}|_W) \bmod (2)$, where $\mathrm{Arf}(\widetilde{W})$ is the Arf invariant in Theorem 10.2, and this is independent of the choice of the complex parallelism on M. Hence the spin cobordism class of $\widetilde{W}$ is independent of β.*

(ii) *In particular $\mathrm{Arf}(K)$ is the reduction $\bmod (2)$ of the geometric genus $\rho_g(V,0)$.*

(iii) *The Rochlin invariant of M is independent of the choice of the complex parallelism on M and equals:*

$$\mu(M) = -s - (W^2 + 8\,\mathrm{Arf}(W)) \qquad \bmod (16),$$

where s is the number of irreducible components in the exceptional divisor of $\widetilde{V}$.

(iv) *If the singularity $(V,0)$ is smoothable and V_t is a smoothing of V, then*

$$\mu(M) = \sigma(V_t) \qquad \bmod (16).$$

The proof of statement (i) is an exercise (using (4.9) above); statement (ii) follows from (i) and Theorem 8.1.i. Statement (iii) is obvious from the previous discussion and the last statement follows from Proposition 7.4.

It is worth remarking that the Rochlin invariant was studied by several authors for singularities whose link is a homology sphere, specially in [187]. By (ii) above this is essentially an invariant of the singularity $(V,0)$, independent of the various choices. This invariant was computed in [231] for all quasi-homogeneous singularities and for the cusps of Hirzebruch [104] (and also for all 3-manifolds of the form $\Gamma \backslash G$, where G is a 3-dimensional Lie group and Γ a discrete subgroup). As a corollary one gets an improvement of [225, 4.5]:

10.4 Corollary. *Let (V_1, P_1) and (V_2, P_2) be normal, Gorenstein surface singularities with orientation preserving links, and let $\mu(V_i)$ denote the corresponding Rochlin invariant. Then:*

(i) *$\mu(V_1) - \mu(V_2) \equiv 8\left(\rho_g(V_1) - \rho_g(V_2)\right) \bmod (16)$, where ρ_g is the geometric genus.*

(ii) *If the singularities are smoothable and $V_1^{\#}, V_2^{\#}$ are smoothings of them, then:*

$$\sigma(V_1^{\#}) - \sigma(V_2^{\#}) \equiv 8\left(\rho_g(V_1) - \rho_g(V_2)\right) \bmod (16);$$

and

$$\chi(V_1^{\#}) - \chi(V_2^{\#}) \equiv 12\left(\rho_g(V_1) - \rho_g(V_2)\right) \bmod (24).$$

The first statement above follows from Theorem 10.3 and the fact that, by [184], the topology of the link determines the topology of the minimal resolution, which gives the number of irreducible components in the exceptional divisor and determines the canonical class K numerically. The second statement follows from the first one, together with Corollary 9.4 above.

This motivates the following question, whose answer I do not know.

10.5 Question. Is there an analogous statement to Corollary 10.4 for complex singularities in higher dimensions? In particular, does the fact that two isolated hypersurface singularities have orientation preserving links imply anything about their Milnor numbers? This question was asked of me by V.I. Arnold and I do not know the answer.

For instance, the singularities $z_1^2 + z_2^7 + z_3^{14} = 0$ and $z_1^3 + z_2^4 + z_3^{12} = 0$ have orientation preserving links, but their Milnor numbers are 78 and 66 respectively, so the congruences above are best possible. On the other hand one may ask under what circumstances is the geometric genus a topological invariant? (see [181, 182]).

10.6 Remark. There is an invariant of framed cobordism called *the real Adams e-invariant* $e_{\mathbb{R}}$. For framed 3-manifolds this takes values in $\mathbb{Z}_{24}$. There is also a complex Adams e-invariant, but this is weaker (though easier to compute). We refer to [224, 231] for details on this and for the relation with complex singularities. From the viewpoint of surface singularities as above, the invariant $e_{\mathbb{R}}$ gives essentially the same information as the Rochlin invariant (which takes values in $\mathbb{Z}_{16}$); each of these has some advantages, but $e_{\mathbb{R}}$ is easier to handle.

IV.11   Comments on new 3-manifolds invariants and surface singularities

As pointed out earlier, since the orientation preserving homeomorphism type of the link M of a normal surface singularity $(V, 0)$ depends only on the analytic type $(V, 0)$, it follows that every 3-manifolds invariant is an invariant of singularities. This has been used by several authors in both ways. On the one hand, whatever invariant we want to understand, the links of surface singularities are a great source of examples which we have more chances of being able to put our hands on. Conversely, the various invariants of 3-manifolds give a lot of information about the singularities themselves. The previous sections are a taste of what is known in that direction concerning more classical invariants.

There are however numerous, very important invariants that have been defined in more recent years, and which are just beginning to be studied. Among these are the η-invariants of Atiyah, Patodi and Singer, published in 1975 and 1976 in Math. Proc. Cambridge Philos. Soc. There is one such invariant for each first-order elliptic, self-adjoint operator acting on the sections of a vector bundle over the link of the singularity. Very little is yet known about these (see, for instance, [230, 25, 190, 180]).

Related with the eta-invariants, via the spectral flow, is the Floer homology (also called "instanton homology"). This homology $HF_*(\Sigma)$ was defined by

A. Floer [79] for every integral homology sphere, as the homology of a Z_8-graded complex generated by the critical points of a perturbed Chern-Simons functional (on a certain space of connections). The Casson invariant $\lambda(\Sigma)$ is essentially the alternating sum of the Betti numbers of the Floer homology, and it is an integral lifting of the Rochlin invariant, i.e., $\lambda(\Sigma)$ is an integer whose reduction modulo 16 gives the Rochlin invariant. For homology 3-spheres which are links of surface singularities, the signature of the Milnor fibre provides another integral lifting of the Rochlin invariant, and the so-called *Casson invariant conjecture* [188] states that these two liftings of the Rochlin invariant should coincide, i.e., that for homology 3-spheres which are links of surface singularities, the Casson invariant $\lambda(\Sigma)$ should equal $\frac{1}{8}\sigma(F)$, an eighth of the signature of the corresponding Milnor fibre.

I am grateful to W.D. Neumann for explaining to me that this conjecture was first suggested by J. Wahl to M.F. Atiyah at the 1987 Weyl conference at Duke in response to Atiyah asking if there were connections between the analytic structure of singularities and gauge-theoretic invariants; that led Atiyah to subsequently make a conjecture about Floer homology. Fintushel and Stern studied in [78] the Floer homology of the Brieskorn homology spheres $\Sigma(p, q, r)$ and proved the Casson invariant conjecture for these; some of the calculations in [78] were extended in [84] to Seifert fibred homology 3-spheres in general. In [188] and [83], independently, the authors proved the Casson invariant conjecture for all homology spheres given by weighted homogeneous surface singularities (in [188] they proved it also for other families of singularities). There is also a very interesting article by N. Saveliev [219], where the author relates in a precise way the Floer homology of Brieskorn homology spheres with the topology of the Milnor fibre and with classical invariants of 4-manifolds and knots. But the Casson invariant conjecture remains open. Recently, in a beautiful set of lectures delivered by W.D. Neumann in a workshop at Luminy, France, he spoke about his work in process with J. Wahl in this direction, and about their new *"Milnor fibre conjecture"*, which implies the Casson invariant conjecture.

In fact, since 1982 much of the progress in low-dimensional topology has arisen from applications of gauge theory, pioneered by S.K. Donaldson. In particular, Donaldson's polynomial invariants have been used to prove a variety of results about the topology and geometry of 4-manifolds. Kronheimer and Mrowka showed in [115] that there is a deep structure encoded in the Donaldson invariants which is related to embedded surfaces in 4-manifolds; this was a significant step for understanding these invariants, and it was a motivation for important further developments, most spectacularly for Witten's introduction of the now-called Seiberg-Witten (SW) monopole equations for the study of four-manifolds. These 4-dimensional SW equations yield differentiable invariants that have all the power of the polynomial invariants of Donaldson, but they are much easier to handle; these are known as the Seiberg-Witten invariants of smooth 4-manifolds. There are also the 3-dimensional Seiberg-Witten equations, giving rise to the Seiberg-Witten invariants of 3-manifolds, whose understanding is important also for 4-manifolds theory. These invariants associate a rational number $sw_M^0(\sigma)$ to each spin^c structure σ on a 3-manifold M. For homology spheres there is only one such structure and the Seiberg-Witten invariant coincides with the Casson invariant (see [142]). When M is the link of a surface singularity $(V, 0)$, the complex structure on $V - 0$ determines a canonical spin^c structure on M and the remarkable articles of A. Némethi and L. Nicolaescu [181, 182, 190, 180] are throwing light into the relation of the corresponding SW invariants with other classical invariants of surface singularities, particularly with the geometric genus. The hope is to find topological invariants of singularities, which determine the analytic ones. Their results led them to make a conjecture, *the Seiberg-Witten invariant conjecture*, which implies the Casson invariant conjecture, and they managed to prove it in some cases. However the recent article [77] provides counterexamples to this more general conjecture (but the original Casson invariant conjecture of [188] remains alive!).

Chapter V

On the Geometry and Topology of Quadrics in $\mathbb{CP}^n$

In Chapter II we saw how the action of the group $SU(2)$ on $\mathbb{C}^2$ gives very precise information about certain surface singularities; in particular we described the close relation between $SO(3,\mathbb{R}) = SU(2)/\pm Id$ and the quadric $z_1^2 + z_2^2 + z_3^2 = 0$. In this chapter, which is based on [135], we look at the canonical action of $SO(n+1,\mathbb{R})$ on $\mathbb{C}^{n+1}$ and on $\mathbb{CP}^n$, the complex projective space, in order to get a better understanding of the geometry and topology of the pair $(\mathbb{C}^{n+1}, V)$, where V is the Fermat quadric

$$z_0^2 + z_1^2 + \cdots + z_n^2 = 0.$$

This is of course related to the classical problem studied by Zariski [267] and others, of studying the topology of the complement of an affine algebraic hypersurface $V \subset \mathbb{C}^{n+1}$. We actually look with more care at the projectivized situation. We notice that the complement of a non-singular hyperquadric Q in $\mathbb{CP}^n$ is diffeomorphic to the total space of the tangent bundle of the real projective n-space $\mathbb{R}P^n$, $\mathbb{CP}^n - Q \cong T(\mathbb{R}P^n)$. Then we use this observation to describe $\mathbb{CP}^n$ as the double mapping cylinder of the double fibration:

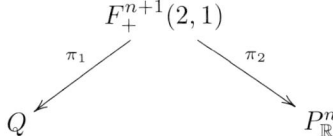

where

$$F_+^{n+1}(2,1) \cong SO(n+1,\mathbb{R})/(SO(n-1,\mathbb{R}) \times (\mathbb{Z}/2\mathbb{Z}))$$

is the partial flag manifold of *oriented* 2-planes in $\mathbb{R}^{n+1}$ and *non-oriented* lines in these planes. The manifold $F_+^{n+1}(2,1)$ is diffeomorphic to the unit sphere normal bundle of Q in $\mathbb{CP}^n$, and it is also diffeomorphic to the unit sphere tangent bundle of $P^n_{\mathbb{R}}$. This decomposition is related to previous work by V. Vassiliev, J. Tits, C.T.C. Wall and others, and we refer to [135] for details and for more on the subject.

In Section 3 we look more carefully at the decomposition of $\mathbb{CP}^n$ arising from the above double fibration. This describes $\mathbb{CP}^n$ as a 1-parameter family of codimension 1 submanifolds $F^{n+1}_+(2,1) \times \{t\}$, for $t \in (0,1)$, together with two "special" fibres: Q and a copy of the real projective space. We prove that these are the orbits of the natural action of $SO(n+1,\mathbb{R})$ on $\mathbb{CP}^n$, regarded as a subgroup of the complex orthogonal group $SO(n+1,\mathbb{C})$. This is a cohomogeneity 1 isometric action with respect to the Fubini-Study metric.

In Section 4 we look at the (now classical) theorem saying that $\mathbb{CP}^2$ modulo conjugation is the sphere $\mathbb{S}^4$. This theorem has a long and remarkable history. The first time this appeared in print (without proof) was in [8, p. 175], where Arnold used it to study real algebraic curves in $\mathbb{R}P^2$. In 1973–74 appeared two independent proofs of this theorem, given by W. Massey and N. Kuiper [116, 159]. Several other proofs of this result have been given by various authors afterward, including important improvements and generalizations (see for instance [10, 11, 12, 22, 153, 175]). We refer particularly to the recent article of M. Atiyah and J. Berndt [17], whose proof is along the same lines of the one we describe here (which is taken from [135]), but they do it in a more general setting. Here, as in [17], we prove an equivariant version of the Arnold-Kuiper-Massey theorem, showing that the equivalence $\mathbb{CP}^2/j \cong \mathbb{S}^4$ can be realized by a real algebraic map Φ which conjugates the natural cohomogeneity 1-actions of $SO(3,\mathbb{R})$ on $\mathbb{CP}^2$ and $\mathbb{S}^4$. Our proof uses only linear algebra.

V.1   The topology of a quadric in $\mathbb{CP}^n$

Let h be a homogeneous polynomial of degree 2 in $n+1$ complex variables with an isolated critical point at $0 \in \mathbb{C}^{n+1}$; let $\widetilde{Q} = h^{-1}(0)$, $K = \widetilde{Q} \cap \mathbb{S}^{2n+1}$ be its link, and let $Q \subset \mathbb{CP}^n$ be its projectivization, $Q = (\widetilde{Q} - 0)/\mathbb{C}^*$, which is non-singular by hypothesis.

1.1 Theorem.

(i) The hypersurface $\widetilde{Q}$ is isotopic to the hypersurface given by the Fermat polynomial equation $z_0^2 + \cdots + z_n^2 = 0$. The analogous statement holds for the projectivized quadrics.

(ii) The Milnor fibre F of h is diffeomorphic to the total space of the tangent bundle of the n-sphere $\mathbb{S}^n$. Hence the link K is diffeomorphic to the total space of the unit sphere tangent bundle of $\mathbb{S}^n$. Therefore K is diffeomorphic to the Stiefel manifold $V_{n+1,2}$ of orthonormal 2-frames in $\mathbb{R}^{n+1}$.

(iii) The quadric $Q \subset \mathbb{CP}^n$ is diffeomorphic to the Grassmannian $G_{n+1,2}$ of oriented 2-planes in $\mathbb{R}^{n+1}$.

iv) *The complement $\mathbb{C}P^n - Q$ is diffeomorphic to the quotient space $F/\mathbb{Z}_2$ of the Milnor fibre F by the monodromy of the Milnor fibration, which is cyclic of order 2. Hence $\mathbb{C}P^n - Q$ is diffeomorphic to the total space of the real projective space $\mathbb{R}P^n$.*

Proof. The first statement in Theorem 1.1 actually holds for homogeneous polynomials of any degree $d > 1$. To prove this, let $\mathcal{P}$ be the projective space of coefficients of homogeneous polynomials of degree d in $n + 1$ complex variables. The general homogeneous equation of degree d in $n + 1$ variables is:

$$\sum_{\alpha_o + \cdots + \alpha_n = d} a_{\alpha_o, \ldots, \alpha_n} z_0^{\alpha_o} \cdots z_n^{\alpha_n} = 0 .$$

This defines a polynomial, and hence a hypersurface $\mathcal{X}$, in $\mathcal{P} \times \mathbb{C}P^n$. The family of projective hypersurfaces of degree d in $\mathbb{C}P^n$ is given by the map:

$$\mathcal{E} : \mathcal{X} \to \mathcal{P} ,$$

induced by the projection of $\mathcal{P} \times \mathbb{C}P^n$ onto $\mathcal{P}$. In $\mathcal{P}$, the polynomials defining singular hypersurfaces in $\mathbb{C}P^n$ form a closed subvariety of complex codimension 1. Hence its complement Ω is connected. Since the map $\mathcal{E}$ is a locally trivial fibration over Ω, by Ehresmann's Lemma (Chapter I), one knows that any nonsingular hypersurface $X \subset \mathbb{C}P^n$ of degree d is ambient isotopic to the hypersurface defined by the Fermat polynomial $\mathcal{F}_d^n := z_0^d + \cdots + z_n^d$. That is, up to isotopy we can assume that X is the projectivization of the affine variety $V := \{z_0^d + \cdots + z_n^d = 0\}$ after removing the singular point $0 \in V$ (c.f. [133, Lemme 2.2]). The corresponding statement for the affine hypersurfaces follows easily from this, proving the generalization of statement (i) for polynomials of degree d.

Also notice that $\mathbb{C}P^n$ is the orbit space of $\mathbb{C}^{n+1} - \{0\}$ by the $\mathbb{C}^*$-action:

$$g_t(z_o, \ldots, z_n) = (tz_o, \ldots, tz_n) , \ t \in \mathbb{C}^* = \mathbb{C} - \{0\} .$$

If V is as above, then V is an invariant set for this $\mathbb{C}^*$-action. It follows that $\mathbb{C}P^n - X$ is the image of $\mathbb{C}^{n+1} - V$. Moreover, $\mathbb{C}^*$ is $\mathbb{S}^1 \times \mathbb{R}^+$ and if we divide $\mathbb{C}^{n+1} - \{0\}$ by the $\mathbb{R}^+$-action we get the sphere $\mathbb{S}^{2n-1}$. Thus $\mathbb{C}P^n - X$ is the quotient of $\mathbb{S}^{2n-1} - (V \cap \mathbb{S}^{2n-1})$ by the corresponding $\mathbb{S}^1$-action. By [168], these $\mathbb{S}^1$-orbits are transversal to the Milnor fibres of the polynomial $\mathcal{F}_d^n(z) = z_0^d + \cdots + z_n^d$, and their action on the fibres is given by the monodromy of the Milnor fibration, which is cyclic of period d. Therefore the Milnor fibre F is a d-fold cyclic cover of $\mathbb{C}P^n - X$, showing that the first statement in (iv) also holds for hypersurfaces of degree $d > 1$.

In the quadratic case $d = 2$ the Milnor fibre is diffeomorphic to the affine variety $z_0^2 + \cdots + z_n^2 = 1$. Let us decompose each vector $Z := (z_0, \ldots, z_n)$ in its real and imaginary parts, $Z = U + iV$; then the Milnor fibre is given as the set $(U, V) \in \mathbb{R}^{n+1} \times \mathbb{R}^{n+1}$ such that $|U|^2 - |V|^2 = 1$ and $U \perp V$. We notice that the map $(U, V) \mapsto (U/\|U\|, V)$ induces an isomorphism of this Milnor fibre F with the tangent bundle of $\mathbb{S}^n$, proving the first statement in (ii). Of course we may think of F as being the unit disc tangent bundle of the sphere. Hence the link K is the unit sphere tangent bundle of $\mathbb{S}^n$, since K can be regarded as the boundary of F. Therefore each point in K corresponds to a point in $\mathbb{S}^n$, i.e., a unit vector in $\mathbb{R}^{n+1}$ together with a unit vector orthogonal to the first one. Hence K is the Stiefel manifold $V_{n+1,2}$, as claimed in (ii).

Statement (iii) follows easily from statement (ii). In fact, it is clear that $(V - 0)/\mathbb{C}^*$ is diffeomorphic to $K/\mathbb{S}^1$. The link K is $V_{n+1,2}$ and the corresponding $\mathbb{S}^1 \cong SO(2)$-action identifies all 2-frames which are in the same 2-plane in $\mathbb{R}^{n+1}$. Hence $K/\mathbb{S}^1$ is the Grassmannian $G_{n+1,2}$.

To complete the proof of Theorem 1.1 it only remains to prove the last statement in (iv). From the previous discussion we know already that $\mathbb{CP}^n - Q$ is diffeomorphic to the quotient of F, the Milnor fibre, by the monodromy, which is cyclic of order 2. We also know that F is the total space of the tangent bundle $T(\mathbb{S}^n)$. We observe that, using the previous notation, the monodromy is given by multiplication by -1, $(U, V) \mapsto (-U, -V)$. Hence the quotient of F by this involution is the tangent bundle of the real projective n-space. $\square$

We notice that part of the argument above is similar to that of Lemmas 2.2 and 2.3 in [133] (see also Libgober in [139, Lemma 1.1]), implying Corollary 1.2 below. We denote by X_0 the projectivization of the affine hypersurface defined by the Fermat polynomial $\mathcal{F}_d^n := z_0^d + \cdots + z_n^d$, and we denote by $C_d^n := \mathbb{CP}^n - X_0$ the complement of X_0.

1.2 Corollary. *Let X be a non-singular hypersurface of $\mathbb{CP}^n$ of degree d. Then:*

 (i) *the pair $(\mathbb{CP}^n, X)$ is isotopic to the pair $(\mathbb{CP}^n, X_0)$; and*

 (ii) *the Milnor fibre F of $\mathcal{F}_d^n$ is a d-fold cyclic cover of C_d^n, the projection map $F \to C_d^n$ being given by the monodromy of the Milnor fibration of $\mathcal{F}_d^n$ (which is cyclic of period d).*

Since by [198, 168] the Milnor fibre has the homotopy type of a bouquet $\bigvee_\mu S^n$ of μ spheres $\mathbb{S}^n$, where $\mu = (d - 1)^{n+1}$ is the Milnor number, one has (as in [139]) that for $n > 1$ the fundamental group $\pi_1(C_d^n)$ is isomorphic to $\mathbb{Z}/d\mathbb{Z}$ and $\pi_j(C_d^n) \cong \pi_j(\bigvee_\mu S^n)$ for $j > 1$. In particular:

$$\pi_j(C_d^n) \cong 0 \ \text{ if } 1 < j < n \quad \text{and} \quad \pi_j(C_d^n) \cong \mathbb{Z}^\mu \text{ if } j = n. \tag{1.3}$$

1.4 Remark. It should be noted that (1.3) is a special case of the results proved by A. Libgober in [139] about the topology of the complement of projective hypersurfaces. There are also interesting results in the recent article [15] about the topology of non-singular complete intersections in complex projective spaces.

V.2   The space $\mathbb{CP}^n$ as a double mapping cylinder

From now on we let $Q \subset \mathbb{CP}^n$ be the non-singular hyperquadric in $\mathbb{CP}^n$ with equation

$$z_0^2 + \cdots + z_n^2 = 0,$$

in homogeneous projective coordinates. Let $j : \mathbb{CP}^n \to \mathbb{CP}^n$ be the involution on $\mathbb{CP}^n$ given by complex conjugation: $j([z_0, \ldots, z_n]) = [\bar{z}_0, \ldots, \bar{z}_n]$, and let Π be the fixed point set of j, so that $\Pi \cong P_{\mathbb{R}}^n$.

Theorem 1.1 says that $\mathbb{CP}^n - Q$ is diffeomorphic to the tangent bundle $T(\Pi)$, and Π is the zero section of this bundle. Hence $\mathbb{CP}^n - (Q \cup \Pi)$ can be regarded as the set of non-zero tangent vectors of Π, so it is diffeomorphic to the cylinder $T_1(\Pi) \times (0,1)$, where $T_1(\Pi)$ is the unit sphere tangent bundle.

In other words, we have that $\mathbb{CP}^n$ is obtained by taking the product $T_1(\Pi) \times (0,1)$ and attaching to it in some way the quadric Q on one end and the real projective space $\Pi = \mathbb{R}P^n$ on the other end. We now explain these "attaching" functions. For this we have:

2.1 Proposition. *The unit sphere tangent bundle $T_1(\Pi)$ is diffeomorphic to the homogeneous space $SO(n+1, \mathbb{R})/(SO(n-1, \mathbb{R}) \times \mathbb{Z}_2)$, and it is therefore diffeomorphic to $F_+^{n+1}(2,1)$, the (partial) flag manifold of oriented 2-planes in $\mathbb{R}^{n+1}$ and unoriented lines in these planes.*

Proof. Notice that the group $SO(n+1, \mathbb{R})$ acts linearly on $\mathbb{C}^{n+1}$ and this action descends to an action on $\mathbb{CP}^n$ that preserves Q. This action also leaves invariant the real projective space Π, where it acts in the usual way (i.e., via the action induced from the linear $SO(n+1, \mathbb{R})$-action on $\mathbb{R}^{n+1}$). This extends, via the differential, to a transitive action of $SO(n+1, \mathbb{R})$ on $T_1(\Pi)$, with isotropy subgroup $SO(n-1, \mathbb{R}) \times \mathbb{Z}_2$. Hence $T_1(\Pi)$ is diffeomorphic to $SO(n+1, \mathbb{R})/(SO(n-1, \mathbb{R}) \times \mathbb{Z}_2)$, as claimed in Proposition 2.1. But $SO(n+1, \mathbb{R})$ also acts transitively on $F_+^{n+1}(2,1)$ with isotropy $SO(n-1, \mathbb{R}) \times \mathbb{Z}_2$. Hence:

$$T_1(\Pi) \cong \frac{SO(n+1, \mathbb{R})}{SO(n-1, \mathbb{R}) \times \mathbb{Z}_2} \cong F_+^{n+1}(2,1),$$

as stated. $\square$

Now observe that each point in the flag manifold $F_+^{n+1}(2,1)$ consists of an oriented plane in $\mathbb{R}^{n+1}$ and a line in this plane. If we forget the 2-plane and keep the line, we get an obvious map from $F_+^{n+1}(2,1)$ into $\mathbb{R}P^n$, which is actually a fibration. Similarly, we may forget the line and keep the 2-plane, getting a projection into the Grassmannian $G_{n+1,2}$, which is diffeomorphic to the quadric Q, by Theorem 1.1. Thus one has a double fibration:

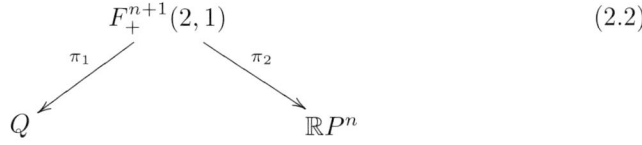

$$F_+^{n+1}(2,1) \qquad\qquad (2.2)$$

$$\pi_1 \swarrow \qquad \searrow \pi_2$$

$$Q \qquad\qquad\qquad \mathbb{R}P^n$$

where π_1 and π_2 are the maps that assign to each flag (P, l) either the 2-plane $P \in G_{n+1,2}$ or the line $l \in \mathbb{R}P^n$.

We form the corresponding double mapping cylinder $\left(F_+^{n+1}(2, 1) \times [0, 1]\right) / \sim$ where $\sim$ identifies a point

$$((P_0, l_0), 0) \in F_+^{n+1}(2, 1) \times \{0\}$$

with the point $\pi_1(P_0, l_0) = P_0$ in $G_{n+1,2} \cong Q$, and a point

$$((P_1, l_1), 1) \in F_+^{n+1}(2, 1) \times \{1\}$$

with the point $\pi_2(P_1, l_1) = l_1 \in \mathbb{R}P^n$. The space we obtain is homeomorphic to $\mathbb{C}P^n$. Furthermore, the double fibration (2.2) splits into two fibrations, corresponding to the maps π_1 and π_2. The open mapping cylinder of π_1 is $\mathbb{C}P^n - \Pi$, while the open mapping cylinder of π_2 is $\mathbb{C}P^n - Q$. One has:

2.3 Theorem. *The projective space $\mathbb{C}P^n$ is the double mapping cylinder of the double fibration (2.2). If we remove Q from $\mathbb{C}P^n$ we obtain a manifold diffeomorphic to the total space of the normal bundle of $\Pi \cong P_{\mathbb{R}}^n$ in $\mathbb{C}P^n$. Reciprocally, if we remove Π from $\mathbb{C}P^n$, what we get is diffeomorphic to the total space of the normal bundle of Q in $\mathbb{C}P^n$. If we remove $Q \cup \Pi$ from $\mathbb{C}P^n$, what we get is diffeomorphic to $F_+^{n+1}(2, 1) \times (0, 1)$.*

Proof. We notice that if we replace in Theorem 2.3 the word *diffeomorphic* by *homeomorphic*, then this theorem follows immediately from the previous discussion. Let us prove that we actually have diffeomorphisms. By Theorem 1.1, this is clear for $\mathbb{C}P^n - Q$. In fact, the fibration of $\mathbb{C}P^n - (Q \cup \Pi)$ given by the manifolds $F_+^{n+1}(2, 1)$ corresponds to the fibration on $T(\Pi) - \Pi$ given by sphere bundles of radius $r > 0$, for some metric on $T(\Pi)$. These correspond to boundaries of tubular neighborhoods $\tilde{\nu}_r(\Pi)$ of $\Pi \subset \mathbb{C}P^n$. In particular $\mathbb{C}P^n - Q$ is a tubular neighborhood of Π, hence $\mathbb{C}P^n - Q$ is diffeomorphic to the total space of the normal bundle of $\Pi \cong P_{\mathbb{R}}^n$ in $\mathbb{C}P^n$. This bundle is isomorphic to $T(\Pi)$.

Let us prove that $\mathbb{C}P^n - \Pi$ is diffeomorphic to the total space of the normal bundle of Q in $\mathbb{C}P^n$. We observe that for all $r > 0$, the interior of $\mathbb{C}P^n - \tilde{\nu}_r(\Pi)$ is diffeomorphic to $\mathbb{C}P^n - \Pi$. Now we prove that $\mathbb{C}P^n - \Pi$ is actually a tubular neighborhood of Q. For this we recall that one has the Fubini-Study metric on $\mathbb{C}P^n$, which can be thought of as being the descent to $\mathbb{C}P^n \cong \mathbb{S}^{2n+1}/\mathbb{S}^1$ of the usual metric on the sphere, which is $\mathbb{S}^1$-invariant. If N is a Riemannian submanifold of $\mathbb{C}P^n$, its *normal map* $\mathcal{N}_N$ is the function that associates to each normal vector v of N the projection to $\mathbb{C}P^n$ (via the exponential map) of the end-point of $v \in T(\mathbb{C}P^n)$ (see, for instance [169], p. 32). Let us denote by $\nu(Q)$ the normal bundle of Q in $\mathbb{C}P^n$ and consider the normal map:

$$\mathcal{N}_Q : \nu(Q) \to \mathbb{C}P^n .$$

We notice that every complex projective line $\mathcal{L}$ in $\mathbb{C}P^n$ orthogonal to Q, for the Fubini-Study metric, is invariant under conjugation, which is an isometry. So $\mathcal{L}$ is defined by equations with real coefficients (c.f. Section 3 below), and it is totally geodesic in $\mathbb{CP}^n$ since it is a complex projective line. Therefore $\mathcal{L}$ intersects Π transversally in a real projective line. This implies that the normal map $\mathcal{N}_Q$ is a diffeomorphism from the open disk bundle in $\nu(Q)$ of radius $\frac{\pi}{2}$ into $\mathbb{CP}^n - \Pi$. The union of all closed geodesic segments normal to Q of length $\frac{\pi}{2}$ fill up all of $\mathbb{CP}^n$. Thus the distance from a point $p \in \mathbb{CP}^n - (\Pi \cup Q)$ to Q is exactly the length of the unique geodesic segment joining p and the unique point $q \in Q$ such that this segment is orthogonal to Q. Hence every tubular neighborhood of Q in $\mathbb{CP}^n$, of diameter less than $\frac{\pi}{2}$, is diffeomorphic to $\mathbb{CP}^n - \Pi$. $\qquad\square$

We remark that one has a construction for the Milnor fibre F of the Fermat polynomial $\mathcal{F}_2^n$ in the spirit of Theorem 2.3, since F can be regarded as the open mapping cylinder of the fibration

$$V_{n+1,2} \cong SO(n+1, \mathbb{R})/SO(n-1, \mathbb{R}) \longrightarrow SO(n+1, \mathbb{R})/SO(n, \mathbb{R}) \cong \mathbb{S}^n \,,$$

where $V_{n+1,2}$ is the aforementioned Stiefel manifold.

V.3 The orthogonal group $SO(n+1, \mathbb{R})$ and the geometry of $\mathbb{CP}^n$

We now look more carefully at the decomposition of $\mathbb{CP}^n$ arising from the double fibration (2.2). It is convenient to look at two other interesting foliations that arise naturally from the double fibration (2.2), and from other considerations too.

3.1 Proposition. *The double fibration (2.2) induces two foliations $\mathcal{F}_1$ and $\mathcal{F}_2$ such that:*

(i) *The first foliation $\mathcal{F}_1$ is defined on $\mathbb{CP}^n - \Pi$; its leaves are embedded copies of $\mathbb{R}^2$ orthogonal to Q, which are the image under the normal map of Q of the fibres of the normal disc bundle of Q of radius less than $\frac{\pi}{2}$. The closure of each such leaf is a closed 2-disc that meets Π orthogonally in a projective line which is a closed geodesic in $\mathbb{CP}^n$. For each pair of conjugate points in Q, the corresponding leaves are naturally glued together along their common limit set in Π, forming a complex projective line defined by real coefficients.*

(ii) *The second foliation $\mathcal{F}_2$ is defined on $\mathbb{CP}^n - Q$; its leaves are embedded n-discs, orthogonal to Π, which are the image under the normal map of Π of the fibres of the normal disc bundle of Π of radius less than $\frac{\pi}{2}$. The closure of each such leaf is a closed n-disc that meets Q orthogonally in an $(n-1)$-sphere, invariant under complex conjugation.*

Here, by *the limit set* of a leaf $\mathcal{L}$ we mean the difference $\overline{\mathcal{L}} - \mathcal{L}$, where $\overline{\mathcal{L}}$ is its topological closure.

Proof. The leaves of the first foliation $\mathcal{F}_1$ on $\mathbb{CP}^n - \Pi$ are the fibres of π_1, which are 2-discs transversal to Q, by Theorem 2.3. By construction, each leaf of $\mathcal{F}_1$ is transversal to all the manifolds $F_+^{n+1}(2,1) \times t \subset \mathbb{CP}^n$ for $t \in (0,1)$, intersecting each in a copy of $\mathbb{R}P^1$ and approaching Π as $t \to 1$. Let us construct this foliation in a different way, which actually gives even more information than what is stated in Proposition 3.1. Recall one has on $\mathbb{CP}^n$ the Fubini-Study metric. From the proof of Theorem 2.3 we know that the normal map $\mathcal{N}_Q$ of Q induces a diffeomorphism between the open disc bundle of radius $\pi/2$ and $\mathbb{CP}^n - \Pi$. The leaves of $\mathcal{F}_1$ are the images of the normal discs. Since the conjugation $j : \mathbb{CP}^n \to \mathbb{CP}^n$ is an isometry, we have that a projective line $\mathcal{L}$ in $\mathbb{CP}^n$ intersects Q at two conjugate points iff it is orthogonal to Q, and this happens iff $\mathcal{L}$ can be defined by equations with real coefficients. So we call these $\mathbb{CR}$-lines. If two distinct $\mathbb{CR}$-lines intersect, they do so in a point in $\Pi \cong P_{\mathbb{R}}^n$. Also, each $\mathbb{CR}$-line $\mathcal{L}$ meets Π in a real projective line, which is an equator of $\mathcal{L}$. Since all complex lines in $\mathbb{CP}^n$ are totally geodesic, the real projective line $\mathcal{L} \cap \Pi$ is a geodesic in $\mathbb{CP}^n$, at equal distance $\pi/2$ from both intersection points in $\mathcal{L} \cap Q$. This divides $\mathcal{L}$ into two round discs of maximal diameter, orthogonal to Q. One can prove that through each point in $\mathbb{CP}^n - \Pi$ passes a unique $\mathbb{CR}$-line, hence these lines foliate this space. Therefore the open discs into which the $\mathbb{CR}$-lines split fill out the whole of $\mathbb{CP}^n - \Pi$, they are totally geodesic in $\mathbb{CP}^n$ and orthogonal to Q, thus providing a fibre bundle decomposition of $\mathbb{CP}^n - \Pi$, equivalent to the open disc bundle of the normal bundle $\nu(Q)$ of Q in $\mathbb{CP}^n$. By construction, the closure of each leaf in $\mathbb{CP}^n$ is obtained by attaching to the leaf a real projective line $\mathbb{R}P^1 \subset \Pi$, which is its boundary (or limit set). This circle (a real projective line in Π) is invariant by conjugation and it is an equator of the unique $\mathbb{CR}$-line, therefore it is also a closed geodesic for the Fubini-Study metric of $\mathbb{CP}^n$.

In the case of the foliation $\mathcal{F}_2$, the leaves are the fibres of π_2, up to isotopy. They are transverse to $F_+^{n+1}(2,1) \times t$, for every $t \in (0,1)$, and these leaves are also transverse to Π. We can describe this foliation, more precisely, as follows. Given $z \in \Pi$, we let $\mathcal{P}_z$ be the pencil of real projective lines in Π passing through z. Note that the tangent vectors at z to the lines of this pencil give the tangent space of Π at z. Let l_z be one of the lines of the pencil $\mathcal{P}_z$. Its complexification is a projective line L_z in $\mathbb{CP}^n$ defined by an equation with real coefficients, invariant under conjugation. This implies that L_z intersects Q at two points w_1 and w_2, which are conjugate; the intersection $L_z \cap Q$ is necessarily orthogonal and l_z is an equator in L_z. Thus, there is a segment $\hat{l}_z$, half of a real projective line (a circle) in L_z, joining the points w_1, z and w_2. This line is orthogonal to Π and to Q, it is geodesic in $\mathbb{CP}^n$ and has length π, by the minimality of L_z. Doing this for all lines in the pencil $\mathcal{P}_z$, we get an open n-disc of radius $\pi/2$ in $\mathbb{CP}^n$, orthogonal to Π at z, filled by geodesics in $\mathbb{CP}^n$ of length $\pi/2$ and intersecting Q orthogonally. Thus the normal map $\mathcal{N}_\Pi$ is regular for vectors of norm $< \pi/2$. The leaves of $\mathcal{F}_2$ are the image under $\mathcal{N}_\Pi$ of the fibres of the open disc normal bundle of $\Pi \subset \mathbb{CP}^n$ of radius $\frac{\pi}{2}$.

There is another interesting way of thinking about this foliation, up to isotopy, which is helpful to understand the way how its leaves approach Q. By Corollary 1.2 we have that $\mathbb{CP}^n - Q$ is the Milnor fibre $F := \{z_0^2 + \cdots + z_n^2 = 1\}$ divided by the monodromy $(z_1, \ldots, z_n) \mapsto (-z_1, \ldots, -z_n)$. The fibre F is the tangent bundle of the n-sphere, so it has a natural foliation by leaves diffeomorphic to n-planes. These planes can be described as follows. Let us decompose each $Z := (z_1, \ldots, z_n)$ in its real and imaginary parts, $Z = U + iV$. The fibre F is the set $(U, V) \in \mathbb{R}^{n+1} \times \mathbb{R}^{n+1}$ such that $\|U\| \geq 1$, $\|U\|^2 - \|V\|^2 = 1$ and $U \perp V$. If $\|U\| = 1$, then we are on the n-sphere and $\|V\| = 0$. Given a fixed $U_o \in \mathbb{S}^n \subset \mathbb{R}^{n+1}$, its "tangent space" is the plane defined as follows: for each $\lambda \in \mathbb{R}$ with $\lambda > 1$, let $\mathbb{S}_\lambda(U_0)$ be the $(n-1)$-sphere in the affine n-plane perpendicular to λU_o, consisting of all vectors V such that the vector $Z = \lambda U_o + iV$ is in F; these must satisfy $\|V\|^2 = \lambda^2 - 1$. The radius of the sphere $\mathbb{S}_\lambda(U_0)$ grows with λ, while for $\lambda = 1$ the corresponding "sphere" is just one point. For a given $U_0 \in \mathbb{S}^n$, let us denote by $\mathcal{L}(U_0)$ the union of all these $(n-1)$-spheres $\mathbb{S}_\lambda(U_0)$, for all $\lambda \geq 1$. Then $\mathcal{L}(U_0)$ is a copy of $\mathbb{R}^n$ embedded in F as a component of the 2-sheeted hyperboloid consisting of $\mathcal{L}(U_0) \cup \mathcal{L}(-U_0)$. The monodromy map interchanges these two sheets of the hyperboloid, so their image in $\mathbb{CP}^n$ is a manifold diffeomorphic to a plane, that we denote by $\mathcal{F}(U_0)$. By the uniqueness of the tubular neighbourhood, these are the leaves of $\mathcal{F}_2$ up to isotopy.

From this description of $\mathcal{F}_2$ one can see the way the leaves approach Q. In fact, let us denote by $\mathbb{S}'_\lambda(U_0)$ the image of the sphere $\mathbb{S}_\lambda(U_0)$ in $\mathbb{CP}^n$. It lies in $\mathcal{F}(U_0)$. Let $\gamma_\lambda(U_0)$ be the intersection of the unit sphere $\mathbb{S}^{2n+1} \subset \mathbb{C}^{n+1}$ with the real half cone over $\mathbb{S}_\lambda(U_0)$ with vertex at 0. The image of $\gamma_\lambda(U_0)$ in $\mathbb{CP}^n$ is also $\mathbb{S}'_\lambda(U_0)$. The sphere $\gamma_\lambda(U_0)$ is the set of vectors $(\frac{\lambda}{\sqrt{2\lambda^2-1}} U_0, \frac{1}{\sqrt{2\lambda^2-1}} V)$ with $(\lambda U_0, V)$ in $\mathbb{S}_\lambda(U_0)$. Therefore the limit of $\gamma_\lambda(U_0)$ is the set of vectors $(\frac{1}{\sqrt{2}} U_0, \frac{1}{\sqrt{2}} v)$ where v is $V/\|V\|$, with V as above. Since the vectors $\frac{1}{\sqrt{2}} U_0$ and $\frac{1}{\sqrt{2}} v$ have equal length, the image $\Lambda(U_0)$ in $\mathbb{CP}^n$ of this limit set is in Q, and it is an $(n-1)$-sphere. By continuity, the limit set of $\mathbb{S}'_\lambda(U_0)$ in $\mathbb{CP}^n$ is also $\Lambda(U_0)$. Since the conjugate of the vector (U, V) is $(U, -V)$, the sets $\gamma_\lambda(U_0)$ and their limit, are invariant under conjugation. Hence $\Lambda(U_0)$ is also invariant by conjugation.

We notice that the previous discussion also proves the following fact, that we state as a proposition. We recall that given a Riemannian submanifold N of $\mathbb{CP}^n$, its *focal points* are the critical values of the normal map of N, see [169].

3.2 Proposition. *The real projective space* $\Pi \cong P_{\mathbb{R}}^n$, *consisting of the points in* $\mathbb{CP}^n$ *with homogeneous real coordinates, is the set of focal points of the quadric Q defined by the Fermat polynomial* $z_0^2 + \cdots + z_n^2 = 0$. *Conversely, the quadric Q is the set of focal points of* Π.

Thus, both manifolds Q and Π can be regarded as *caustics* in $\mathbb{CP}^n$, i.e., they are the critical values of the corresponding co-normal maps of Π and Q, respectively (see [13]).

Let us consider now the action of $SO(n+1, \mathbb{R})$ on $\mathbb{CP}^n$, regarded as a subgroup of the complex orthogonal group $O(n+1, \mathbb{C})$. This action leaves Q invariant and it is by isometries with respect to the Fubini-Study metric. An isometry of $\mathbb{CP}^n$ that leaves Q invariant necessarily carries the set of focal points of Q into itself. Hence Π is also an invariant set for the action of $SO(n+1, \mathbb{R})$. We know already that Q is the Grassmannian $G_{n+1,2} \cong SO(n+1, \mathbb{R})/(SO(n-1, \mathbb{R}) \times SO(2, \mathbb{R}))$, so the action of $SO(n+1, \mathbb{R})$ is transitive on Q. Thus Q is one single orbit, and so is Π. Let us look at the orbit of a point $w \in \mathbb{CP}^n - (Q \cup \Pi)$. We claim that its orbit is the manifold $(F_+^{n+1}(2,1) \times t)$ passing through w. For this we use again the normal map:

$$\mathcal{N}_Q : \nu(Q) \to \mathbb{CP}^n .$$

By the previous discussion, this map is a diffeomorphism from the open disc bundle in $\nu(Q)$ of radius $\frac{\pi}{2}$ into $\mathbb{CP}^n - \Pi$ and the image of the fibres are the leaves of $\mathcal{F}_1$. Hence each point $w \in \mathbb{CP}^n - (Q \cup \Pi)$ is in the image of the normal map $\mathcal{N}_Q$, i.e., there is a (unique) vector $v_w \in \nu(Q)$ normal to Q, such that $w = \mathcal{N}_Q(v_w)$; the norm of v_w equals the distance $d_w = d(w, Q)$ from w to Q, which is > 0 and $< \pi/2$. That is, w corresponds, via $\mathcal{N}_Q$, to a point in the sphere bundle $S_{d_w}(\nu(Q))$ of radius d_w in $\nu(Q)$. We claim that the $SO(n+1, \mathbb{R})$-orbit $\mathcal{O}_w$ of w is the image of this sphere bundle, i.e., $\mathcal{O}_w = \mathcal{N}_Q(S_{d_w}(\nu(Q)))$. For this we notice that the group $SO(n+1, \mathbb{R})$ also acts on the tangent bundle $T(\mathbb{CP}^n)$ via the differential, and this action preserves the (C^∞) splitting $T(\mathbb{CP}^n)|_Q \cong TQ \oplus \nu(Q)$. This induces an action of $SO(n+1, \mathbb{R})$ on the normal bundle $\nu(Q)$ of Q, and this action is isometric and commutes with $\mathcal{N}_Q$, proving the claim. Hence the $SO(n+1, \mathbb{R})$-orbits are all manifolds $(F_+^{n+1}(2,1) \times t)$, for some $t \in (0,1)$, with two exceptional orbits which are Q and Π, corresponding to $t = 0$ and $t = 1$. By [106, 1.1], this implies that Q and Π are minimal submanifolds of $\mathbb{CP}^n$, which is obvious for Q, being a complex submanifold. The orbits of maximal dimension, which in this case are diffeomorphic to $F_+^{n+1}(2,1)$, are called *principal orbits*.

The previous arguments also show that each $SO(n+1, \mathbb{R})$-orbit in $\mathbb{CP}^n$ is at constant distance from Q, and also from Π, and these distances go from 0 to $\frac{\pi}{2}$. This proves that the space of $SO(n+1, \mathbb{R})$-orbits in $\mathbb{CP}^n$ is the interval $[0, \frac{\pi}{2}]$, with the two special orbits corresponding to the end-points of the interval. But one can actually be more precise about this statement. Let us consider again the geodesic $\hat{l}_z$ described above, in the construction of the foliation $\mathcal{F}_2$. In fact we are interested in half of this geodesic segment. To construct this "half geodesic segment", that we shall denote by $\check{l}$, we can start with any complex projective $\mathbb{CR}$-line $\mathcal{L}$. This line intersects Π in a real projective line, and it meets Q orthogonally at two conjugate points, say w and $\bar{w}$. Now we choose a point $z_0 \in \Pi \cap \mathcal{L}$. Then $\check{l}$ is the geodesic (of length $\frac{\pi}{2}$) in $\mathcal{L}$ joining the points z_0 and w, and it is a geodesic in $\mathbb{CP}^n$ because $\mathcal{L}$ is totally geodesic. This geodesic $\check{l}$ starts at $z_0 \in \Pi$ and finishes at $w \in Q$. Hence it meets each $SO(n+1, \mathbb{R})$-orbit orthogonally in exactly one point, since the orbits are the level sets of the function distance to Π. Hence $\check{l}$ parametrizes the orbits of $SO(n+1, \mathbb{R})$. This shows that the $SO(n+1, \mathbb{R})$-action on $\mathbb{CP}^n$ is a hyperpolar isometric action of cohomogeneity 1, which is already well known (see for instance [98]). In fact, *cohomogeneity 1* means that the principal orbits have codimension 1, and we know that this happens in our case (see the following section). An isometric action is said to be *polar* if there exists a closed, connected submanifold Σ that meets all orbits orthogonally. In our case this can be, for instance, the complete geodesic in $\mathcal{L}$ determined by $\check{l}$. Such a manifold is called *a section*. If one can chose such a section to be also flat, one says that the action is *hyperpolar*. This is obviously satisfied in our case since the section is a geodesic.

We have thus proved the following:

3.3 Theorem.

(i) *The natural $SO(n+1, \mathbb{R})$-action on $\mathbb{CP}^n$ is isometric, hyperpolar of cohomogeneity 1, with space of orbits the interval $[0, \pi/2]$. A section for this action (i.e., a submanifold that intersects transversally each orbit at exactly one point) can be constructed by considering some (any) $\mathbb{CR}$-line $\mathcal{L}$, choosing a point $z \in \mathcal{L} \cap \Pi$ and taking the geodesic (a circle) in $\mathcal{L}$ that passes through z and the two points where $\mathcal{L}$ meets Q.*

(ii) *There are three orbit types: two special orbits, Q and Π, which correspond to the endpoints $\{0, \pi/2\}$, and the principal orbits, which are copies of the partial flag manifold,*

$$F_+^{n+1}(2, 1) \cong SO(n+1, \mathbb{R})/(SO(n-1, \mathbb{R}) \times \mathbb{Z}_2),$$

of oriented 2-planes in $\mathbb{R}^{n+1}$ and lines in these planes.
The manifold $F_+^{n+1}(2, 1)$ is diffeomorphic to the unit sphere normal bundle of Q in $\mathbb{CP}^n$, and also diffeomorphic to the unit sphere tangent bundle of $P_{\mathbb{R}}^n$. Each of the two special orbits is the set of focal points of the other, and they are minimal submanifolds of $\mathbb{CP}^n$.

(iii) *The complex projective lines in $\mathbb{CP}^n$ whose homogeneous coordinates are real, i.e., the $\mathbb{CR}$-lines, foliate $\mathbb{CP}^n - \Pi$ and they are everywhere transversal to the orbits of $SO(n+1, \mathbb{R})$ (away from Π). In particular, they are orthogonal to Q.*

(iv) *The real projective space $\Pi \cong P_{\mathbb{R}}^n$ is embedded in $\mathbb{CP}^n$ so that its normal bundle is isomorphic to its tangent bundle. Its "tangent spaces" naturally define a foliation of $\mathbb{CP}^n - Q$ by embedded copies of $\mathbb{R}^n$, which are everywhere transversal to the orbits of $SO(n+1, \mathbb{R})$ (away from Q). In particular, they are orthogonal to Π.*

3.4 Remark. Let $q : \mathbb{CP}^n \to [0, \pi/2] \subset \mathbb{R}$ be the function $q(Z) = [d(Z, Q)]^2$, i.e., q is the square of the distance to Q. It is clear that q is constant along the $SO(n+1, \mathbb{R})$-orbits, which are its level sets. Hence q has the two special orbits Q and Π as critical set. It is clear that if Σ is a small disc in $\mathbb{CP}^n$ orthogonal to Q (or to Π), then the restriction of q to Σ is the ordinary quadratic map, so it is a Morse function on Σ. This means, by definition, that q is a Morse-Bott function (c.f. [33, 61]).

V.4 Cohomogeneity 1-actions of $SO(3)$ on $\mathbb{CP}^2$ and $\mathbb{S}^4$

We recall that a smooth manifold is said to be a *homogeneous space* if there is a Lie group that acts transitively on it. More generally, given a connected Lie group G, a smooth, connected manifold M and an action $\Phi : G \times M \to M$, one has that the G-orbits of highest dimension form a dense open set in M; these are called *principal orbits* while the others are the *exceptional orbits* (see for instance [106]). If the principal orbits have codimension k we say that the action has *cohomogeneity k*. If G and M are compact, cohomogeneity 1-actions have a simple global structure: either there are no exceptional orbits and the space of orbits is a circle, or else there are just two exceptional orbits and the quotient is the closed unit interval, with the exceptional orbits corresponding to the end-points of the interval. See for instance [175, 98].

In the previous sections we studied an example of such a cohomogeneity 1-action of $SO(n + 1, \mathbb{R})$ on $\mathbb{CP}^n$, whose principal orbits were the boundaries of tubular neighborhoods of the quadric Q, or equivalently of the real projective space $\mathbb{R}P^n \hookrightarrow \mathbb{CP}^n$. This was a case where we had two exceptional orbits: the quadric and $\mathbb{R}P^n$. When $n = 2$ the principal orbits are copies of the flag manifold $F_+^3(2, 1)$ and the quadric Q is diffeomorphic to the 2-sphere, since it is the Grassmannian of oriented 2-planes in $\mathbb{R}^3$ (by Theorem 1.1 above), and every such plane corresponds uniquely to a point in $\mathbb{S}^2$.

We notice that this action is very similar to a classical cohomogeneity 1-action of $SO(3)$ on the sphere $\mathbb{S}^4$, studied by Hsiang and Lawson in [106, Example 1.4], and by several other authors. To recall this action it is convenient to think of the sphere in a non-canonical way that we describe now.

Let $\mathcal{S}$ be the vector space of real 3×3, traceless and symmetric matrices. As a real vector space $\mathcal{S}$ is $\mathbb{R}^5$, and it can be equipped with a metric given by the inner product $(A, B) \mapsto trace(AB)$. Let $\mathbb{S}^{(4)}$ be the space of matrices in $\mathcal{S}$ with norm 1. One has an obvious diffeomorphism $\mathbb{S}^4 \cong \mathbb{S}^{(4)}$, which becomes isometric if we endow $\mathbb{S}^4$ with its usual round metric and $\mathbb{S}^{(4)}$ with the metric given by the inner product in $\mathcal{S}$. We shall identify these two spaces in the sequel, denoting both of them by $\mathbb{S}^4$ indistinctly. The group $SO(3, \mathbb{R})$ acts on $\mathcal{S}$ by $A \mapsto O^t A O$, where O^t is the transposed matrix (which is equal, in our case, to O^{-1}). This induces an isometric action Γ of $SO(3, \mathbb{R})$ on $\mathbb{S}^4$. This action on $\mathbb{S}^4$ has two disjoint copies of $\mathbb{R}P^2$ as special fibres. The space of orbits is the interval $[0, 1]$, with the endpoints giving the special orbits. Each principal orbit (i.e., the orbits of highest dimension) is a flag manifold:

$$F^3(2, 1) \cong SO(3, \mathbb{R}) / (\mathbb{Z}_2 \times \mathbb{Z}_2) \cong L(4, 1) / \mathbb{Z}_2,$$

of pairs (P, l) with P a plane in $\mathbb{R}^3$ and l line in P, where $L(4, 1)$ is the lens space $\mathbb{S}^3 / (\mathbb{Z}/4\mathbb{Z}) \cong SO(3, \mathbb{R}) / \mathbb{Z}_2$.

We notice that this decomposition of the sphere $\mathbb{S}^4$ into the orbits of the $SO(3)$-action is very similar to that of $\mathbb{CP}^2$ described previously. In fact this similarity is even closer than we have seen so far. To explain this, let us give a description of $\mathbb{CP}^2$ in the spirit of the one we just gave of $\mathbb{S}^4$.

Let

$$\mathfrak{H}(3, \mathbb{C}) = \{H \in M(3, \mathbb{C}) \mid H = H^*\}$$

be the space of complex 3×3 Hermitian matrices, where $H^* = \bar{H}^t$ is the adjoint matrix of H, obtained by conjugating first each entry of H and then transposing the matrix. We equip $\mathfrak{H}(3, \mathbb{C})$ with the Hermitian inner product:

$$\langle H_1, H_2 \rangle = \frac{1}{2} \text{ trace } (H_1 H_2). \tag{4.1}$$

As a vector space, with this inner product, $\mathfrak{H}(3, \mathbb{C})$ is the ordinary Euclidean space $\mathbb{E}^9$. Consider the subset $P(2)$ of $\mathfrak{H}(3, \mathbb{C})$ defined by:

$$P(2) = \{H \in \mathfrak{H}(3, \mathbb{C}) \mid H^2 = H, \text{ and } \text{trace}(H) = 1\}. \tag{4.2}$$

4.3 Lemma. *The set $P(2)$ is a manifold, diffeomorphic to $\mathbb{CP}^2$. Moreover, if we endow $P(2)$ with the metric defined by (4.1), then $P(2)$ is isometric to $\mathbb{CP}^2$ equipped with the Fubini-Study metric (of constant holomorphic sectional curvature 4).*

We remark that it is possible to describe $\mathbb{CP}^n$ in a similar way, but we restrict to $n = 2$ because this is all we need.

Proof. We claim that if H is in $P(2)$, then it is an orthogonal projection over a complex line. In fact, if H is in $P(2)$, then it is diagonalizable by a unitary matrix and its eigenvalues are 0 or 1, because $H^2 = H$. Since the trace is 1, two eigenvalues must be 0 and the other is 1. Hence H is a surjection of $\mathbb{C}^3$ over a complex line, and this map has to be an orthogonal projection because H is Hermitian. Conversely, it is clear that each line $L \in \mathbb{C}^3$ determines a unique orthogonal projection of $\mathbb{C}^3$, and this is given by a matrix in $P(2)$. The diffeomorphism in Lemma 4.3 is achieved by the map that carries H into the corresponding line in $\mathbb{C}^3$. To prove that this map gives a metric equivalence, we notice that the unitary group $U(3)$ acts on $\mathfrak{H}(3, \mathbb{C})$ by: $H \mapsto U^* H U$, and $P(2)$ is an orbit of this action, with isotropy $\big(U(2) \times U(1)\big)$. Thus,

$$P(2) \cong U(3)/\big(U(2) \times U(1)\big) \cong \mathbb{CP}^2,$$

and the metric on $P(2)$ is obviously $U(3)$-invariant. Hence the induced metric on $\mathbb{CP}^2$ is also $U(3)$-invariant, and this characterizes the Fubini-Study metric, up to scaling. $\qquad\square$

Now we notice that if L_H is the complex line in $\mathbb{C}^3$ that corresponds to a point of $H \in P(2)$ under the diffeomorphism of Lemma 4.3, and if we pick a non-zero point $(z_1, z_2, z_3) \in L_H$, then H is the point in $\mathbb{CP}^2$ with projective coordinates $[z_1, z_2, z_3]$. To the matrix $\bar{H}$ corresponds the line with projective coordinates $[\bar{z}_1, \bar{z}_2, \bar{z}_3]$. Therefore we have:

4.4 Lemma. *The involution $j*$ of $P(2)$ defined by $j * (H) = \bar{H}$, coincides with the involution j of $\mathbb{CP}^2$ given by complex conjugation, $[z_1, z_2, z_3] \overset{j}{\mapsto} [\bar{z}_1, \bar{z}_2, \bar{z}_3]$.*

Let us denote by j the complex conjugation on both $P(2)$ or $\mathbb{CP}^2$. It is easy to see that the $SO(3)$-action commutes with complex conjugation on $P(2)$ and therefore descends to an $SO(3)$-action on the orbit space $P(2)/j$. Each principal orbit becomes a copy of the flag manifold $F^3(2,1)$; the projective space $\mathbb{R}P^2$ corresponds to the matrices in $P(2)$ with real entries, so it is the fixed point set of j. The other exceptional orbit is the quadric $Q \cong \mathbb{S}^2$, and complex conjugation on it corresponds to identifying opposite points, hence its quotient by j is another copy of $\mathbb{R}P^2$. Summarizing, the quotient $P(2)/j$ admits an $SO(3)$-action with two exceptional fibres, both copies of $\mathbb{R}P^2$ and the principal orbits are copies of $F^3(2,1)$. The space of orbits of this action on $P(2)/j$ is the interval and the two exceptional orbits correspond to the end-points of the interval. As noted by Atiyah-Witten in [22], this is enough to deduce the classical Arnold-Kuiper-Massey theorem (see [8, 10, 116, 159]), saying that $\mathbb{CP}^2$ modulo conjugation is the 4-sphere, using the classification of closed 4-manifolds that admit a cohomogeneity 1-action of $SO(3)$. However we prefer, just as in [135, 17], to give a direct proof of this theorem. This is what we do in the following section.

V.5　The Arnold-Kuiper-Massey theorem.

The previous discussion motivates an equivariant version of the Arnold-Kuiper-Massey theorem In this section we give (following [135]) a proof of this theorem. We construct an explicit algebraic map $\Phi : \mathbb{CP}^2 \to \mathbb{S}^4$ which is equivariant with respect to the cohomogeneity 1 isometric actions of $SO(3, \mathbb{R})$ on $\mathbb{CP}^2$ and $\mathbb{S}^4$ and induces a diffeomorphism $\mathbb{CP}^2/\text{conjugation} \cong \mathbb{S}^4$. This method for proving the Arnold-Kuiper-Massey theorem is exactly the same method used by Atiyah and Berndt in [17] and has the advantage of generalizing to other situations. In this way they managed to prove that not only has one $\mathbb{CP}^2/\text{conjugation} \cong \mathbb{S}^4$, but also one has the theorem of Arnold $\mathcal{H}^2/SO(2) \cong \mathbb{S}^7$, where $\mathcal{H}^2$ is the quaternionic projective plane, and also $\mathcal{C}^2/Sp(1) \cong \mathbb{S}^{13}$, where $\mathcal{C}^2$ is now the projective Cayley plane; this last result of was not known before. Furthermore, their diffeomorphisms are all equivariant with respect to certain cohomogeneity 1 group actions.

To begin with, we notice that complex conjugation j on $\mathbb{CP}^2$ has fixed points, so it is not at all obvious that the quotient $\mathbb{CP}^2/j$ is even a smooth manifold. This is carefully explained in [153], so we only sketch here a few ideas. Away from the fixed point set $\Pi \cong \mathbb{R}P^2$, the involution j is free, so the quotient is a smooth manifold. The problem is on Π. A tubular neighborhood of Π in $\mathbb{CP}^2$ can be regarded as a normal open disc bundle, and conjugation carries each normal fibre into itself. Since the quotient of each normal 2-disc by the involution is again a 2-disc, it follows that the quotient $\mathbb{CP}^2/j$ is a topological manifold. Making this argument more carefully one gets that $\mathbb{CP}^2/j$ is in fact a PL-manifold, as noticed in [116], and therefore it is smooth, since every piecewise linear 4-manifold is smooth. In [153] Marin defines the smooth structure on $\mathbb{CP}^2/j$ directly, without using PL-structures. An important point is that the smooth structure on $\mathbb{CP}^2/j$ is such that the obvious projection $\mathbb{CP}^2 \to \mathbb{CP}^2/j$ is differentiable.

Let us denote by Γ the aforementioned isometric action of $SO(3,\mathbb{R})$ on $\mathbb{S}^4$, and denote by $\widetilde{\Gamma}$ the standard action of $SO(3,\mathbb{R})$ on $\mathbb{CP}^2$, which is by isometries with respect to the Fubini-Study metric. This action is defined either by considering $SO(3,\mathbb{R})$ as a subgroup of $O(3,\mathbb{C})$, acting on the space of lines in $\mathbb{C}^3$, or via the action of $SO(3,\mathbb{R})$ on the space of matrices $P(2) \subset H(3,\mathbb{C})$ given by

$$(O, A) \mapsto O^t A O.$$

By Lemma 4.3, both metrics on $\mathbb{CP}^2$ are equivalent; also for every $O \in SO(3,\mathbb{R})$, $H \in P(2)$ and $v \in \mathbb{C}^3$ such that $H(v) = v$, one has: $O^t H O(O^{-1}(v)) = O^{-1}(v)$, because $O^{-1} = O^t$. Hence both actions on $\mathbb{CP}^2 \cong P(2)$ are equivalent. Similarly, given the $SO(3,\mathbb{R})$-action $\widetilde{\Gamma}$ on $\mathbb{CP}^2$ and Γ on $\mathbb{S}^4$, we say that these actions are equivariant if there exists a map $\Phi : \mathbb{CP}^2 \to \mathbb{S}^4$, making commutative the following diagram:

$$
\begin{array}{ccc}
SO(3,\mathbb{R}) \times \mathbb{CP}^2 & \xrightarrow{\ \widetilde{\Gamma}\ } & \mathbb{CP}^2 \\
{\scriptstyle Id \times \Phi}\Big\downarrow & & \Big\downarrow{\scriptstyle \Phi} \\
SO(3,\mathbb{R}) \times \mathbb{S}^4 & \xrightarrow{\ \Gamma\ } & \mathbb{S}^4
\end{array}
$$

In this case we say that Φ *conjugates* the actions Γ and $\widetilde{\Gamma}$. The map Φ carries orbits into orbits, i.e., the decompositions of $\mathbb{CP}^2$ and $\mathbb{S}^4$ into orbits are (smoothly) equivalent.

Let us denote by Id_3 the 3×3 identity matrix and by $\|A\| = \mathrm{trace} A^2$ the norm of a real symmetric matrix. The real part of a Hermitian matrix H is denoted $\mathrm{Re}(H)$.

5.1 Theorem. *Identify $\mathbb{S}^4$ with the real 3×3 symmetric matrices A with $\mathrm{trace}(A) = \mathrm{trace}(A^2) = 0$, and identify $\mathbb{CP}^2$ with the complex 3×3 Hermitian matrices H with $\mathrm{trace}\, 1$ and $H^2 = H$. Then the map $\Phi : \mathbb{CP}^2 \to \mathbb{S}^4$,*

$$\Phi(H) = \frac{\frac{1}{3} Id_3 - \mathrm{Re}(H)}{\|\frac{1}{3} Id_3 - \mathrm{Re}(H)\|}$$

is a well-defined 2-to-1 branched covering, ramified over a copy of $\mathbb{R}P^2$ embedded in $\mathbb{S}^4$. This map is $SO(3,\mathbb{R})$-equivariant and also invariant by the complex conjugation j, inducing an equivariant diffeomorphism $\mathbb{CP}^2/j \cong \mathbb{S}^4$.

We notice that Theorem 5.1, together with [106], imply that the image of $\mathbb{R}P^2 \subset \mathbb{CP}^2$ under the above map is the image of $\mathbb{R}P^2$ by the classical Veronese embedding $(\mathbb{CP}^2, \mathbb{R}P^2) \hookrightarrow (\mathbb{CP}^5, \mathbb{S}^4)$.

The proof of Theorem 5.1 follows from several lemmas below.

5.2 Lemma. *Let A be a real (3×3)-matrix. Then A is the real part of a matrix H in $P(2)$ if and only if it satisfies:*

(i) *A is symmetric with trace 1;*

(ii) *A has 0 as an eigenvalue and the other two eigenvalues λ_i and λ_j are roots of an equation of the form:*

$$\lambda^2 - \lambda + k = 0,$$

for some constant $k \in \mathbb{R}$ with $0 \le k \le \frac{1}{4}$.

If A and H are as above, and if $O \in SO(3, \mathbb{R})$ is such that $O^t A O$ is a diagonal matrix, then the imaginary part B of H, taken into its canonical form $O^t B O$, has only two possible non-zero entries, which are $\pm\sqrt{k}$. In particular, if $k = 0$, then $H = A$.

Proof. Let us consider a matrix $H \in P(2)$ and we decompose it in its real and imaginary parts: $H = A + iB$. Then one has: $\bar{H}^t = A^t - iB^t$. Also $H = \bar{H}^t$ because H is Hermitian. Hence $A = A^t$ and $B = -B^t$, i.e., A is symmetric and B is anti-symmetric. Thus the trace of A is 1, proving statement (i). One also has:

$$H^2 = A^2 - B^2 + i(AB + BA),$$

and $H^2 = H$ because H is in $P(2)$. Therefore: $A = A^2 - B^2$ and, $B = AB + BA$.

Now, A is symmetric, and so is A^2; these two matrices obviously commute, so they can be diagonalized simultaneously by a matrix $O \in SO(3, \mathbb{R})$. Since $B^2 = A^2 - A$, one knows that $O^t B^2 O$ is also diagonal:

$$O^t B^2 O = \begin{pmatrix} \mu_1 & 0 & 0 \\ 0 & \mu_2 & 0 \\ 0 & 0 & \mu_3 \end{pmatrix},$$

with $\mu_i = \lambda_i^2 - \lambda_i$, for each $i = 1, 2, 3$, where the λ_i are the eigenvalues of A. But B is antisymmetric and commutes with B^2, which is symmetric. Hence the same matrix O takes B to its canonical form:

$$O^t B O = \begin{pmatrix} 0 & a & c \\ -a & 0 & b \\ -c & -b & 0 \end{pmatrix},$$

for some $a, b, c \in \mathbb{C}$. This implies:

$$O^t B^2 O = (O^t B O)(O^t B O) = \begin{pmatrix} -a^2 - c^2 & -bc & ab \\ -bc & -a^2 - b^2 & -ac \\ ab & -ac & -b^2 - c^2 \end{pmatrix},$$

which we know is a diagonal matrix. Therefore two of the numbers a, b, c must be zero. Assume for instance that a and b are 0, then both eigenvalues λ_1 and λ_3 are roots of the polynomial:

$$\lambda^2 - \lambda + c^2 = 0.$$

This implies:
$$\lambda_1 + \lambda_3 = 1, \text{ and } \lambda_1 \cdot \lambda_3 = c^2 \geq 0.$$

Hence $\lambda_2 = 0$ (because the trace of A is 1), so 0 is an eigenvalue of A. The other eigenvalues λ_1 and λ_3 must be both ≥ 0 and ≤ 1, because their product is a non-negative number and their sum is 1. Moreover the roots must be real, therefore $k = c^2 \leq \frac{1}{4}$, proving statement (ii).

Also, in this case the eigenvalues of A determine the imaginary part B of H up to sign:

$$B = \pm O \begin{pmatrix} 0 & 0 & c \\ 0 & 0 & 0 \\ -c & 0 & 0 \end{pmatrix} O^t,$$

with $c^2 = \lambda_1 - \lambda_3^2 = \lambda_3 - \lambda_3^2$, proving in this case the last statement of Lemma 5.2. The other cases, when either $\{a, c\}$ or $\{b, c\}$ are both zero, are similar to the previous one. This proves that if $A = \text{Re}(H)$ for some matrix $H \in P(2)$, then A is as stated in Lemma 5.2. Conversely, given A satisfying these conditions, the arguments above tell us how to construct B so that these matrices are the real and imaginary parts of some H in $P(2)$. □

Now, given $H \in P(2)$, its real part is $\text{Re}(H) = \frac{1}{2}(H + \bar{H})$. Define

$$\psi : P(2) \to M(3, \mathbb{R}),$$

the space $M(3, \mathbb{R})$ being the space of real (3×3)-matrices, by the formula:

$$\psi(H) = \frac{1}{3} Id_3 - \text{Re}(H) \in M(3, \mathbb{R}), \tag{5.3}$$

where Id_3 is the (3×3)-identity matrix. In other words, $\psi(H)$ is the real part of the matrix $(\frac{1}{3} Id_3 - H)$. Since $H \in P(2)$, it follows that $\psi(H)$ is actually contained in $\mathcal{S}$.

It is clear that the above action of $SO(3, \mathbb{R})$ on $P(2)$ given by conjugation is equivalent, via the above diffeomorphism $P(2) \cong \mathbb{CP}^2$, with the standard action studied above. It is also clear that, for every $O \in SO(3, \mathbb{R})$, one has:

$$\psi(O^t H O) = \frac{1}{3} Id_3 - \frac{1}{2}(O^t(H + \bar{H})O) = O^t \left(\frac{1}{3} Id_3 - \frac{1}{2}(H + \bar{H}) \right) O = O^t \psi(H) O.$$

Hence we have:

5.4 Lemma. *The map ψ is equivariant. That is, for every $O \in SO(3, \mathbb{R})$ and $H \in P(2)$ one has: $\psi(O^t H O) = O^t \psi(H) O$.*

5.5 Lemma. *Given $S \in \mathcal{S} - \{0\}$, there exists a unique positive $t \in \mathbb{R}$, such that the matrix $(\frac{1}{3} Id_3 - tS)$ is the real part of some matrix $H \in P(2)$.*

Proof. By Lemma 5.4, we can assume S is diagonal. Hence the matrix $\widehat{S}_t = (\frac{1}{3}Id_3 - tS)$ is also diagonal, say

$$\widehat{S}_t = \begin{pmatrix} \lambda_1(t) & 0 & 0 \\ 0 & \lambda_2(t) & 0 \\ 0 & 0 & \lambda_3(t) \end{pmatrix}$$

with $\lambda_i(t) = \frac{1}{3} - t\mu_i$, where the μ_i are the eigenvalues of S. We notice that for all $t \in \mathbb{R}$, one has

$$\text{trace } \widehat{S}_t = 1 - t\,(\text{trace } S) = 1\,,$$

because S has trace 0. Hence all these matrices satisfy condition (i) of Lemma 5.2.

Let us look for the possible values of t that give solutions of Lemma 5.2. That is, we want $t > 0$ for which one eigenvalue $\lambda_i(t)$ is 0 and the others satisfy that their sum is 1 and their product is ≥ 0 and $\leq \frac{1}{4}$.

Let us number the eigenvalues of S so that $\mu_1 \leq \mu_2 \leq \mu_3$. Since their sum is 0 and S is not the zero matrix, one must have $\mu_1 < 0$ and $\mu_3 > 0$. If we want t as above, we must have that one $\lambda_i(t)$ must vanish. Let us look for solutions with $\lambda_1(t) = 0$. This means $t = \frac{1}{3\mu_1} < 0$, and we want $t > 0$. Hence, there are no solutions with $\lambda_1(t) = 0$.

Now let us look for solutions with $\lambda_2(t) = 0$. This implies $t = \frac{1}{3\mu_2}$; for this to be possible we must have $\mu_2 \neq 0$. If $\mu_2 < 0$, then $t < 0$ and we want t to be positive. Thus, we only care about $\mu_2 > 0$. We have:

$$\lambda_1(t) = \frac{1}{3}(1 - \frac{\mu_1}{\mu_2}) \quad \text{and} \quad \lambda_3(t) = \frac{1}{3}(1 - \frac{\mu_3}{\mu_2})\,.$$

We have $\mu_1 < 0 < \mu_2$, so $\lambda_1(t) > 0$. If $\mu_2 < \mu_3$, then $\lambda_3(t) < 0$, thus the product $\lambda_1(t)\lambda_3(t)$ is < 0, so there are no solutions like this that satisfy Lemma 5.5. The other possibility is $\mu_2 = \mu_3$; this implies $\lambda_3(t) = 0$ too. In this case one has $\lambda_1(t) = 1$ and $\lambda_2(t) = \lambda_3(t) = 0$, and $t = \frac{1}{3\mu_2}$ is positive. Hence we have a solution, and this is unique because $\mu_2 = \mu_3$. If $\mu_2 = 0$, then $\lambda_2(t)$ cannot be 0 and we cannot find solutions like this.

Summarizing, so far we have seen that: (i) there are no solutions satisfying Lemma 5.5 for which $\lambda_1(t) = 0$; (ii) if $\mu_2 \leq 0$, there are no such solutions for which $\lambda_2(t) = 0$; and (iii) if $\mu_2 = \mu_3$, then there is a unique solution satisfying Lemma 5.5, for which one has $\lambda_2(t) = \lambda_3(t) = 0$ and $\lambda_1(t) = 1$.

Finally, let us look for solutions with $\lambda_3(t) = 0$, i.e., with $t = \frac{1}{3\mu_3}$. We know, by hypothesis, that $\mu_2 \leq \mu_3$ and $\mu_3 > 0$. If $\mu_2 = \mu_3$, then we are in the previous case and there is a unique positive t giving a solution satisfying Lemma 5.5. Let us assume now that $\mu_2 < \mu_3$. Then we have:

$$\lambda_1(t) = \frac{1}{3}\left(1 - \frac{\mu_1}{\mu_3}\right) \quad \text{and} \quad \lambda_2(t) = \frac{1}{3}\left(1 - \frac{\mu_2}{\mu_3}\right)\,,$$

which are both ≥ 0. Since their sum is 1, it follows that each $\lambda_i(t)$ is also ≤ 1.

The product of $\lambda_1(t)$ and $\lambda_2(t)$ satisfies:

$$0 \leq \lambda_1(t) \cdot \lambda_2(t) = \frac{1}{9}\left(1 - \frac{\mu_1 + \mu_2}{\mu_3} + \frac{\mu_1\mu_2}{\mu_3}\right) = \frac{1}{9}\left(2 + \frac{\mu_1\mu_2}{\mu_3^2}\right)$$

$$= \frac{1}{9}\left(2 + \frac{\mu_1\mu_2}{(\mu_1 + \mu_2)^2}\right) \leq \frac{1}{4},$$

since $\mu_1 + \mu_2 + \mu_3 = 0$ and $\frac{\mu_1\mu_2}{(\mu_1+\mu_2)^2} \leq \frac{1}{4}$ because the inequality $\frac{1}{4}(a+b)^2 \geq ab$ is valid for any pair of real numbers a and b (and equality occurs iff $a = b$). Hence $t = \frac{1}{3\mu_3}$ is the unique solution satisfying the conditions of Lemma 5.5. $\quad\square$

We now "normalize" the map ψ so that its image is contained in $\mathbb{S}^4 \subset \mathcal{S}$. For this we define a function

$$\alpha(H) = [\text{trace}\,(\psi(H)^2)]^{-\frac{1}{2}},$$

i.e., $\alpha(H)$ is the inverse of the norm of $\psi(H)$ in $\mathcal{S}$, and we set:

$$\Phi(H) = \alpha(H)\,\psi(H).$$

One has:

$$\text{trace}\,[\psi(H)^2] = \text{trace}\left[\left(\frac{1}{3}Id_3 - \frac{1}{2}(H + \bar{H})\right)^2\right]$$

$$= \text{trace}\left[\frac{1}{9}Id_3 - \frac{1}{3}(H + \bar{H}) + \frac{1}{4}(H^2 + \bar{H}^2 + H\bar{H} + \bar{H}H)\right]$$

$$= \frac{1}{6} + \frac{1}{4}\text{trace}\,(H\bar{H} + \bar{H}H),$$

which is always positive since the matrix $(H\bar{H} + \bar{H}H)$ is positive semi-definite, so its trace is ≥ 0. Hence the maps α and Φ are well defined. It is clear that the image of Φ is contained in $\mathbb{S}^4 \subset \mathcal{S}$, because the linearity of the trace implies:

$$[\text{trace}\,(\Phi(H))]^2 = \alpha^2(H)\,[\text{trace}\,\psi(H)]^2 = 1.$$

It is also clear that Φ is $SO(3, \mathbb{R})$-equivariant, since the trace is invariant under conjugation and ψ is equivariant by Lemma 5.4. These considerations imply both Lemma 5.5 and the following lemma.

5.6 Lemma. *The map Φ is an equivariant surjection from $P(2)$ over $\mathbb{S}^4 \subset \mathcal{S}$, and it is two-to-one, except over the image of the real matrices in $P(2)$ where it is one-to-one.*

This gives the map in Theorem 5.1 that determines an equivariant diffeomorphism between $\mathbb{S}^4$ and $\mathbb{CP}^2$ modulo the involution given by conjugation.

Then Φ is invariant under this involution, since $\text{Re}(H) = \text{Re}(\bar{H})$, proving Theorem 5.1. $\quad\square$

To complete the proof of Theorem 5.1 we notice that Φ is invariant under the involution of $P(2)$ that corresponds to complex conjugation in $\mathbb{CP}^2$ (see Lemma 4.4 above).

Chapter VI

Real Singularities and Complex Geometry

The topology of isolated complex singularities has long been studied by many authors, as we know already, and there is a beautiful and well-developed theory in this respect, though there are still many things to be understood. The previous chapters are a sample of this. This chapter is a turning point for us, as from now on we are concerned with the real counterpart of this theory, largely inspired by [168]. The theme here is the interplay between complex geometry and real analytic singularities. We consider several classes of real hypersurface and complete intersection singularities that arise naturally from complex geometry. The motivation for this is the pioneering work of S. López de Medrano, L. Meersseman and A. Verjovsky on new constructions of complex manifolds via holomorphic dynamics. These are now called LVM-manifolds. The first step they take is to construct real analytic complete intersection singularities associated to appropriate actions of $\mathbb{C}^m$ on $\mathbb{C}^n$, $n > 2m$. We briefly describe this in Sections 1 and 2 below. In Section 3 we give a general method for constructing real analytic singularities using holomorphic vector fields in general, which is motivated by the constructions in Sections 1 and 2. In Section 4 we look at a very simple construction of real singularities, motivated by Section 3 and closely related to the material in Chapter VII of this text: the real hypersurfaces that arise when we consider a $\mathbb{C}$-valued holomorphic function H on an open set $\mathcal{U}$ in $\mathbb{C}^n$ and we compose it with the projection onto a real line through the origin in $\mathbb{C}$. As a byproduct of this we get explicit real analytic embeddings in $\mathbb{R}^3$ of closed oriented 2-manifolds of all genera.

VI.1 The space of Siegel leaves of a linear flow

This section is based on [50, 149, 150], which are our basic references. Let us consider now a holomorphic vector field in $\mathbb{C}^n$ of the form $F(z) = (\lambda_1 z_1, \ldots, \lambda_n z_n)$, where the λ_i are non-zero complex numbers. The origin $0 \in \mathbb{C}^n$ is the only zero (or singularity) of this vector field. The solutions of F, other than 0, are Riemann surfaces (non-compact), immersed in $\mathbb{C}^n$. These are given by $\{z = e^{AT} w\}$, with $T \in \mathbb{C}$, $w \in \mathbb{C}^n$ and A the diagonal matrix with $(\lambda_1, \cdots, \lambda_n)$ in its diagonal. The solutions of F give a 1-dimensional holomorphic foliation $\mathcal{F}$ with singular set $0 \in \mathbb{C}^n$.

We want to look at the set of points in $\mathbb{C}^n - 0$ where the foliation $\mathcal{F}$ has contacts with the codimension 1 real foliation $\mathcal{S}$ given by all the spheres in $\mathbb{C}^n$ with centre at 0. That is, we want to look at the points where these two foliations are tangent. Given $z \in \mathbb{C}^n - 0$, we denote by $\mathcal{L}_z$ the leaf of $\mathcal{F}$ that passes through z. It is clear that $\mathcal{L}_z$ is tangent at z to the sphere $\mathbb{S}(0, \|z\|)$, with centre at 0 and radius the norm of z, iff the complex line spanned by the vector $F(z)$ is contained in the tangent space $T_z(\mathbb{S}(0, \|z\|))$; this is equivalent to saying that the Hermitian product of $F(z)$ and $z = (z_1, \ldots, z_n)$ vanishes, i.e.,

$$\langle F(z), z \rangle = \sum_{i=1}^{n} \lambda_i |z_i|^2 = 0 \, .$$

In other words, the set of contacts of the two foliations $\mathcal{F}$ and $\mathcal{S}$ on $\mathbb{C}^n - 0$ is the real analytic variety $M^* = M - 0$ defined by the intersection of the two quadrics:

$$V_F = \left\{ z \in \mathbb{C}^n \mid \operatorname{Re}\left(\sum_{i=1}^{n} \lambda_i |z_i|^2\right) = 0 \right\} \cap \left\{ z \in \mathbb{C}^n \mid \operatorname{Im}\left(\sum_{i=1}^{n} \lambda_i |z_i|^2\right) = 0 \right\} \, .$$

There are two drastically different cases according as *the convex hull*,

$$\mathcal{H}(\lambda_1, \ldots, \lambda_n) \subset \mathbb{C} \, ,$$

of the eigenvalues contains or does not contain the origin. To see why this fact makes such a big difference, observe that the analytic set V_F, being the set of contacts of the two foliations, does not depend on the actual vector field F but only on the holomorphic foliation $\mathcal{F}$ that it defines. If we multiply F by a unit complex number $e^{i\theta}$, the spectrum of F is rotated by an angle θ, but the foliation remains unchanged. Hence, if $0 \notin \mathcal{H}(\lambda_1, \ldots, \lambda_n)$ then, multiplying F by an appropriate unit complex number if necessary, we can assume that all the eigenvalues have positive real part. This implies that the equation:

$$\operatorname{Re}\left(\sum_{i=1}^{n} \lambda_i |z_i|^2\right) = 0 \, ,$$

does not have non-trivial solutions and therefore V_F consists of 0 alone. This means that every leaf of $\mathcal{F}$ in $\mathbb{C}^n - 0$ is transversal to all the spheres around 0. We will see that in the other case, when $0 \in \mathcal{H}(\lambda_1, \ldots, \lambda_n)$, the variety V_F has very rich and interesting geometry and topology.

1.1 Definition. If $0 \in \mathcal{H}(\lambda_1, \ldots, \lambda_n)$ we say that the linear vector field F is *in the Siegel domain*. Otherwise we say that F is *in the Poincaré domain*.

The analogous situation in the real case, which we can draw, would be a sink (or a source) and a saddle point (see Figure 11).

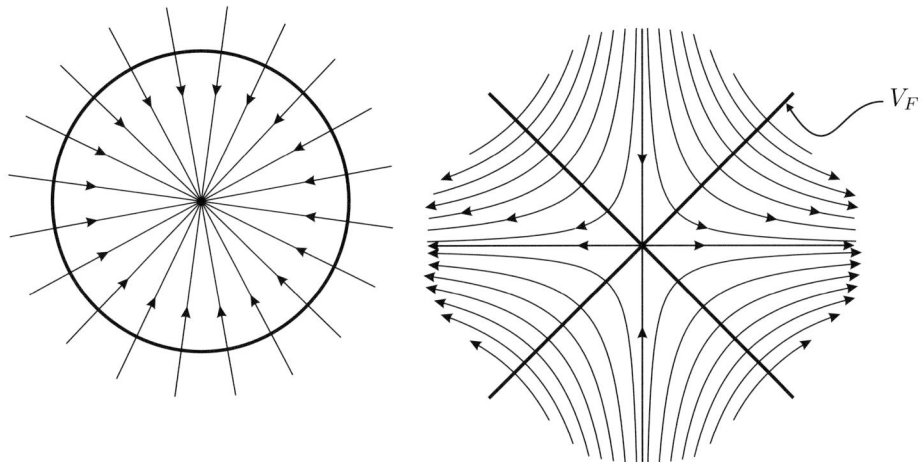

Figure 11: In the Poincaré domain   in the Siegel domain.

The topology of the foliation $\mathcal{F}$ was studied in [50]. The case of linear flows in the Poincaré domain is very interesting from the viewpoint of holomorphic dynamics, but it is not relevant for this text, whose purpose is to study the topology and geometry of analytic spaces, and for these vector fields the analytic space we are looking at consists of the origin alone. So we will restrict from now on to linear vector fields in the Siegel domain.

It is noted in [50] that the equality, and even real dependence, of two eigenvalues of F complicates the topology of $\mathcal{F}$ and V_F very much. Therefore one usually assumes the following generic *hyperbolicity hypothesis*: any two eigenvalues are independent over $\mathbb{R}$:

$$i \neq j \Rightarrow \lambda_i \notin \mathbb{R}\lambda_j, \quad i, j = 1, \ldots, n, \tag{$*$}$$

So we now let F be a linear vector field in the Siegel domain satisfying $(*)$. It is clear that a point $z \in \mathbb{C}^n - 0$ is in V_F iff the restriction of the real function $d(z) = \|z\|^2 = \sum_{i=1}^n |z_i|^2$ to the leaf $\mathcal{L}_z$ through z has a critical point at z. Furthermore, as noted in [50, §3], the fact that the solutions of F are parametrized by exponential maps implies that the *leaves of F are concave*. Thus, if a leaf $\mathcal{L}$ meets V_F^*, then it has a unique point in V_F and it is the point in $\mathcal{L}$ of minimal distance to $0 \in \mathbb{C}^n$. Such a leaf is called *a Siegel leaf*. It is a copy of $\mathbb{C}$ embedded in $\mathbb{C}^n$ and can be characterized by its unique point in V_F. Furthermore, the fact that the intersection $\mathcal{L} \cap V_F$ of each leaf that meets V_F is at a local minimal point in $\mathcal{L}$, implies that $\mathcal{L} \cap V_F$ is a transverse intersection. By the flow-box theorem for complex differential equations, this means that we have at each $z \in V_F^*$ a neighborhood of the form $U_z \times \mathbb{D}^2$, where U_z is a disc of real dimension $2n-2$ and the second factor denotes small discs in the leaves. It follows that V_F^* is a smooth real submanifold of $\mathbb{C}^n$ of codimension 2. That is, V_F is a real analytic complete intersection in $\mathbb{C}^n$ and the union $W = V_F^* \times \mathbb{C}$ of all the Siegel leaves of F is an open subset of $\mathbb{C}^n$ that can be identified with the total space of the normal bundle of V_F^*.

It is shown in [50] that W is actually dense in $\mathbb{C}^n$. Notice that the n coordinate axes of $\mathbb{C}^n$ are contained in the complement of W: it is also shown in [50] that the complement $\mathbb{C}^n - W$ consists of the leaves of $\mathcal{F}$ that contain 0 in their closure; these are called *Poincaré leaves*. It is an exercise to see that V_F is globally embedded as a cone with vertex at 0 and base the intersection $M = V_F \cap \mathbb{S}^{2n-1}$ with the unit sphere, which is the link of the corresponding singularity.

We summarize part of the above discussion in the following theorem, which re-phrases results in [50]:

1.2 Theorem. *Let $F(z) = (\lambda_1 z_1, \ldots, \lambda_n z_n)$ be a linear vector field in the Siegel domain, satisfying the hyperbolicity hypothesis. Then the real analytic variety:*

$$V_F = \left\{ z \in \mathbb{C}^n \mid \mathrm{Re}\left(\sum_{i=1}^n \lambda_i |z_i|^2\right) = 0 \right\} \cap \left\{ z \in \mathbb{C}^n \mid \mathrm{Im}\left(\sum_{i=1}^n \lambda_i |z_i|^2\right) = 0 \right\},$$

is a complete intersection of real codimension 2 with an isolated singularity at 0, and the regular points of V_F parametrize the Siegel leaves of F.

So now the natural problem from the viewpoint of singularities is to determine the topology of the link M. This question was first addressed in general, and beautifully answered, by S. López de Medrano in [149, 150], though there are already a few remarks in this respect in [50]. This is closely related to a problem studied by C.T.C. Wall [260] in relation to the topological stability of smooth mappings. Let us describe briefly some of the results in [149, 150].

We begin by considering (as in [50]) the open set of unordered n-tuples of points $\Lambda = \{\lambda_1, \ldots, \lambda_n\}$ in $\mathbb{C}^*$ satisfying the hyperbolicity hypothesis. This set has a connected component which is the Poincaré domain, i.e., when 0 is not in the convex hull $\mathcal{H}(\lambda)$; the other components form the Siegel domain, and this is the one we are now interested in. It is clear that the topology of $M = V_F \cap \mathbb{S}^{2n-1}$, being a compact smooth manifold, does not change within each component of the Siegel domain. It is pointed out in [50, p. 9] that the Siegel domain is connected for $n = 3$, 4 and it has three components for $n = 5$. This means that for $n = 3$, 4 the diffeomorphism type of the link is constant and for $n = 5$ there are at most three different diffeomorphism types. This was the starting point in [149]. Writing

$\lambda_j = \alpha_j + i\beta_j$ we see that the equations that determine V_F are:

$$\begin{cases} \alpha_1(x_1^2 + y_1^2) + \cdots + \alpha_n(x_n^2 + y_n^2) = 0 \,, \\ \beta_1(x_1^2 + y_1^2) + \cdots + \beta_n(x_n^2 + y_n^2) = 0 \,, \end{cases} \tag{1.3}$$

and one can consider, more generally, the variety V given by the equations:

$$\begin{cases} a_1 x_1^2 + \cdots + a_n x_n^2 = 0 \,, \\ b_1 x_1^2 + \cdots + b_n x_n^2 = 0 \,, \end{cases} \tag{1.4}$$

where the $A_i = (a_i, b_i)$ are pairs of real numbers. One would like to know what are the "degenerate" cases and what is the topology of the link in the "non-degenerate cases". Obviously the first thing to look at is the case of a single quadric $\sum a_i x_i^2 = 0$, where the answer is classical: non-degenerate means no a_i is zero, and the topology of the link is then given by the Morse index (or signature), i.e., by the number q of negative a_i and the number p of positive a_i. The link is then a product $\mathbb{S}^{p-1} \times \mathbb{S}^{q-1}$.

Let us look now at the case given by (1.4). We consider (following [149]) the points $A_i = (a_i, b_i) \in \mathbb{R}^2$. The natural non-degeneracy condition is that the gradient vector fields of these two equations be linearly independent at each point of V, i.e., that V is a complete intersection of codimension 2. It is easy to see that this condition is equivalent to saying that the origin is not in the convex hull of any pair A_i, A_j. This is called in [149] *the weak hyperbolicity condition* (w.h.c. for short), a notation taken from work by M. Chaperon. Under this assumption, the link M is a smooth submanifold of $\mathbb{R}^n$ of codimension 3.

To describe the topology of the link, consider the configurations of n points $A_i = (a_i, b_i)$ as an unordered set of n vectors in $\mathbb{R}^2$, some of which may coincide. If one configuration satisfying the w.h.c. is deformed into another without breaking this condition, then the corresponding links are diffeomorphic. Thus one needs to describe the connected components of the space of configurations satisfying the w.h.c., and in each such component one may deform the A_i to take them into a "normal form". It is noted that the points A_i come in "bunches" that can be collapsed into one multiple point without breaking the w.h.c., and this multiple point can be pushed into the unit circle. Furthermore, one can move these points in the unit circle, preserving the w.h.c., so that they become the vertices of a regular polygon (see [150]), and a standard graph theory argument shows that this polygon has an odd number of vertices.

Thus one can associate to each system of equations as in (1.4) and satisfying the w.h.c. a normal form determined by $k = 2r + 1$ vertices of a regular k-gon, each having a certain multiplicity n_i (the number of points A_i concentrated at this vertex), and $n = n_1 + n_2 + \cdots + n_k$. This gives an odd cyclic partition of the positive integer n, and one may define two such partitions to be equivalent iff they differ by a rotation of the circle. Then one has ([150, Thm. 1]):

1.5 Theorem. *This association establishes a one-to-one correspondence between the set of connected components of the space of configurations of points in $\mathbb{R}^2$ satisfying the w.h.c. and the set of odd cyclic partitions of n.*

Thus the partition $n = n_1 + n_2 + \cdots + n_k$ plays for the system of equations (1.4) the analogous role as the Morse signature $n = p + q$ of a single non-degenerate quadric; the diffeomorphism types of the corresponding link of the singularity are given in terms of this partition [150, Thm. 2]:

1.6 Theorem. *Let M be the link of the singularity determined by the partition $n = n_1 + n_2 + \cdots + n_k$, $k = 2r + 1$. Then (up to diffeomorphism):*

(i) *If $k = 1$, then $M = \emptyset$.*

(ii) *If $k = 3$, then M is a product of spheres: $M \cong \mathbb{S}^{n_1-1} \times \mathbb{S}^{n_2-1} \times \mathbb{S}^{n_3-1}$.*

(iii) *For $k \geq 5$, M is "always" a connected sum of products of spheres. More precisely, for each n_i set $d_i = n_i + n_{i+1} + \cdots + n_{i+r-1}$, one has: if $k = 5$ and all $d_i > 2$, or if $k > 5$ and $n \neq 7$, then M is the connected sum:*

$$M \cong \#_{i=1}^{k} \left(\mathbb{S}^{d_i-1} \times \mathbb{S}^{n-d_i-2} \right).$$

The original systems of equations as in (1.3), corresponding to linear vector fields in the Siegel domain with the hyperbolicity condition, correspond to systems as in (1.4) with n and all n_i even. In particular, for vector fields in $\mathbb{C}^3$ the link M is always the 3-torus $\mathbb{S}^1 \times \mathbb{S}^1 \times \mathbb{S}^1$. One has in general [149, Theorems 1 & 2]:

1.7 Theorem. *The components of the Siegel domain are in one-to-one correspondence with the odd cyclic partitions $n = n_1 + \cdots + n_k$ of n into $k > 1$ positive integers, and the link M of the corresponding singularity is:*

(i) *If $k = 3$, then M is a product of spheres: $M \cong \mathbb{S}^{2n_1-1} \times \mathbb{S}^{2n_2-1} \times \mathbb{S}^{2n_3-1}$.*

(ii) *If $k = 2r + 1 > 3$, then*

$$M \cong \#_{i=1}^{k} \left(\mathbb{S}^{2d_i-1} \times \mathbb{S}^{2n-2d_i-2} \right).$$

An important and difficult question is to determine how the topology of the singularities defined by equation (1.3) varies as we pass from one component in the Siegel domain to another one, i.e., as we break the weak hyperbolicity condition. This is beautifully answered, in a more general setting, in the recent article of F. Bosio and L. Meersseman [32]. This is a "wall-crossing" problem as they explain, and they show that crossing a wall means performing a precise surgery, which they describe. From the viewpoint of singularities, what we do when crossing a wall is to put two of the above complete intersection singularities in a 1-parameter family of singularities which are all complete intersections, except that they bifurcate when crossing the wall.

VI.2 Real singularities and the López de Medrano-Verjovsky-Meersseman manifolds

We now follow [160, 161, 162, 32] and extend the previous constructions to sets of m commuting vector fields on $\mathbb{C}^n$. This type of situations were previously envisaged by N. Kuiper in [116], but his interest was more in the dynamics rather than in the geometry involved. In [160, 161, 162, 32], and also in [151], their interest is focused essentially on constructing and studying complex manifolds which arise as quotients of the links of the singularities that we describe here.

Let m and n be positive integers such that $n > 2m$. Let $\Lambda = (\Lambda_1, \ldots, \Lambda_n)$ be an n-tuple of vectors in $\mathbb{C}^m$, $\Lambda_i = (\lambda_i^1, \ldots, \lambda_i^m)$ for $i = 1, \ldots, n$, and let $\mathcal{H}(\Lambda_1, \ldots, \Lambda_n)$ be the convex hull of Λ in $\mathbb{C}^m$.

2.1 Definition. The n-tuple $\Lambda = (\Lambda_1, \ldots, \Lambda_n)$ is an *admissible configuration* if it satisfies:

(i) *the Siegel condition*: $0 \in \mathcal{H}(\Lambda)$; and

(ii) *the weak hyperbolicity condition (w.h.c.)*: for every $2m$-tuple of integers $(i_1, \ldots, i_{2m})$, $1 \le i_1 < \cdots < i_{2m} \le n$, we have $0 \notin \mathcal{H}(\Lambda_{i_1}, \ldots, \Lambda_{i_{2m}})$.

Geometrically this means that 0 is in the convex polytope $\mathcal{H}(\Lambda)$, but 0 is not contained in any hyperplane passing through $2m$ of its vertices. Note that conditions (i) and (ii) together form an open condition, so a small perturbation of an admissible configuration is again admissible.

To an admissible configuration $(\Lambda_1, \ldots, \Lambda_n)$ we associate the holomorphic foliation $\mathcal{F}$ of $\mathbb{C}^n$ generated by the m commuting vector fields $(1 \le j \le m)$

$$F_j : (z_1, \ldots, z_n) \in \mathbb{C}^n \longmapsto \sum_{i=1}^{n} \lambda_i^j z_i \frac{\partial}{\partial z_i} ;$$

in other words, each vector field F_j has components $(\lambda_1^j z_1, \ldots, \lambda_n^j z_n)$. The fact that these linear vector fields commute implies that they define together a holomorphic action of $\mathbb{C}^m$ on $\mathbb{C}^n$ and the orbits of this action are the leaves of $\mathcal{F}$. This foliation is singular at the origin 0. In analogy with the previous case of $m = 1$, one has:

2.2 Definition. Let $\mathcal{L}$ be a leaf of $\mathcal{F}$. If 0 is the closure of $\mathcal{L}$, we say this is a *Poincaré leaf*. Otherwise $\mathcal{L}$ is a *Siegel leaf*.

Let us consider the Siegel leaves of $\mathcal{F}$ and consider the set V_Λ of contacts of this foliation with the spheres around the origin. As in the previous case, this is defined by the real analytic equations:

$$V_\Lambda = \left\{ z \in \mathbb{C}^n \mid \sum_{i=1}^{n} \Lambda_i |z_i|^2 = 0 \right\} = V_{F_1} \cap \cdots \cap V_{F_m} ,$$

where each V_{F_j} is the variety defined in the previous section, i.e., $V_{F_j} - 0$ is the space of Siegel leaves of the vector field F_j. It is an exercise to show that the weak hyperbolicity condition implies that each V_{F_j} is a real codimension 2 complete intersection in $\mathbb{C}^n$, and they all meet transversally away from 0; the variety V_Λ is a global complete intersection in $\mathbb{C}^n$ of real codimension $2m$ and singular only at 0. (This is proved in [160, I.1] and it is analogous to Hamm's result in [96] giving necessary and sufficient conditions for a set of Brieskorn varieties to define a complete intersection, c.f. II.7 above.)

Again, the fact that the vector fields are linear implies that we can parametrize the leaves of $\mathcal{F}$ by exponential maps, and this implies that each Siegel leaf has a unique point in $V_\Lambda - 0$, which is the point in the leaf of minimal distance to the origin. One has:

2.3 Theorem. *Let $(\Lambda_1, \ldots, \Lambda_n)$ be an admissible configuration as above, let $\mathcal{F}$ be the associated holomorphic foliation of $\mathbb{C}^n$ and let V_Λ be the real analytic variety defined by $V_\Lambda = \{ z \in \mathbb{C}^n \mid \sum_{i=1}^n \Lambda_i |z_i|^2 = 0 \}$. Then:*

(i) *V_Λ is a complete intersection in $\mathbb{C}^n$ of dimension $2n - 2m$, and it is singular only at 0.*

(ii) *The smooth manifold $V_\Lambda^* = V_\Lambda - 0$ parametrizes the Siegel leaves of $\mathcal{F}$: each such leaf has a unique point in V_Λ^*; this is the point in the leaf of minimal distance to the origin.*

(iii) *The foliation $\mathcal{F}$ is everywhere transversal to V_Λ^* and the leaves of $\mathcal{F}$ can be identified with the fibres of the normal bundle of V_Λ^*. Hence V_Λ^* is, canonically, a complex manifold of dimension $n - m$.*

The last statement in this theorem, claiming that V_Λ^* has a canonical complex structure comes from the observation of A. Haefliger in [94], that if a smooth real submanifold N of a complex manifold X is everywhere transversal to a holomorphic foliation of complementary dimension, then the holomorphic foliated atlas of the foliation determines a holomorphic atlas for N. This does not mean however that V_Λ^* is embedded as a complex submanifold of $\mathbb{C}^n$. Neither does this mean that the germ of V_Λ at 0 is complex analytic, which is false in general. For instance, if $m = 1$ and $n = 3$, then we know from the previous section that the link is the 3-torus, and we know from [246] (see Chapter III above) that the 3-torus is not the link of an isolated surface singularity. It would be interesting to know whether or not these singularities are ever complex analytic, and I believe they are not (c.f. Chapter VII below).

The results of [32] show that the topology of these singularities can be rather complicated and interesting, as is already shown in the case $m = 1$ of the previous section. These manifolds admit a natural action of the real torus $\mathbb{T}^n = (\mathbb{S}^1)^n$ given by:

$$((u_1, \ldots, u_n), (z_1, \ldots, z_n)) \longmapsto (u_1 z_1, \ldots, u_n z_n) \in \mathbb{C}^n,$$

for $(u_1, \ldots, u_n) \in \mathbb{T}^n$ and $(z_1, \ldots, z_n) \in \mathbb{C}^n$. This action leaves invariant the link $M_\mathcal{F}$ and they show in [32] (in analogy with [149, 150] for $m = 1$) that this action can be used to associate a natural convex polytope to $M_\mathcal{F}/\mathbb{T}^n$, which captures the topology of the link.

The principal goal of [151, 160, 124] was to use the previous constructions to obtain new families of compact, complex non-symplectic manifolds. These are now called the *López de Medrano-Verjovsky-Meersseman (LVM for short) manifolds*. A motivation for this comes from the generalized Hopf manifolds of [46, 94]. Recall that the Hopf manifolds are constructed as follows: given a real number $r > 1$, we can take in $\mathbb{C}^n - 0$ the action of $\mathbb{Z}$ given by $m \cdot z = r^m z$. The quotient is a manifold diffeomorphic to $\mathbb{S}^{2n-1} \times \mathbb{S}^1$. Since the action is holomorphic and totally discontinuous, the quotient inherits a natural complex structure. This construction was generalized by Haefliger taking quotients of $\mathbb{C}^n - 0$ by other holomorphic and totally discontinuous actions of $\mathbb{Z}$, obtaining complex manifolds diffeomorphic to $\mathbb{S}^{2n-1} \times \mathbb{S}^1$ but with different complex structures. For this he used the observation just mentioned above, that a smooth manifold transversal to a holomorphic foliation inherits a complex structure.

The construction in [46] of compact manifolds that generalizes those of Hopf is given by considering the Brieskorn singularities

$$V_{(a_0,\dots,a_n)} = \{z_0^{a_0} + \cdots + z_n^{a_n} = 0\},$$

and taking the quotient of $V^* = V_{(a_0,\dots,a_n)} - 0$ by the analogous $\mathbb{Z}$-action: given a real number $r > 1$, define a $\mathbb{Z}$-action on V^* by

$$m \cdot (z_1, \dots, z_n) = (r^{m/a_0} z_0, \dots, r^{m/a_n} z_n).$$

The quotient $V^*/\mathbb{Z}$ is a compact complex manifold diffeomorphic to $M \times \mathbb{S}^1$, where M is the link. Even when the link is the standard sphere, some of these complex structures are different to those obtained in [94].

Another motivation for the LVM-manifolds comes from [49, 268]. Calabi and Eckmann [49] generalized Hopf's construction to give complex structures on all products of spheres $\mathbb{S}^{2n-1} \times \mathbb{S}^{2m-1}$, $n, m > 1$. Each of these manifolds is the total space of a principal bundle

$$\mathbb{S}^{2n-1} \times \mathbb{S}^{2m-1} \to \mathbb{CP}^{n-1} \times \mathbb{CP}^{m-1},$$

with fibre the 2-torus; endowing the torus with the structure of an elliptic curve one turns $\mathbb{S}^{2n-1} \times \mathbb{S}^{2m-1}$ into a complex manifold. The Calabi-Eckmann manifolds were generalized by Loeb-Nicolau in [268], inspired in Haefliger's work. They consider a linear vector field F in $\mathbb{C}^{n+m}$ in the Poincaré domain, satisfying a weak hyperbolicity hypothesis (weaker than the one in §1 above). Then they prove that there is an embedding of $\mathbb{S}^{2n-1} \times \mathbb{S}^{2m-1}$ in $\mathbb{C}^{n+m}$ which is transversal to the foliation $\mathcal{F}$ of F, and therefore (by [94]) inherits a complex structure from that of $\mathcal{F}$. In this way they get a large class of complex structures on these products of spheres, including those of [49].

The LVM-manifolds combine all these ideas to give a new construction of compact complex manifolds. This class contains "most" of the manifolds constructed in [94, 268], and many more, where "most" means that in both cases the LVM-manifolds leave out certain complex structures that arise by considering vector fields with a certain type of resonances which are valid in [94, 268] but not in [151, 160, 124].

To construct the LVM-manifolds we start with an admissible configuration $\Lambda = (\Lambda_1, \ldots, \Lambda_n)$ as above, let $\mathcal{F}$ be the associated holomorphic foliation of $\mathbb{C}^n$ and let V_Λ be the real analytic variety defined by

$$V_\Lambda = \{ z \in \mathbb{C}^n \mid \sum_{i=1}^n \Lambda_i |z_i|^2 = 0 \},$$

so that $V_\Lambda^* = V_\Lambda - 0$ is the space of Siegel leaves of the $\mathbb{C}^m$-action. Let $\mathcal{S} \subset \mathbb{C}^n$ be the union of all the Siegel leaves, so that $\mathcal{S}$ is V_Λ^* saturated by the $\mathbb{C}^m$-action. Observe that the usual $\mathbb{C}^*$-action:

$$t \cdot (z_1, \ldots, z_n) \mapsto (tz_1, \ldots, tz_n)$$

commutes with the $\mathbb{C}^m$-action and both actions have $\mathcal{S}$ as an invariant set. Therefore they give together a free action of $\mathbb{C}^m \times \mathbb{C}^*$ on $\mathcal{S}$ by holomorphic transformations. The quotient

$$N = N_\Lambda = \mathcal{S}/(\mathbb{C}^m \times \mathbb{C}^*)$$

is a complex manifold. To prove that it is compact we observe that the quotient $\mathcal{S}/\mathbb{C}^m$ is the manifold V_Λ^*, hence N is nothing but the projectivization of V_Λ^*. This can also be regarded as the quotient of the link M of 0 in V_Λ divided by the $\mathbb{S}^1$-action induced from that of $\mathbb{C}^*$. We also observe that one can follow the method of [46] described above and construct $\mathbb{Z}$-actions on V_Λ^* so that the orbit space is a compact, complex manifold which is a principal bundle over N with fibre an elliptic curve.

The manifold N is, by definition, an *LVM-manifold*. These manifolds have fascinating geometry and topology. Their topology is beautifully described in [32]. In [160] it is shown (using [143]) that the m vector fields defined by the admissible configuration Λ give rise to m holomorphic vector fields on N which define a (non-singular) holomorphic foliation $\mathcal{G}$ of dimension m (recall N has complex dimension $(n-m-1) \geq m$). And if the configuration Λ satisfies a certain "rational condition (K)", then [162, Theorem A] this foliation is by compact, complex tori. Furthermore, in this case the quotient $N/\mathcal{G}$ is a projective (quasi-regular) toric variety equipped with an orbifold structure. The projection $N \to N/\mathcal{G}$ is a *generalized Calabi-Eckmann* fibration, in the notation of [162] Corollary C, and they also show that every LVM manifold has an arbitrary small deformation of its complex structure so that the deformed manifold is the total space of a generalized Calabi-Eckmann fibration. Conversely, Theorem G and Corollary H in [162] tell us that every projective toric manifold, and every projective quasi-regular toric variety, can be realized as a quotient $N/\mathcal{G}$, for an appropriate LVM-manifold N.

VI.3   Real singularities and holomorphic vector fields

Let us now give a method for constructing codimension 2 real analytic complete intersection singularities in $\mathbb{C}^n$ with a very rich geometry and topology; this is motivated by the constructions in Section 1 and is reminiscent of the geometry of complex singularities. This construction, taken from [226], is actually inspired by [88], where the goal is to study the topology of holomorphic vector fields near an isolated singularity, in a similar way to the one used by Milnor to study the topology of an analytic space near an isolated singularity. It would be interesting to make a similar study for singularities of higher codimension, using ideas of §2 above.

Let $\chi(\mathbb{C}^n, 0)$ denote the space of all germs of continuous vector fields at $0 \in \mathbb{C}^n$, and let F, X be elements in $\chi(\mathbb{C}^n, 0)$. One has a continuous map,

$$\psi_{F,X} : \mathbb{C}^n \cong \mathbb{R}^{2n} \to \mathbb{C} \cong \mathbb{R}^2 \,,$$

defined by $\psi_{F,X}(z) = \langle F(z), X(z) \rangle$, where

$$\langle F(z), X(z) \rangle = \sum_{i=1}^n F_i(z) \cdot \overline{X}_i(z)$$

is the usual Hermitian product. We note that if F and X are both differentiable of class C^r, then $\psi_{F,X}$ is of class C^r; if F and X are both real analytic, then $\psi_{F,X}$ is also real analytic, but even if both F and X are complex analytic, $\psi_{F,X}$ is not complex analytic, unless X is constant.

3.1 Examples.

(i) Let F be a linear vector field in $\mathbb{C}^n$, $F(z) = (\lambda_1 z_1, \ldots, \lambda_n z_n)$, and let X be the radial vector field $X(z) = (z_1, \ldots, z_n)$. Then

$$\psi_{F,X}(z) = \sum_{i=1}^n \lambda_i |z_i|^2$$

is the map envisaged in §1 and its zero-set $V_{F,X} = \psi_{F,X}^{-1}(0)$ consists of the origin union the set of contacts of the foliation $\mathcal{F}$ of F with the foliation given by all the spheres around 0. Thus, if F is in the Poincaré domain, then $V_{F,X} = \{0\}$, and if F is in the Siegel domain and satisfies the weak hyperbolicity condition, then $V_{F,X} - 0$ is the space of Siegel leaves of F.

(ii) Let $f : \mathbb{C}^2 \to \mathbb{C}$ be the Pham-Brieskorn polynomial $f(z) = z_1^p + z_2^q$, with $p, q > 2$, let F be the Hamiltonian vector field $F = \left(\frac{\partial f}{\partial z_2}, -\frac{\partial f}{\partial z_1} \right)$, and let $X = (a, b)$ be constant. Then

$$\psi_{F,X} = a \frac{\partial f}{\partial z_2} - b \frac{\partial f}{\partial z_1}$$

and $V_{F,X}$ is the polar curve of f and a linear form (much studied by B. Teissier, Lê Dũng Tráng and others).

(iii) Let F be holomorphic and let X be the gradient vector field of a real analytic function $f : \mathbb{C}^n \to \mathbb{R}$. Then $V_{F,X}$ is the union of the zeroes of F, the zeroes of X and the *polar variety* of the foliations $\mathcal{F}$ (defined by F) and the one given by the level surfaces of f, i.e., the set of points in $\mathbb{C}^n$ where the two foliations are tangent (c.f. [251]). If f is the square of the function distance to the origin, then this is example i).

Although some of the things we say below hold in general for F and X continuous vector fields, we restrict ourselves to the case of this last example, F holomorphic and X the gradient vector field of a real analytic function $f : \mathbb{C}^n \to \mathbb{R}$, where the geometry is very rich and interesting.

Let $V_{F,X} = \psi_{F,X}^{-1}(0)$ be the polar variety of $\mathcal{F}$ and the level surfaces of f, that we denote by $\mathcal{S}$. So $V_{F,X}$ is defined by the real analytic equations:

$$\mathrm{Re}\,\langle F(z), X(z) \rangle = 0 \qquad \text{and} \qquad \mathrm{Im}\,\langle F(z), X(z) \rangle = 0.$$

Away from $V_{F,X}$ these two foliations meet transversally, defining a foliation Γ by real curves which are tangent to both foliations.

In order to study the geometry of $V_{F,X}$ it is convenient to modify the function that determines it by composing $\psi_{F,X}$ with the automorphism $\iota : \mathbb{C} \to \mathbb{C}$ given by $\iota(z) = i\,\bar{z}$. The composition $\widehat{\psi}_{F,X} = \iota \circ \psi$ is the map $z \mapsto i\,\overline{\langle F(z), X(z) \rangle}$. Clearly one has $V_{F,X} = \widehat{\psi}_{F,X}^{-1}(0)$.

3.2 Lemma. *The foliation Γ is given by the integrals of the real analytic vector field*

$$\tau(z) = \widehat{\psi}_{F,X}(z) \cdot F(z),$$

whose zero locus is $V_{F,X}$.

Proof. It is clear that $\tau(z)$ is always tangent to $\mathcal{F}$, because at each point $z \in \mathbb{C}^n$, $\tau(z)$ is $F(z)$ multiplied by a complex number. To prove that $\tau(z)$ is tangent to $\mathcal{S}$ we must prove that $\tau(z)$ is normal to $X(z)$. One has:

$$\langle \tau(z),\, X(z) \rangle = \langle\, ((i\,\overline{\langle F(z), X(z) \rangle}\,) \cdot F(z),\, X(z) \,\rangle$$
$$= (i\,\overline{\langle F(z), X(z) \rangle}) \cdot \langle\, F(z), X(z) \,\rangle = i\,\|\langle F(z), X(z) \rangle\|^2.$$

Hence $\mathrm{Re}\,\langle \tau(z), X(z) \rangle = 0$, so $\tau(z)$ is normal to $X(z)$ because the real part of the Hermitian product is the usual inner product in $\mathbb{R}^{2n} \cong \mathbb{C}^n$. $\qquad\square$

This vector field τ describes the behavior of F in the direction determined by the level surfaces of f. For instance, if $X(z) = z$ and F is a linear vector field in the Poincaré domain with generic eigenvalues, then one knows by work of Guckenheimer that τ is Morse-Smale, a fact used in [50] to prove that the vector field F is structurally stable.

The following immediate consequence of Lemma 3.2 gives a geometric interpretation of the map $\widehat{\psi}_{F,X}$. For simplicity set $\widehat{\psi} = \widehat{\psi}_{F,X}$.

3.3 Proposition. *For each $z \in \mathbb{C}^n - V_{F,X}$, the argument of the complex number $\widehat{\psi}(z)$ is the angle by which we must rotate the vector $F(z)$ in its complex line (counter-clock-wise) to make it tangent to the level surface of f at z.*

Let us now look at the map $\phi = \phi_{F,X} : \mathbb{C}^n - V_{F,X} \to \mathbb{S}^1 \subset \mathbb{C}$ given by

$$z \xmapsto{\phi} \arg(\widehat{\psi}(z)) = \frac{\widehat{\psi}(z)}{|\widehat{\psi}(z)|} = \frac{i\,\overline{\langle F(z), X(z) \rangle}}{|\langle F(z), X(z) \rangle|}$$

and set $E_\theta = \phi^{-1}(e^{i\theta})$. For each $\theta \in [0, \pi) \subset \mathbb{R}$ we define a map $\mathbb{C}^n \xrightarrow{\psi_\theta} \mathbb{R}$ by

$$\psi_\theta(z) = \mathrm{Re}\,\langle e^{i\theta}\, F(z), X(z) \rangle \,,$$

and we set $V_\theta = \psi_\theta^{-1}(0)$. One has the following decomposition theorem, which is reminiscent of Milnor's fibration theorem, but so far we are not claiming anything in this respect. We will return to that point in Chapter VII.

3.4 Theorem. *The union of all V_θ covers $\mathbb{C}^n$; their intersection is the polar variety $V_{F,X}$ and each pair of antipodal fibres E_θ and $E_{\theta+\pi}$ are naturally glued along $V_{F,X}$ forming the variety V_θ. That is, for $\theta \in [0, \pi)$ one has:*

$$\mathbb{C}^n = \cup V_\theta \;;\quad V_{F,X} = \cap \cup V_\theta \;;\quad \text{and}\quad M_\theta = E_\theta \cup V_{F,X} \cup E_{\theta+\pi}\,.$$

The proof of this result is exactly the same as the proof of 4.1 below.

We now recall that X is the gradient ∇f of a function f. By restricting f to the leaves of $\mathcal{F}$ one gets a vector field $X_\mathcal{F}$ whose solutions are contained in the leaves of $\mathcal{F}$; locally, for $z \notin V_{F,X}$, $X_\mathcal{F}(z)$ is obtained by projecting $X(z)$ to the complex line spanned by F. The zeroes (or singularities) of $X_\mathcal{F}$ form precisely the variety $V_{F,X}$. We say (following Thom [251]) that a point $z \in V_{F,X}$ is a *generic contact* of the two foliations if none of them is singular at z and the corresponding zero of $X_\mathcal{F}$ is non-degenerate. This means that the restriction of f to the leaf has a Morse singularity at z. Hence these contacts are either non-degenerate local minimal points in the leaves of $\mathcal{F}$ (with respect to the level surfaces of f), or local maximal points, or saddle points. Just as in the case of the varieties of Siegel leaves in Sections 1 and 2, one has that in either case of a generic contact, each leaf of $\mathcal{F}$, which is (real) 2-dimensional, is locally contributing to $V_{F,X}$ with one point. It follows that at a generic contact the polar variety $V_{F,X}$ is smooth of real codimension 2. Furthermore, with a little more work one can prove that the converse also holds and one has the following theorem (we refer to [88, 2.3] for details).

3.5 Theorem. *Assume that the function f and the vector field F vanish only at 0. Then the variety of contacts $V_{F,X}$ is a real analytic geometric complete intersection in $\mathbb{C}^n$ of real codimension 2 and smooth away from 0 if and only if all the contacts of $\mathcal{F}$ and the level surfaces of f are generic.*

Of course in all this discussion we can consider F and X to be defined only on an open set U of $\mathbb{C}^n$ and not on the whole space.

VI.4   On the topology of certain real hypersurface singularities

This section is taken from [228] and it is motivated by [168, 197] and by the previous section. Consider now a connected open set $\mathbb{D} \subset \mathbb{C}^{n+1}$, $0 \in \mathbb{D}$, and let

$$H : (\mathbb{D}, 0) \to (\mathbb{C}, 0)$$

be a continuous function. For each real line $\mathcal{L}_\theta \subset \mathbb{C}$ passing through the origin with an angle θ, $\theta \in [0, \pi)$, we let $\pi_\theta \colon \mathbb{C} \to \mathcal{L}_\theta$ be the orthogonal projection; set $h_\theta = \pi_\theta \circ H$, so that h_0 and $h_{\frac{\pi}{2}}$ are, respectively, the real and the imaginary parts of H. We set $M_\theta = h_\theta^{-1}(0)$ and $V = H^{-1}(0)$. We define the map $\tilde{\phi} \colon \mathbb{D} - V \to \mathbb{S}^1$ by $\tilde{\phi}(z) = \frac{iH(z)}{\|H(z)\|}$. For each $e^{i\theta} \in \mathbb{S}^1$, we set $E_\theta = \tilde{\phi}^{-1}(e^{i\theta})$.

The following lemma establishes that one has in general a decomposition similar to the one in Theorem 3.4 above:

4.1 Lemma. *One has:* $\mathbb{D} = \cup M_\theta$ *and* $V = \cap M_\theta$, *for* $\theta \in [0, \pi)$. *Also, for each* $\theta \in [0, \pi)$ *one has:* $M_\theta = E_\theta \cup V \cup E_{\theta+\pi}$. *Similarly, if* $\mathbb{S}_\varepsilon$ *is a sphere embedded in* $\mathbb{D}$ *with centre at* 0, *one has,*

$$\mathbb{S}_\varepsilon = \cup(M_\theta \cap \mathbb{S}_\varepsilon) \quad V \cap \mathbb{S}_\varepsilon = \cap(M_\theta \cap \mathbb{S}_\varepsilon); \quad and$$
$$(M \cap \mathbb{S}_\varepsilon) = (E_\theta \cap \mathbb{S}_\varepsilon) \cup (V \cap \mathbb{S}_\varepsilon) \cup (E_{\theta+\pi} \cap \mathbb{S}_\varepsilon) .$$

Proof. If $z \in M_\theta$, then $H(z)$ is contained in the line passing through 0 with an angle $\theta \pm \frac{\pi}{2}$. Hence $z \in M_{\theta_1} \cap M_{\theta_2}$ if and only if $z \in V$, where $\theta_1 = \theta_2 + k\pi$, $k \in \mathbb{Z}$. We prove that $V = \cap M_\theta$ for each $\theta \in [0, \pi)$. It is clear that each point in $\mathbb{D}$ is either in V itself or else it is in a certain E_θ. So the claim $\mathbb{D} = \cup M_\theta$ follows from the claim $M_\theta = E_\theta \cup V \cup E_{\theta+\pi}$. We prove this last claim. For this we notice that M_θ is the set of points $z \in \mathbb{C}^{n+1}$ such that $e^{i\theta} H(z)$ is a purely imaginary number. One has:

$$M_\theta = \{z \in \mathbb{C}^{n+1} \,|\, \mathrm{Re}\, e^{i\theta} H(z) = 0\} .$$

It is now clear that $V \subset M_\theta$. Let us prove that $E_\theta \subset M_\theta$. If $z \in E_\theta$, then

$$\frac{iH(z)}{\|H(z)\|} = e^{i\theta}, \qquad \text{which implies} \qquad \frac{e^{-i\theta} iH(z)}{\|H(z)\|} = 1 ,$$

hence

$$\mathrm{Re}\, \frac{e^{i\theta} H(z)}{\|H(z)\|} = 0 .$$

Thus one has $z \in M_\theta$. Similarly, if $z \in E_{\theta+\pi}$, then

$$\frac{iH(z)}{\|H(z)\|} = e^{i(\theta+\pi)} = -e^{i\theta}, \qquad \text{so} \qquad e^{-i\theta} \frac{iH(z)}{\|H(z)\|} = -1 ,$$

which implies Re $e^{-i\theta} H(z) = 0$. Hence one has $E_{\theta+\pi} \subset M_\theta$. Conversely, if $z \in M_\theta$, then Re $e^{-i\theta} H(z) = 0$. If $H(z) = 0$, one has $z \in V$ and there is nothing to prove. If $H(z) \neq 0$, then

$$e^{-i\theta} \frac{iH(z)}{\|H(z)\|} = \pm 1 \,,$$

so z is in E_θ or in $E_{\theta+\pi}$, according to its sign, and we arrive at the formula above. To prove the second statement in the lemma we just restrict the previous discussion to the sphere $\mathbb{S}_\varepsilon$. $\qquad\square$

4.2 Theorem. *Assume now that H is holomorpic; let $h_\theta = \pi_\theta \circ H$ be as before and let*

$$\phi = \frac{H}{\|H\|} : \mathbb{S}_\varepsilon - K \to \mathbb{S}^1$$

be the usual Milnor fibration, where $K = V \cap \mathbb{S}_\varepsilon$ is the link of 0. Then:

(i) *Each $M_\theta = h_\theta^{-1}(0)$ is a real analytic hypersurface in $\mathbb{D}$, whose singular set is the singular set of V.*

(ii) *Each pair of antipodal fibres F_θ and $F_{\theta+\pi}$ of ϕ are naturally glued together along K, forming a real analytic variety isomorphic to $N_\theta = M_\theta \cap \mathbb{S}_\varepsilon$.*

Proof. The second statement in Theorem 4.2 is a consequence of the lemma above, because the map $\tilde{\phi}$ is ϕ followed by the diffeomorphism $z \to iz$ of $\mathbb{C}$. Hence the intersection of each M_θ with each sphere $\mathbb{S}_\varepsilon$ is a real analytic variety of dimension $2n$; since M_θ is a cone near 0, by [168], we know that M_θ is a real hypersurface. Its singular set consists of the critical points of h_θ, i.e., the points where all the partial derivatives of h_θ vanish. This set does not change if we multiply h_θ by the number $e^{-i\theta}$. Thus, to prove the claim about the singularities of M_θ, it is enough to consider the case $\theta = 0$, i.e., for the real part of H, $f = \text{Re } H$. One has,

$$f = \frac{1}{2}(H + \overline{H}) \,,$$

where $\overline{H}$ is the complex conjugate of H. Therefore,

$$2\Delta f = \left(\frac{\partial H}{\partial z_1}, \frac{\overline{\partial H}}{\partial z_1}, \dots, \frac{\partial H}{\partial z_{n+1}}, \frac{\overline{\partial H}}{\partial z_{n+1}} \right),$$

because the partial derivatives of H with respect to the $\overline{z}_i$ are all 0, the partial derivatives of $\overline{H}$ with respect to the z_i are all 0 and $\frac{\overline{\partial H}}{\partial z_i} = \frac{\partial \overline{H}}{\partial z_i}$. Thus, the critical points of H are the critical points of f. $\qquad\square$

This theorem tells us that the link N_θ of M_θ is, in some sense, the double of the Milnor fibre of H, but N_θ is singular if K is singular. We recall (see Ch. I.6 above) that even in this case there is a Milnor fibration

$$H : H^{-1}(\mathbb{S}^1_\delta) \cap B_\varepsilon \to \mathbb{S}^1_\delta \,,$$

where $\mathbb{S}_\delta^1$ is a small circle centred at $0 \in \mathbb{C}$ and B_ε is a small open ball in $\mathbb{C}^{n+1}$. The closure of each fibre $E_\theta = H^{-1}(e^{i\theta})$ in this fibration is a compact non-singular variety with boundary $\widetilde{K}$, while the closure of the Milnor fibre F is the union of F with the link K, which may be singular. I thank Lê Dũng Tráng for explaining to me that there is a natural contraction function from $\widetilde{K}$ onto K; this can be constructed by lifting to $\mathbb{C}^{n+1}$ a radial vector field in $\mathbb{C}$ around 0. It would be interesting to study the relation of this function to our construction (c.f. [163]). Equivalently, one may study the relation between the double of the Milnor fibre, which is a closed manifold, and the link N_θ of the real analytic variety $M_\theta = h_\theta^{-1}(0)$.

We notice that M_0 and $M_{\frac{\pi}{2}}$ are, respectively, the sets of points where Re $H(z) = 0$ and Im $H(z) = 0$.

If the singular set of V consists of an isolated point, one has the following theorem:

4.3 Theorem. *If $0 \in \mathbb{D}$ is an isolated critical point of H, then the M_θ's are non-singular away from 0, and they are all homeomorphic. The link $N_\theta = M_\theta \cap \mathbb{S}_\varepsilon$ is diffeomorphic to the double of the Milnor fibre F of H, hence N_θ is $(n-1)$-connected; N_θ is always stably parallelizable, and it is actually parallelizable if and only if n is odd and the Milnor number μ of H is 1.*

We recall that a manifold M is parallelizable if its tangent bundle TM is trivial; M is stably parallelizable if $TM \oplus (k)$ is trivial, where (k) is a trivial bundle over M, or equivalently, if M can be embedded in some sphere $\mathbb{S}^N$ with trivial normal bundle.

Since the topology of the Milnor fibre of H and the link K are well understood, these theorems determine the topology of the real hypersurfaces M_θ.

Proof. The claims that the M_θ are regular away from 0, that they are all homeomorphic (actually diffeomorphic away from 0) and the link is diffeomorphic to the double of the Milnor fibre of f, are all consequences of Theorem 4.2. That N_θ is connected follows because F is connected, by [168]. Furthermore, if $n > 1$, then K is connected and F is simply connected, by [168]. Hence Van Kampen's Theorem [239, p. 151] implies that N_θ is simply connected. Moreover, by [168], F is a wedge of n-spheres and K is $(n-2)$-connected, thus Mayer-Vietoris (reduced if n=2) implies that one has

$$H_1(N_\theta; \mathbb{Z}) \cong \cdots \cong H_{n-1}(N_\theta; \mathbb{Z}) \cong 0 \,.$$

Therefore Hurewicz's isomorphism [239, p. 397] implies that N_θ is $(n-1)$-connected. Finally, N_θ is stably parallelizable because it is a codimension 1, oriented submanifold of $\mathbb{S}_\varepsilon$. Thus, [112, Th. IX], N_θ is parallelizable if and only if its Euler-Poincaré characteristic $\chi(N_\theta)$ vanishes. One has:

$$\chi(N_\theta) = 2\chi(F) = 2 + 2(-1)^n \mu \,,$$

where $\mu > 0$ is the Milnor number of H, since N_θ is the double of F, and F is a wedge of μ n-spheres. So $\chi(N_\theta) = 0$ if and only if n is odd and $\mu = 1$. $\square$

The following corollary is actually a special case of the theorem above, but we think it deserves to be stated on its own.

4.4 Corollary. *Let* $f : (\mathbb{C}^{n+1}, 0) \to (\mathbb{C}, 0)$ *be a holomorphic map with an isolated critical point at* $0 \in \mathbb{C}^{n+1}$ *and let* F *be the Milnor fibre of* f, *so* $F = f^{-1}(t) \cap B_\varepsilon$ *for some small ball* B_ε *and some* t *with* $|t|$ *sufficiently small. Let* M_0 *be the link of the real part of* f, *i.e., of the function* $\mathrm{Re}\, f : (\mathbb{C}^{n+1}, 0) \to (\mathbb{R}, 0)$. *Then* M_0 *is diffeomorphic to the double of* F.

4.5 Example. Consider again the polynomial $f(z_1, z_2) = z_1^p + z_2^q$, with $p, q > 1$. The topology of this singularity is well understood (see Ch. I). Its link K is a torus link (or knot if p, q are relatively prime) and its Milnor number μ is $(p-1)(q-1)$; we recall that the Milnor fibre F is in this case an oriented surface in the 3-sphere, with boundary K and the homotopy type of μ circles. Now consider the real and the imaginary parts of f,

$$f_1 = \mathrm{Re}\, f = z_1^p + z_2^q + \bar{z}_1^p + \bar{z}_2^q,$$
$$f_2 = \mathrm{Im}\, f = z_1^p + z_2^q - \bar{z}_1^p - \bar{z}_2^q.$$

Both define real hypersurfaces in $\mathbb{C}^2$, which are cones over their links L_1, L_2. By the theorem above, L_1 and L_2 are homeomorphic, actually diffeomorphic, and they are the double of the Milnor fibre F. Hence L_1 and L_2 are closed, oriented surfaces of genus $(p-1)(q-1)$ in the 3-sphere $\mathbb{S}^3 \subset \mathbb{C}^2$.

This can be interesting because it provides explicit analytic embeddings of surfaces of all genera in $\mathbb{S}^3 \subset \mathbb{C}^2$. For instance, if $p = 2 = q$, then K is the Hopf link, F is a cylinder $\mathbb{S}^1 \times I$ and L_1, L_2 are tori $\mathbb{S}^1 \times \mathbb{S}^1$, obtained by taking two copies of F and gluing them along their boundary. The link K is the intersection of the tori L_1, L_2.

Chapter VII

Real Singularities with a Milnor Fibration

A map $f : (\mathbb{R}^{n+k}, 0) \to (\mathbb{R}^k, 0)$, $n, k > 0$, satisfies the Milnor condition at 0 if it is a submersion at every point in a punctured neighborhood of $0 \in \mathbb{R}^{n+k}$. Milnor's fibration theorem says that every such map determines a locally trivial fibre bundle $\mathbb{S}_\varepsilon - M \xrightarrow{\phi} \mathbb{S}^{k-1}$ for every sufficiently small sphere $\mathbb{S}_\varepsilon \subset \mathbb{R}^n$ centred at 0, where M is the link. However this projection ϕ in general may not be given by the obvious map $\frac{f}{|f|}$. The goal of this chapter is to construct real analytic singularities for which Milnor's condition holds and one actually has that the projection map is $\frac{f}{|f|}$, just as it is for complex singularities. In this case we say that the singularity satisfies the strong Milnor condition.

The results in this chapter are based on [226, 227, 214]. For completeness we present also related results of [111, 218] where the authors discuss the difference between the Milnor condition and the strong Milnor condition, and they give sufficient conditions for a map f as above to satisfy the strong Milnor condition. In particular, for weighted homogeneous maps both conditions are equivalent.

The constructions in this chapter are inspired by those in Chapter VI and the singularities we get are reminiscent of the classical Pham-Brieskorn singularities, so we call them twisted Pham-Brieskorn singularities.

VII.1 Milnor's fibration theorem revisited

Let us consider real analytic functions $f : (\mathbb{R}^n, 0) \to (\mathbb{R}^p, 0)$, $n \geq p$.

1.1 Definition. We say that the map f *satisfies the Milnor condition at* 0 if the derivative $Df(x)$ has rank p at every point $x \in U - 0$, where U is an open neighborhood of $0 \in \mathbb{R}^n$, i.e., if f is a local submersion at every point in a punctured neighborhood of $0 \in \mathbb{R}^n$.

One has the following theorem of Milnor, that we proved in Chapter I:

1.2 Theorem. *If f satisfies the condition of Definition 1.1, then for every $\varepsilon > 0$ sufficiently small, and for every $\delta > 0$ sufficiently small with respect to ε, one has that*

$$f|_{f^{-1}(C_\delta) \cap B_\varepsilon} : f^{-1}(C_\delta) \cap B_\varepsilon \longrightarrow C_\delta \,,$$

is a fibre bundle, where $C_\delta \subset \mathbb{C}$ is the circle in $\mathbb{C}$ of radius δ and centre at 0, and B_ε is the closed ball in $\mathbb{R}^n$ of radius ε and centre 0.

Of course every complex-valued holomorphic function with an isolated critical point in its domain satisfies these conditions, and so does if we compose such function with a real analytic diffeomorphism of the target $\mathbb{C}$. The interesting point here is to find examples which are honestly real analytic. Milnor exhibited the following examples in his book, suggested to him by N. Kuiper, of functions satisfying the condition of Definition 1.1, and therefore giving rise to fibrations. Let A denote either the complex numbers, the quaternions or the Cayley numbers, and define

$$h : A \times A \to A \times \mathbb{R} \,,$$

by $h(x, y) = (2x\bar{y}, |y|^2 - |x|^2)$. Milnor first proves ([168, 11.6]) that this mapping carries the unit sphere of $A \times A$ to the unit sphere of $A \times \mathbb{R}$ by a *Hopf fibration* (see [242, p. 109]). Then he defines, more generally,

$$f : A^n \times A^n \to A \times \mathbb{R} \,,$$

by

$$f(x, y) = (2\langle x, y \rangle, \|y\|^2 - \|x\|^2) \,,$$

where $\langle \cdot, \cdot \rangle$ is the Hermitian inner product in A. One has that this map is a local submersion on a punctured neighbourhood of $(0, 0) \in A^n \times A^n$. The link M of the corresponding singularity is the Stiefel manifold of 2-frames in A^n and the Milnor fibre in this case is a disc bundle over the unit sphere of A^n.

Milnor asked a number of questions regarding his theorem for real singularities. As Milnor pointed out in his book, the hypothesis of Df having maximal rank everywhere near 0 is too strong and it is difficult to find examples, even for $p = 2$. There are in fact pairs (n, p) as above for which no such examples exist, see [53, 208]. For $p = 2$, which is the most relevant case for this work, what is generic is to have real curves in $\mathbb{R}^2$ converging to $(0, 0)$, whose inverse image consists of points where the Jacobian matrix has rank less than 2. Consider for instance the maps $\psi : \mathbb{C}^n \cong \mathbb{R}^{2n} \to \mathbb{C} \cong \mathbb{R}^2$ of Chapter VI:

$$\psi(z) = \langle F(z), z \rangle = \sum_{i=1}^{n} \lambda_i |z_i|^2 \,,$$

where $F = (\lambda_1 z_1, \ldots, \lambda_n z_n)$ is a linear vector field. We know from the previous chapter that if F is in the Siegel domain and satisfies the weak hyperbolicity condition, then the zeroes of ψ define a codimension 2 real complete intersection in $\mathbb{R}^{2n}$ with an isolated singularity at 0, and it has a very rich geometry and topology. These maps are reminiscent of Milnor's examples above and could be natural examples for maps satisfying the Milnor condition at 0. However it is an exercise to see that, regardless of what the eigenvalues of F are, the derivative of these maps has rank less than 2 on the n axis of $\mathbb{C}^n$.

Milnor actually asked whether there exist "non-trivial" examples satisfying the condition of Definition 1.1 when $p = 2$. This question was answered positively by Looijenga [145] for n even and by Church and Lamotke [53] for n odd, using Looijenga's technique. However, no explicit examples of such singularities are given in those articles. Roughly speaking, their proof consists in showing that on one hand, if Df has maximal rank everywhere near 0, then the pair $(\mathbb{S}_\varepsilon, M)$ is a *Neuwirth-Stallings pair* (c.f. [188, 240]), where $M = \mathbb{S}_\varepsilon \cap f^{-1}(0)$ is the link of the singularity; and conversely, a Neuwirth-Stallings pair with appropriate conditions determines an isolated singularity as above. Then they use an inductive process to show that for $p = 2$ one can always construct such a Neuwirth-Stallings pair, proving the existence of non-trivial examples for all $n > 1$.

The first explicit non-trivial example of a real analytic singularity with target $\mathbb{R}^2$ satisfying the Milnor condition at 0, other than those of Milnor, was given by A'Campo [2]. This is given by the map $\mathbb{C}^{m+2} \to \mathbb{C}$ defined by

$$(u, v, z_1, \ldots, z_m) \longmapsto uv(\bar{u} + \bar{v}) + z_1^2 + \cdots + z_m^2,$$

which is not holomorphic due to the presence of complex conjugation. In §3 below we construct, following [227, 214], infinite families of singularities satisfying Milnor's condition, which are in the same vein as that of A'Campo.

VII.2   The strong Milnor condition

The problem of studying real singularities satisfying the Milnor condition at 0, for which the map $\frac{f}{|f|}$ extends to all of $\mathbb{S}_\varepsilon - M \to \mathbb{S}^{p-1}$ as the projection of a fibre bundle (as in the case of holomorphic maps), was first studied in [111], and this turns out to be quite subtle. When $p = 2$ such singularities define open book decompositions on the spheres with the link M as binding; M is a fibred knot (or maybe a link), with the Milnor fibre $f^{-1}(0) \cap \mathbb{D}_\varepsilon$ being a Seifert surface for M.

We recall (Theorem 1.2 above) that given an analytic map $f : \mathbb{R}^n \to \mathbb{R}^2$ satisfying the Milnor condition at 0, for each small circle $C_\delta \subset \mathbb{R}^2$ around the origin, the restriction of f to $f^{-1}(C_\delta) \cap \mathbb{B}_\varepsilon$, where $\mathbb{B}_\varepsilon$ is a small closed disc around $0 \in \mathbb{R}^n$, is a fibre bundle over the circle with projection map f. Then using a vector field we may inflate the tube $f^{-1}(C_\delta) \cap \mathbb{B}_\varepsilon$ to make it become the complement (of a tubular neighborhood) of the link M in the sphere $\mathbb{S}_\varepsilon^{n-1}$ and in this way we have $\mathbb{S}_\varepsilon^{n-1} - M$ fibreing over $C_\delta \cong \mathbb{S}^1$. The point is that when we inflate the tube to carry it into the sphere we lose control, in general, of what map we end up having on the sphere. It is also hard to keep control of the behavior of the fibres as we get closer to the link. When n is even and f is holomorphic, then Milnor showed that you can keep control of this process, so that you end up having the map f as projection (or $f/|f|$ if we want the image to be the unit circle $\mathbb{S}^1 \subset \mathbb{C}$).

2.1 Definition. Let $f = (f_1, f_2) : (\mathbb{R}^n, 0) \to (\mathbb{R}^2, 0)$ be analytic and satisfy the Milnor condition at 0. Let M be the link; f satisfies *the strong Milnor condition at* 0 if for every sufficiently small sphere $\mathbb{S}_\varepsilon$ around 0, the map

$$\frac{f}{|f|} : \mathbb{S}_\varepsilon - M \to \mathbb{S}^1$$

is the projection of a fibre bundle.

The strong Milnor condition is not always satisfied by maps that satisfy the Milnor condition, as noticed by Milnor in [168, p. 99]. Jacquemard gave two conditions that were sufficient to guarantee that $\frac{f}{|f|}$ actually extends to all of $\mathbb{S}_\varepsilon - M$ as the projection map of a fibre bundle. The first condition (A) is geometric: that the angle between the gradient vector fields of these two functions be bounded; the second condition (B) is algebraic:

2.2 Theorem. (Jacquemard) *Let $f : (\mathbb{R}^n, 0) \to (\mathbb{R}^2, 0)$ be an analytic map-germ. If the component functions f_1 and f_2 of f satisfy the following two conditions, then f satisfies the strong Milnor condition. These conditions are:*

A) *there exists a neighborhood U of the origin in $\mathbb{R}^n$ and a real number $0 < \rho < 1$ such that for all $x \in U - 0$ one has:*

$$\frac{|\langle \operatorname{grad} f_1(x), \operatorname{grad} f_2(x) \rangle|}{\| \operatorname{grad} f_1(x) \| \cdot \| \operatorname{grad} f_2(x) \|} \leq 1 - \rho,$$

where $\langle \cdot, \cdot \rangle$ is the usual inner product in $\mathbb{R}^n$; and,

B) *if ε_n denotes the local ring of analytic map-germs at the origin in $\mathbb{R}^n$, then the integral closures in ε_n of the ideals generated by the partial derivatives*

$$\left(\frac{\partial f_1}{\partial x_1}, \frac{\partial f_1}{\partial x_2}, \dots, \frac{\partial f_1}{\partial x_n} \right) \quad and \quad \left(\frac{\partial f_2}{\partial x_1}, \frac{\partial f_2}{\partial x_2}, \dots, \frac{\partial f_2}{\partial x_n} \right) \qquad (2.3)$$

coincide.

Again, no explicit examples of such maps were given in [111] (other than complex polynomials with an isolated critical point).

It was noted in [214] that the above condition (B) can be relaxed and still have sufficient conditions to guarantee the strong Milnor condition. For this we recall the notion of the real integral closure of an ideal as given in [86].

2.4 Definition. Let I be an ideal in the ring ε_m. *The real integral closure* of I, denoted by $\overline{I}_{\mathbb{R}}$, is the set of $h \in \varepsilon_m$ such that for all analytic $\varphi : (\mathbb{R}, 0) \to (\mathbb{R}^m, 0)$, we have $h \circ \varphi \in (\varphi^*(I)) \varepsilon_1$.

Given $f : (\mathbb{R}^n, 0) \to (\mathbb{R}^2, 0)$ as above, let us set:

Condition $B_\mathbb{R}$: The real integral closures of the Jacobian ideals (2.3) coincide.

For complex analytic germs both conditions (B) and $(B_\mathbb{R})$ are equivalent (see [248], [86]). As pointed out in [214], essentially the same proof of Jacquemard in [111] gives:

2.5 Theorem. *Let $f : (\mathbb{R}^n, 0) \to (\mathbb{R}^2, 0)$ be an analytic map-germ that satisfies the Milnor condition. If its components f_1, f_2 satisfy the condition (A) of Theorem 2.2 and condition $(B_\mathbb{R})$, then f satisfies the strong Milnor condition.*

This improvement of Theorem 2.2 was used in [214] to prove a stability theorem for real singularities with the strong Milnor condition (see §6 below). This was also used in [218] to find a theorem in the spirit of Theorem 2.5 but using "regularity conditions" instead of Jacquemard's conditions. This is inspired in [226, 227] and related to VI.4 above. To explain this we need to introduce some concepts from [26, 27].

Let M be a smooth n-manifold, and let X, Y be submanifolds of M such that $Y \subset \overline{X}$.

2.6 Definition. The pair (X, Y) is (a)-regular at $y_0 \in Y$ if for each sequence of points $\{x_i\} \to y_0$ such that the corresponding sequence of tangent spaces $\{T_{x_i}X\}$ of X at x_i converges to some T in the Grassmannian of $(\dim X)$-planes in $\mathbb{R}^n$, then one has $T_{y_0}Y \subset T$. We say that (X, Y) is (a)-regular if it is (a)-regular at every y_0 in Y.

2.7 Definition. Let $\rho : M \to \mathbb{R}$ be a smooth non-negative function such that $\rho^{-1}(0) = Y$. The pair (X, Y) is C-regular at $y_0 \in Y$ with respect to the control function ρ if for each sequence of points $\{x_i\} \to y_0$ such that the sequence of planes $\{\ker D\rho(x_i) \cap T_{x_i}X\}$ converges to a plane T in the Grassmannian of $(\dim X - 1)$-planes in $\mathbb{R}^n$, then $T_{y_0}Y \subset T$. The pair (X, Y) is C-regular with respect to the control function ρ if it is C-regular for every point $y_0 \in Y$ with respect to the function ρ.

It is easy to see that C-regularity implies (a)-regularity.

Now consider a map-germ $f : (\mathbb{R}^n, 0) \to (\mathbb{R}^2, 0)$ as above, satisfying the Milnor condition at 0. Just as in the previous chapter, let $\pi_\theta : \mathbb{C} \to L_{-\theta}$ be the linear projection to the line $L_{-\theta}$ forming angle $-\theta$ with the horizontal axis in $\mathbb{C} = \mathbb{R}^2$, and consider the family $f_\theta = \pi_\theta \circ f$. The projections π_θ are all submersions, hence the $\{f_\theta\}$ are all real-valued maps with an isolated critical point at 0. For each $\theta \in [0, \pi)$, set $V_\theta^* = f_\theta^{-1}(0) - \{0\} \subset \mathbb{R}^n$. Now consider the map $F : (\mathbb{R}^n \times [0, \pi)) \to \mathbb{R} \times \mathbb{R}$ defined by $F(x, \theta) = f_\theta(x)$, and set $X_f = F^{-1}(0) - (0 \times [0, \pi))$, $Y_f = 0 \times [0, \pi)$. Thus $Y_f \subset \overline{X}_f$ and $X_f = \bigcup V_\theta^*$ for all $\theta \in [0, \pi)$.

2.8 Definition. The family $\{f_\theta\}$ is C-regular at $y_0 \in Y$ with respect to a control function ρ if the pair (X_f, Y_f) is C-regular at y_0 with respect to ρ.

The main result of [218] is:

2.9 Theorem. *Let $f : (\mathbb{R}^n, 0) \to (\mathbb{R}^2, 0)$ be a real analytic map-germ with an isolated critical point at the origin, such that f satisfies the Milnor condition at 0 and the family $f_\theta : (\mathbb{R}^n, 0) \to (\mathbb{R}, 0)$ satisfies the C-regularity condition with respect to the function $\rho(x, \theta) = \sum_i x_i^2$. Then, f satisfies the strong Milnor condition.*

Let us give an outline of their proof. First they notice (Lemma 2.6 in [214]) that one has a decomposition of the ambient space as given in VI.§4 above: there is a neighborhood $U \subset \mathbb{R}^n$ of 0 such that for every $x \in U - \{0\}$:

(i) $U = \cup_\theta (V_\theta \cap U)$, $0 \leq \theta < \pi$; $V_\theta = f_\theta^{-1}(0)$.

(ii) $V = \cap_\theta V_\theta = V_{\theta_1} \cap V_{\theta_2}$, where $V = f^{-1}(0)$, $\theta_1 \neq \theta_2$; $\theta_1, \theta_2 \in [0, \pi)$.

(iii) For each $\theta \in [0, \pi)$, $V_\theta = E_\theta \cup V \cup E_{\theta+\pi}$, where $E_\alpha = \widetilde{\phi}^{-1}(e^{i\alpha})$ with $\widetilde{\phi} :$ $U - V \to \mathbb{S}^1$, $\widetilde{\phi}(x) = i\frac{f(x)}{|f(x)|}$.

Furthermore, one also has:

(iv) If f satisfies the Milnor condition at 0, then for each $\theta \in [0, \pi)$, $V_\theta^* = V_\theta - \{0\}$ is a real smooth submanifold of real codimension 1 of $U - \{0\}$, given by the union of E_θ, $E_{\theta+\frac{\pi}{2}}$ and $V - \{0\}$.

Now, by [26, 255], the C-regularity condition of the family f_θ implies that there exists a family of germs of homeomorphisms $h_\theta : (\mathbb{R}^n, 0) \to (\mathbb{R}^n, 0)$, with $h_0 = Id$ and $\|h_\theta(x)\| = \|x\|$, $\forall x \in f_\theta^{-1}(0)$ (preserving small spheres centred at the origin), such that $f_\theta \circ h_\theta(x) = f_0(x)$. Hence the hypersurfaces V_θ are homeomorphic. It then follows from [26] that there exists a sufficiently small $\varepsilon > 0$ such that each V_θ meets the sphere $\mathbb{S}_\varepsilon^{n-1}$ transversally for all θ. Each E_θ is a submanifold of U, and $V_\theta \cap \mathbb{S}_\varepsilon^{n-1} = (E_\theta \cap \mathbb{S}_\varepsilon^{n-1}) \cup (V \cap \mathbb{S}_\varepsilon^{n-1}) \cup (E_{\theta+\frac{\pi}{2}} \cap \mathbb{S}_\varepsilon^{n-1})$. One can thus get a decomposition of the sphere $\mathbb{S}^{n-1}$ satisfying properties similar to (i)–(iv) above. Let $F_\theta := E_\theta \cap \mathbb{S}_\varepsilon^{n-1}$. It follows from (iii) that letting $M_\varepsilon = V \cap \mathbb{S}_\varepsilon$ one has:

$$V_\theta \cap \mathbb{S}_\varepsilon^{n-1} = F_\theta \cup M_\varepsilon \cup F_{\theta+\pi} .$$

For each θ_1, θ_2 one has $(f_{\theta_1} \circ h_{\theta_1})^{-1}(0) = f_0^{-1}(0) = (f_{\theta_2} \circ h_{\theta_2})^{-1}(0)$. Then, $h_{\theta_1}^{-1}(V_{\theta_1}) = h_{\theta_2}^{-1}(V_{\theta_2})$. Hence $h_{\theta_2} \circ h_{\theta_1}^{-1}(V_{\theta_1} \cap \mathbb{S}_\varepsilon^{n-1}) = V_{\theta_2} \cap \mathbb{S}_\varepsilon^{n-1}$.

Using this one can define an $\mathbb{R}$-action on the sphere, $\Gamma_\alpha : \mathbb{R} \times \mathbb{S}_\varepsilon^{n-1} \to \mathbb{S}_\varepsilon^{n-1}$, by $x \longmapsto h_\theta \circ h_\alpha^{-1}(x)$ for $x \in M_\alpha$. Since $\|h_\alpha(x)\| = \|x\|$ one has that Γ_α is well defined and for all $\alpha \in \mathbb{R}$ and $x \in F_\alpha$ one has:

$$h_\theta \circ h_\alpha^{-1}(F_\alpha) = h_\theta \circ h_\alpha^{-1}(E_\alpha \cap \mathbb{S}_\varepsilon^{n-1}) = E_\theta \cap \mathbb{S}_\varepsilon^{n-1} = F_\theta .$$

Moreover, the action Γ_α is transversal to the fibres and satisfies

$$\Gamma_\alpha(K_\varepsilon) = \Gamma_\alpha(M \cap \mathbb{S}_\varepsilon^{n-1}) = \Gamma_\alpha(M) \cap \mathbb{S}_\varepsilon^{n-1} = M \cap \mathbb{S}_\varepsilon^{n-1} = K_\varepsilon .$$

Hence it leaves the link invariant. Now, proceeding as in [168] Theorem 4.8, we get that $\frac{f}{|f|}$ is the projection map of a locally trivial fibre bundle. $\qquad \square$

As a corollary they obtain a generalization of Theorem 4.0.2 of [214]. For this we recall (c.f. III.1 above) that a polynomial h in m real variables is *weighted homogeneous* if there exist non-zero integers $(q_1, \ldots, q_m)$ and a positive integer d, such that:

$$h(t^{q_1} x_1, \ldots, t^{q_n} x_m) = t^d h(x_1, \ldots, x_m) \, .$$

Equivalently, we demand that there exist non-zero rational numbers $(w_1, \ldots, w_m)$, called *the weights* of h, for which h is a sum of monomials $z_1^{i_1} \cdots z_n^{i_m}$ such that

$$\frac{i_1}{w_1} + \cdots + \frac{i_n}{w_m} = 1 \, .$$

A singular variety V (or a map-germ) is said to be *quasi-homogeneous* if it is defined by polynomials in m real variables which are *weighted homogeneous* of the same weights.

2.10 Corollary. *Let $f : (\mathbb{R}^n, 0) \to (\mathbb{R}^2, 0)$ be a quasi-homogeneous polynomial map-germ. Then f satisfies the Milnor condition at 0 iff it satisfies the strong Milnor condition at 0.*

This follows from Theorem 2.9 since it is well known that in this case the family f_θ is C-regular (see [26]).

2.11 Remarks.

(i) It is worth saying that the proof of Theorem 2.9 is similar in spirit to that of Theorem 4.2 below, which is more elementary since in that case the $\mathbb{R}$-action that gives the local triviality of the fibres can be given explicitly, so one does not need to use Bekka's regularity. This has the advantage that one gets for free the monodromy map of the corresponding Milnor fibrations.

(ii) M.A. Ruas and R.N.A. Santos in [218, Theorems 4.4 and 4.8] prove that Theorem 2.9 is stronger than Theorem 2.5 and 2.2: if a map-germ satisfies Jacquemard's conditions (A) and (B), or $(B_\mathbb{R})$, then it satisfies C-regularity, but not conversely.

VII.3   Real singularities of the Pham-Brieskorn type

We recall that the Pham-Brieskorn singularities are defined by

$$\{ z \in \mathbb{C}^{n+1} \, | \, z_1^{a_1} + \cdots + z_n^{a_n} = 0 \, ; \, a_i \geq 2 \; \forall \, i = 1, \ldots, n \} \, ,$$

and one may consider more generally the variety defined by the complex polynomial

$$f(z) = \lambda_1 z_1^{a_1} + \cdots + \lambda_n z_n^{a_n} \, ,$$

where the λ_i are non-zero complex numbers, which is essentially the same. We have seen in Chapter I of this book, and in almost every other chapter, several remarkable geometric and topological features of these singularities. Let us consider now the following real analytic analogues of them:

$$\psi(z) = \lambda_1 z_1^{a_1} \overline{z}_1 + \cdots + \lambda_n z_n^{a_n} \overline{z}_n \ ,$$

or more generally,

$$\psi(z) = \lambda_1 z_1^{a_1} \overline{z}_{\sigma_1} + \cdots + \lambda_n z_n^{a_n} \overline{z}_{\sigma_n} \ ,$$

where $\sigma = \{\sigma_1, \ldots, \sigma_n\}$ is a permutation of the set $\{1, \ldots, n\}$.

3.1 Definition. A *twisted Pham-Brieskorn* real singularity of class $\{a_1, \ldots, a_n; \sigma\}$ is a singularity in $\mathbb{C}^n$ defined by a polynomial map

$$\psi(z) = \lambda_1 z_1^{a_1} \overline{z}_{\sigma_1} + \cdots + \lambda_n z_n^{a_n} \overline{z}_{\sigma_n} \ ,$$

where each $a_i \geq 2$, $i = 1, \ldots, n$, the λ_i are non-zero complex numbers and $\sigma = \{\sigma_1, \ldots, \sigma_n\}$ is a permutation of the set $\{1, \ldots, n\}$. We call *weights* the exponents $\{a_1, \ldots, a_n\}$, and the permutation σ is *the twisting*.

So for instance, if $n = 2$ then the maps in question are of two classes:

$$\lambda_1 z_1^{a_1} \overline{z}_1 + \lambda_2 z_2^{a_2} \overline{z}_2 \qquad \text{or} \qquad \lambda_1 z_1^{a_1} \overline{z}_2 + \lambda_2 z_2^{a_2} \overline{z}_1 \ .$$

For $n = 3$, one has three essentially different classes:

$$\lambda_1 z_1^{a_1} \overline{z}_1 + \lambda_2 z_2^{a_2} \overline{z}_2 + \lambda_3 z_3^{a_3} \overline{z}_3 \ ; \quad \lambda_1 z_1^{a_1} \overline{z}_2 + \lambda_2 z_2^{a_2} \overline{z}_1 + \lambda_3 z_3^{a_3} \overline{z}_3 \quad \text{or}$$

$$\lambda_1 z_1^{a_1} \overline{z}_2 + \lambda_2 z_2^{a_2} \overline{z}_3 + \lambda_3 z_3^{a_3} \overline{z}_1 \ ,$$

and so on.

In Section 5 below we prove that if the twisting σ is the identity, then the corresponding singularity is topologically equivalent to the Pham-Brieskorn singularity $z_1^{a_1-1} + \cdots + z_n^{a_n-1} = 0$, where the a_i are the weights.

In this section we prove that for all twisted Pham-Brieskorn singularities, the corresponding variety $V = \psi^{-1}(0)$ is a codimension 2 complete intersection in $\mathbb{C}^n$ with a unique singularity at 0, and the corresponding maps satisfy the Milnor condition at 0. In the following section we prove that they further satisfy the strong Milnor condition at 0.

We observe that if we let F, X be holomorphic vector fields of the form:

$$F = (k_1 z_{\sigma_1}^{a_1}, \ldots, k_n z_{\sigma_n}^{a_n}) \ , \quad X = (t_1 z_1^{b_1}, \ldots, t_n z_n^{b_n}) \ ,$$

where the k_i and the t_i are non-zero complex numbers, the a_i and b_i are positive integers and $\{\sigma_1, \ldots, \sigma_n\}$ is a permutation of the set $\{1, \ldots, n\}$, and if we let

$$\psi_{F,X}(z) : \mathbb{C}^n \to \mathbb{C}$$

be the real analytic function defined (as in Chapter VI) by

$$\psi_{F,X}(z) = \langle F(z), X(z) \rangle = \sum_{i=1}^{n} k_i \cdot \overline{t}_i \cdot z_{\sigma_i}^{a_i} \cdot \overline{z}_i^{b_i} \ ,$$

then the twisted Pham-Brieskorn singularities correspond to the case when the exponents b_i are all 1.

The purpose of this section is to classify the vector fields F and X as above, for which the corresponding map $\psi_{F,X}$ satisfies the Milnor condition at $0 \in \mathbb{C}^n \cong \mathbb{R}^{2n}$. This has the advantage of giving us more singularities that satisfy the Milnor condition, other than the twisted Pham-Brieskorn singularities. For this we think of $\psi := \psi_{F,X}$ as a map into $\mathbb{R}^2$ and we let (ψ_1, ψ_2) be its components, which correspond to the real and imaginary parts of ψ. We want to determine the rank of its derivative. Suppose first that F is of the form $F(z) = (k_1 z_1^{a_1}, \ldots, k_n z_n^{a_n})$ and $X = (z_1^{b_1}, \ldots, z_n^{b_n})$, i.e., the permutation σ is the identity. In this case one has:

$$\psi_1 = \frac{1}{2} \sum_{i=1}^{n} (k_i z_i^{a_i} \bar{z}_i^{b_i} + \bar{k}_i \bar{z}_i^{a_i} z_i^{b_i}) \ , \ \ \psi_2 = \frac{1}{2} \sum_{i=1}^{n} (k_i z_i^{a_i} \bar{z}_i^{b_i} - \bar{k}_i \bar{z}_i^{a_i} z_i^{b_i}).$$

Let us consider the 2×2 minor of ψ corresponding to the partial derivatives of (ψ_1, ψ_2) with respect to z_i and $\bar{z}_i$. Its determinant (multiplied by -2) is:

$$||k_i||^2 a_i^2 ||z_i||^{2a_i - 2} ||z_i||^{2b_i} - ||k_i||^2 b_i^2 ||z_i||^{2a_i} ||z_i||^{2b_i - 2},$$

so this is 0 iff $z_i = 0$ or $a_i = b_i$. Hence, if $z_i \neq 0$ and $a_i \neq b_i$ for all i, then $\psi_{F,X}$ is a submersion. Conversely, if $a_i = b_i$ for some i, then the rank of $\psi(z)$ is less than 2 whenever we are on the z_i-axis, proving that in this case the function $\psi_{F,X}$ does not satisfy the Milnor condition at 0.

Given F and X as above, let us set: $\widehat{F} = (k_1 \bar{t}_1 z_{\sigma_1}^{a_1}, \ldots, k_n \bar{t}_n z_{\sigma_n}^{a_n})$, and $\widehat{X} = (z_1^{b_1}, \ldots, z_n^{b_n})$, then $k_i \bar{t}_i$ is in $\mathbb{C}^*$ for each i, and one has:

$$\psi_{F,X} = \psi_{\widehat{F},\widehat{X}}.$$

Thus we can assume, with no loss of generality, that each t_i is 1.

Now, given $F = (k_1 z_{\sigma_1}^{a_1}, \ldots, k_n z_{\sigma_n}^{a_n})$, where σ is a permutation of $\{1, \ldots, n\}$, we can split σ into cycles of length $r \geq 1$. Consider such a cycle and re-label the components so that the corresponding components of F are $(k_1 z_2^{a_1}, \ldots, k_{r-1} z_r^{a_{r-1}}, k_r z_1^{a_r})$. In this case ψ is of the form:

$$\psi(z) = (k_1 z_2^{a_1} \bar{z}_1^{b_1} + \cdots + k_r z_1^{a_r} \bar{z}_r^{b_r}) + (\text{terms in other variables}).$$

Hence, to determine whether or not ψ satisfies the Milnor condition at 0, we only need to look at the minors of the Jacobian matrix determined by the various cycles of σ. In other words, it is enough to consider the case when F is a vector field in $\mathbb{C}^r$ of the form $F(z) = (k_1 z_2^{a_1}, k_2 z_3^{a_2}, \ldots, k_r z_1^{a_r})$, and $X = (z_1^{b_1}, \ldots, z_r^{b_r})$. The case $r = 1$ was discussed above, so we assume $r > 1$. In this case the determinant of the minor of ψ corresponding to the partial derivatives of (ψ_1, ψ_2) with respect to $z_i, \bar{z}_i$, multiplied by -2, is:

$$||k_{i-1}||^2 a_{i-1}^2 ||z_i||^{2a_{i-1} - 2} ||z_{i-1}||^{2b_{i-1}} - ||k_i||^2 b_i^2 ||z_{i+1}||^{2a_i} ||z_i||^{2b_i - 2}.$$

If the rank of $D\psi$ at a point $z = (z_1, \ldots, z_r)$ is less than 2, then all these determinants must vanish; making $\lambda_i = ||k_i||^2$ and $\nu_i = ||z_i||^2$ for all $i = 1, \ldots, r$, we get the system of equations:

$$
\begin{aligned}
\textbf{(3.2)} \qquad \lambda_1 a_1^2 \nu_2^{a_1-1} \nu_1^{b_1} &= \lambda_2 b_2^2 \nu_3^{a_2} \nu_2^{b_2-1}, \\
\lambda_2 a_2^2 \nu_3^{a_2-1} \nu_2^{b_2} &= \lambda_3 b_3^2 \nu_4^{a_3} \nu_3^{b_3-1}, \\
&\;\;\vdots \\
\lambda_r a_r^2 \nu_1^{a_r-1} \nu_r^{b_r} &= \lambda_1 b_1^2 \nu_2^{a_1} \nu_1^{b_1-1}.
\end{aligned}
$$

Suppose now that each a_i is greater than 1, and also that one b_i is greater than 1, say $b_j > 1$. Let us assume first that $r \geq 5$. Considering the indices as integers modulo r, there are exactly three equations that involve ν_{j+1}, these are:

$$
\lambda_{i-1} a_{i-1}^2 \nu_i^{a_{i-1}-1} \nu_{i-1}^{b_{i-1}} = \lambda_i b_i^2 \nu_{i+1}^{a_i} \nu_i^{b_i-1},
$$

for $i = j, j+1, j+2$. It is then clear that the rank of $D\psi$ is less than 2 at all points in $\mathbb{C}^r$ satisfying $z_{j-1} = z_j = z_{j+2} = z_{j+3} = 0$. Hence, if some b_i is more than 1, then ψ is not a submersion on a punctured neighborhood of 0. The cases $r = 2, 3, 4$ can be easily worked out case by case, using the same technique as above. Conversely, we claim that ψ is in fact a submersion near 0 when $b_i = 1$ for all i (in fact this statement also holds if each b_i is either 0 or 1). This is proved in [227] and it is a straightforward computation using the above system of equations (3.2): first we note that if $(\nu_1, \ldots, \nu_r)$ satisfies the system (3.2) and some $\nu_i = 0$, then $\nu_i = 0$ for all i, so the solution is trivial. Thus we can assume $\nu_i > 0$ for all i. Multiplying the left-hand side and the right-hand side of all the equations in (3.2) we get:

$$
\prod_{i=1}^{n} [\lambda_i a_i^2 \nu_i^{a_i-1}] = \prod_{i=1}^{n} [\lambda_i \nu_i^{a_i-1}].
$$

This implies $a_i = 1$ for all i, because λ_i and ν_i are both non-zero for all i, which is a contradiction. Thus we arrive at Theorem 3.3 below; we recall that F, X are vector fields of the form:

$$
F = (k_1 z_{\sigma_1}^{a_1}, \ldots, k_n z_{\sigma_n}^{a_n}), \quad X = (t_1 z_1^{b_1}, \ldots, t_n z_n^{b_n}),
$$

where the k_i and the t_i are non-zero complex numbers, all a_i and b_i are positive integers, $\{\sigma_1, \ldots, \sigma_n\}$ is some permutation of the set $\{1, \ldots, n\}$ and $\psi_{F,X}(z) = \langle F(z), X(z) \rangle$.

3.3 Theorem. *Assume $a_i > 1$ for all $i = 1, \ldots, n$. Then $\psi_{F,X}$ satisfies the Milnor condition at $0 \in \mathbb{C}^n$ if and only if either one of the following two conditions is satisfied:*

(i) *The permutation σ is the identity, so that $F(z) = (k_1 z_1^{a_1}, \ldots, k_n z_n^{a_n})$, and one has $a_i \neq b_i$ for each $i = 1, \ldots, n$.*

(ii) *Split σ into cycles $\sigma_1, \ldots, \sigma_m$, where each σ_j is a permutation of length o_j. Then: for each $j = 1, \ldots, m$ such that $o_j = 1$, one must have that the corresponding exponents satisfy $a_i \neq b_i$; for each $j = 1, \ldots, m$ such that $o_j > 1$, the exponents b_i corresponding to these components are all 1.*

3.4 Remark. We notice that in the cases above, whenever $\psi_{F,X}$ is a submersion away from 0, one actually has that $\psi_{F,X}$ is a submersion on all of $\mathbb{C}^n - \{0\}$. This is not always the case. For instance, if $F = (k_1 z_2^a, k_2 z_1)$, $a > 1$, and $X = (z_1, z_2)$, then $\psi_{F,X}$ is a submersion near $0 \in \mathbb{C}^2$, but the rank of $D\psi$ drops at every point $(0, z_2)$ such that $||z_2||^{a-1} = ||k_2||/||k_1||$. As we will see, this is related to the fact that the singularities in Theorem 3.3 are all quasi-homogeneous.

Let us motivate the following theorem with an example:

3.5 Example. Let $F(z) = (k_1 z_2^{a_1}, k_2 z_3^{a_2}, k_3 z_4, k_4 z_1)$, $a_1 \geq 1$, $a_2 \geq 1$, and $X = (z_1, z_2, z_3, z_4)$. The system of equations (3.2) becomes

$$\lambda_1 a_1^2 \nu_2^{a_1 - 1} \nu_1 = \lambda_2 \nu_3^{a_2} \ ; \ \lambda_2 a_2^2 \nu_3^{a_2 - 1} \nu_2 = \lambda_3 \nu_4 \ ; \ \lambda_3 \nu_3 = \lambda_4 \nu_1 \ ; \ \lambda_4 \nu_4 = \lambda_1 \nu_2^{a_1} ,$$

where $\nu_i = ||z_i||^2$. If $a_2 = 1$, then every $z \in \mathbb{C}^4$ of the form $z = (0, z_2, 0, z_4)$, with

$$\nu_4 = \frac{\lambda_2}{\lambda_3} \nu_2 \ , \ \nu_2^{a_1 - 1} = \frac{\lambda_2 \lambda_4}{\lambda_1 \lambda_3} ,$$

is a point where $\psi_{F,X}$ is not a submersion. If $a_1 = 1$, this implies $|k_1 k_3| = |k_2 k_4|$ or else z is the origin.

3.6 Theorem. *We now let F be as above, but we allow the a_i to be ≥ 1, and $X = (z_1, \ldots, z_n)$. Then the function $\psi_{F,X}$ satisfies the Milnor condition at 0 iff it satisfies the following conditions:*

(i) *For each permutation of σ of odd length, there is at least one a_i strictly bigger than 1.*

(ii) *For each permutation of σ of even length, either there is at least one a_i strictly bigger than 1, or else the corresponding coefficients $k_i, \ldots, k_{i+2m+1}$ satisfy:*

$$||k_i \, k_{i+2} \ \cdots \ k_{i+2m}|| \neq ||k_{i+1} \, k_{i+3} \ \cdots \ k_{i+2m+1}|| .$$

Proof. As in the proof of Theorem 3.3 above, we can assume σ has only one cycle. In this case the system of equations (3.2) becomes:

$$\textbf{(3.7)} \qquad \begin{aligned} \lambda_1 a_1^2 \nu_2^{a_1 - 1} \nu_1 &= \lambda_2 \nu_3^{a_2} , \\ \lambda_2 a_2^2 \nu_3^{a_2 - 1} \nu_2 &= \lambda_3 \nu_4^{a_3} , \\ &\vdots \\ \lambda_{n-1} a_{n-1}^2 \nu_n^{a_{n-1} - 1} \nu_{n-1} &= \lambda_n \nu_1^{a_n} , \\ \lambda_n a_n^2 \nu_1^{a_n - 1} \nu_n &= \lambda_1 \nu_2^{a_1} . \end{aligned}$$

If n is odd, one has that if $\nu_i = 0$ for some i, then $\nu_i = 0$ for all i, so the solution is trivial. Hence we can assume ν_i is non-zero for all i. Then, as before, multiplying the left-hand side of these equations, and the right-hand side, we obtain:

$$\prod_{i=1}^{n}[\lambda_i a_i^2 \nu_i^{a_i-1}] = \prod_{i=1}^{n}[\lambda_i \nu_i^{a_i-1}].$$

Therefore $a_i = 1$ for all i, or else $\nu_i = 0$ for all i and we have the trivial solution. Hence, if we assume that at least one a_i is greater than 1, then $\psi_{F,X}$ is a submersion on $\mathbb{C}^n - 0$. Conversely, if $n = 2m+1$ and $a_i = 1$ for all i, then one can easily verify that the system of equations (3.7) has non-trivial solutions given by:

$$\nu_{2m+1} = \frac{\lambda_{2m-1}}{\lambda_{2m}}\nu_{2m-1} = \cdots = \frac{\lambda_{2m-1}\lambda_{2m-3}\ldots\lambda_3\lambda_1}{\lambda_{2m}\lambda_{2m-2}\ldots\lambda_2}\nu_1 ;$$

$$\nu_{2m} = \frac{\lambda_{2m-2}}{\lambda_{2m-1}}\nu_{2m-2} = \cdots = \frac{\lambda_{2m-2}\lambda_{2m-4}\ldots\lambda_2}{\lambda_{2m-1}\lambda_{2m-3}\ldots\lambda_3}\nu_2 ;$$

$$\nu_2 = \frac{\lambda_{2m+1}\lambda_{2m-1}\ldots\lambda_3}{\lambda_{2m}\lambda_{2m-2}\ldots\lambda_2}\nu_1 ;$$

where $\nu_1 = ||z_1||^2$ can be taken as near to 0 as we please, proving statement (i) in the theorem.

Assume now $n = 2m$ is even and at least one $a_i > 1$; for simplicity we assume $a_1 > 1$. It is easy to see that if one ν_{even} is 0 then ν_i is 0 for all i. However, if a ν_{odd} is zero, this only implies that all the ν_{odd} are 0. Hence, as in the example above, one has in this case that a point $z \in \mathbb{C}^n$ of the form $z = (0, z_1, 0, z_2, \ldots, 0, z_{2m})$ is a solution of (3.7) iff it satisfies:

$$\nu_{2m} = \frac{\lambda_1}{\lambda_{2m}}\nu_2^{a_1} ; \quad \nu_{2m-2} = \frac{\lambda_{2m-1}}{\lambda_{2m-2}}\nu_{2m}^{a_{2m-1}} , \ldots, \quad \nu_2 = \frac{\lambda_3}{\lambda_2}\nu_4^{a_3} ,$$

and

$$\nu_2^{a_1 a_3 \ldots a_{2m-1}-1} = \frac{1}{\frac{\lambda_3}{\lambda_2}\left(\frac{\lambda_5}{\lambda_4}\right)^{a_3}\left(\frac{\lambda_7}{\lambda_6}\right)^{a_3 a_5}\ldots\left(\frac{\lambda_1}{\lambda_n}\right)^{a_3 a_5 \ldots a_{n-1}}} .$$

Hence every non-trivial solution is at a bounded distance away from $0 \in \mathbb{C}^n$, thus $\psi_{F,X}$ is a submersion on a punctured neighborhood of 0. Finally assume $n = 2m$ and $a_i = 1$ for all i. Then the system of equations (3.7) splits into the two independent systems of m-equations given by: $\lambda_1\nu_1 = \lambda_2\nu_3$, $\lambda_3\nu_3 = \lambda_4\nu_5$, $\ldots$, and $\lambda_2\nu_2 = \lambda_3\nu_4$, $\lambda_4\nu_4 = \lambda_5\nu_6$, $\ldots$; in either case, the corresponding system of equations has a non-trivial solution iff:

$$||\lambda_i \lambda_{i+2} \ldots \lambda_{i+2m}|| = ||\lambda_{i+1} \lambda_{i+3} \ldots \lambda_{i+2m+1}||,$$

proving the theorem. $\square$

VII.4 Twisted Pham-Brieskorn singularities and the strong Milnor condition

In this section we prove that all the twisted Pham-Brieskorn singularities satisfy the strong Milnor condition. We also prove that they are all quasi-homogeneous singularities (which is not obvious). Hence the fact that they satisfy the strong Milnor condition is a consequence of Corollary 2.10 and Theorem 3.3 above. However we prove this fact directly, using only Theorem 3.3 and not Corollary 2.10, because this is almost as hard as proving that they are quasi-homogeneous (which we must prove anyhow), and it throws light into the geometry of these singularities, since we obtain the monodromy map of the corresponding Milnor fibrations.

More generally, in this section we classify the vector fields F and X as above,

$$F = (k_1 z_{\sigma_1}^{a_1}, \ldots, k_n z_{\sigma_n}^{a_n}) \quad \text{and} \quad X = (t_1 z_1^{b_1}, \ldots, t_n z_n^{b_n}),$$

such that the function $\psi := \psi_{F,X}$ satisfies the strong Milnor condition at 0.

For each $\theta \in \mathbb{R}$, define a map $\psi_\theta : \mathbb{C}^n \to \mathbb{R}$ by:

$$\psi_\theta(z) = \operatorname{Re}\langle e^{i\theta} F(z), X(z)\rangle = \frac{1}{2} \sum_{i=1}^n (e^{i\theta} F_i(z)\overline{X}_i(z) + e^{-i\theta}\overline{F}_i(z)X_i(z)),$$

and set $V_\theta := \psi_\theta^{-1}(0)$. It is clear that $V_\theta = V_{\theta+\pi r}$ for all $r \in \mathbb{Z}$. One also has $\psi_\theta(z) = \operatorname{Re} e^{i\theta}\psi(z)$, so that ψ_θ is ψ followed by the projection from $\mathbb{C}$ onto the line $\mathcal{L}_{-\theta}$ that passes through the origin in $\mathbb{C}$ with an inclination of $-\theta$ radians. Thus ψ_θ is a surjection whenever ψ is a surjection.

We recall that, just as in VI.4 and in the proof of 2.8, one has a decomposition of $\mathbb{C}^n$ in the V_θ satisfying:

(i) $\mathbb{C}^n$ is the union of all V_θ, $\theta \in [0, \pi)$.

(ii) $V = \cap_{\theta \in [0,\pi)} V_\theta = V_{\theta_1} \cap V_{\theta_2}$, for any distinct $\theta_1 \theta_2 \in [0, \pi)$.

(iii) For each $\theta \in [0, \pi)$, one has

$$V_\theta = E_\theta \cup V \cup E_{\theta+\pi},$$

where $E_\alpha = \widetilde{\phi}^{-1}(e^{i\alpha})$ and $\widetilde{\phi} : \mathbb{C}^n - V \to \mathbb{S}^1$ is the function

$$\widetilde{\phi}(z) = \arg(i\overline{\langle F(z), z\rangle}).$$

(iv) If ψ satisfies the Milnor condition at 0, then each $V_\theta^* := V_\theta - \{0\}$ is a smooth real submanifold of $\mathbb{C}^n$ of codimension 1.

For every sufficiently small sphere $\mathbb{S}_\varepsilon$ in $\mathbb{C}^n$ with centre at 0, we let $M_\varepsilon = V \cap \mathbb{S}_\varepsilon$ be the link of V and we set

$$\phi_\varepsilon = \phi = \frac{\psi}{|\psi|} : \mathbb{S}_\varepsilon^{2n-1} - M_\varepsilon \to \mathbb{S}^1. \tag{4.1}$$

4.2 Theorem. *Let* $F = (k_1 z_{\sigma_1}^{a_1}, \ldots, k_n z_{\sigma_n}^{a_n})$ *and* $X = (t_1 z_1^{b_1}, \ldots, t_n z_n^{b_n})$ *be vector fields as above, and assume the following conditions are satisfied:*

(i) *for each cycle of* σ *of length* 1, *the corresponding exponents satisfy* $a_i \neq b_i$;

(ii) *for each cycle of length* $r > 1$, r *odd, one has* $a_i \geq 1$, $b_i = 1$ *and at least one* a_i *is strictly bigger than* 1, *for each* $i = 1, \ldots, r$.

(iii) *for each cycle of* σ *of length* $r > 1$, r *even, for each* $i = 1, \ldots, r$, *one has* $a_i \geq 1$, $b_i = 1$ *and at least two* a_i, *say* a_{i_1} *and* a_{i_2}, *are strictly bigger than* 1, *with* i_1 *being odd and* i_2 *even in the cycle.*

Then $\psi_{F,X}$ *satisfies the strong Milnor condition at* 0. *That is, for every sufficiently small sphere* $\mathbb{S}_\varepsilon^{2n-1}$, (4.1) *is a* C^∞ *fibre bundle. Each pair of antipodal fibres* $\phi^{-1}(e^{i\theta})$, $\phi^{-1}(e^{i(\theta+\pi)})$, *are glued together along* M_ε *forming the smooth* $(2n-1)$-*manifold:*

$$V_\theta \cap \mathbb{S}_\varepsilon^{2n-1} = \{ z \in \mathbb{S}_\varepsilon^{2n-1} \subset \mathbb{C}^n \mid \operatorname{Re} \langle e^{i\theta} F(z), z \rangle = 0 \},$$

and the monodromy of this bundle is the first return map of the $\mathbb{S}^1$-*action in Lemma 4.3 below.*

Notice that the twisted Pham-Brieskorn singularities satisfy the hypothesis of Theorem 4.2 and therefore the statements of this theorem hold for them.

From now on we assume F, X are vector fields satisfying the conditions of Theorem 4.2, so we know already, by Section 3, that $\psi_{F,X}$ satisfies the Milnor condition at 0. Theorem 4.2 will be a consequence of the previous results in §3 and the following two lemmas:

4.3 Lemma. *There exists a norm-preserving, smooth action* γ_λ *of* $\mathbb{S}^1$ *on* $\mathbb{C}^n$, *permuting the* V_θ. *More precisely, if we think of this action as an* $\mathbb{R}$-*action, via the identification* $\mathbb{S}^1 \cong \mathbb{R} \bmod (\pi)$, *then for every* $\lambda \in \mathbb{R}$, γ_λ *carries* V_θ *into* $V_{[\theta+\lambda]}$, *where* $[\theta+\lambda]$ *means the residue class of* $(\theta+\lambda)$ *modulo* π. *Hence,* V *is an invariant set for this action.*

This lemma is proved by constructing explicitly the $\mathbb{S}^1$-action. Before we do this, let us observe that Lemma 4.3 implies that the V_θ^* of Theorem 4.2 are all diffeomorphic and the map $\widetilde{\phi} : \mathbb{C}^n - V \to \mathbb{S}^1$ is the projection map of a locally trivial C^∞-fibre bundle. It is clear that $\widetilde{\phi}(z)$ is ϕ followed by the diffeomorphism of $\widetilde{\mathbb{C}}$ given by $z \mapsto iz$. Therefore one has that $\phi : \mathbb{C}^n - V \to \mathbb{S}^1$ is also a locally trivial fibre bundle. To prove Theorem 4.2 we shall prove that the restriction of ϕ to every sphere around $0 \in \mathbb{C}^n$, is also the projection map of a locally trivial fibre bundle (notice this is more than we need: it would be sufficient to prove this claim for small spheres). This is an immediate consequence of the previous discussion and the following lemma:

4.4 Lemma. *With the above hypotheses, there exists a flow* $\{f_t\}$ *on* $\mathbb{C}^n$, *whose orbits are transversal to every sphere around* $0 \in \mathbb{C}^n$, *except for* 0 *itself, which is a fixed point, and they converge to* $0 \in \mathbb{C}^n$ *when the time tends to* $-\infty$. *This flow leaves invariant each* V_θ, *thus it also leaves* V *invariant.*

Of course Lemma 4.4 implies that each V_θ is embedded (globally) as a cone in $\mathbb{C}^n$, intersecting transversally each sphere around $0 \in \mathbb{C}^n$, and so does V.

Before proving these lemmas, let us state a result which is a corollary of the proof of Lemma 4.4.

4.5 Corollary. *If F and X satisfy the conditions of Theorem 4.2, then the polynomial in $\mathbb{R}^{2n} \cong \mathbb{C}^n$ defined by $\psi_{F,X}$, is weighted homogeneous.*

To prove Lemmas 4.3 and 4.4, let us split $\mathbb{C}^n$ into direct summands, according to the cycles of σ, and construct the corresponding flows on each of these direct summands.

Proof of Lemma 4.4. We shall construct an action $\Gamma : \mathbb{R}^+ \times \mathbb{C}^n \to \mathbb{C}^n$ of the form $\Gamma(t,(z_1,\ldots,z_n)) = (t^{m_1} z_1,\ldots,t^{m_n} z_n)$, $m_i \in \mathbb{Q}$, $m_i > 0$, such that $\psi(\Gamma(t,z)) = t\psi(z)$. It is clear that this implies Lemma 4.4 and Corollary 4.5. Consider first a 1-cycle (if there is any). This corresponds to a monomial in $\psi := \psi_{F,X}$ of the form $k_i z_i^{a_i} \bar{z}_i^{b_i}$. Define $w := \Gamma(t, z_i) = (t^{\frac{1}{(a_i+b_i)}} z_i)$. Then $k_i w_i^{a_i} \bar{w}_i^{b_i} = t(k_i z_i^{a_i} \bar{z}_i^{b_i})$, as we wanted.

Now consider an r-cycle, $r > 1$, and let $\psi_{(r)}$ be the polynomial consisting of the monomials in ψ containing the variables in this cycle. We re-label the components so that this cycle is $(k_1 z_2^{a_1}, \ldots, k_{r-1} z_r^{a_{r-1}}, k_r z_1^{a_r})$. Let B be the $r \times r$-matrix,

$$\begin{pmatrix} a_2 & 1 & 0 & 0 & \cdots & \cdots & 0 \\ 0 & a_3 & 1 & 0 & \cdots & \cdots & 0 \\ \vdots & \vdots & \vdots & \vdots & & & \vdots \\ \vdots & \vdots & \vdots & \vdots & & & \vdots \\ 0 & 0 & 0 & 0 & \cdots & a_r & 1 \\ 1 & 0 & 0 & 0 & \cdots & 0 & a_1 \end{pmatrix}.$$

Observe that its determinant is $a_1 \ldots a_r + (-1)^{r+1} \neq 0$, since $a_i > 0$ for all i and at least one of them is > 1. Define $\Gamma(t,(z_1,\ldots,z_r)) = (t^{m_1} z_1,\ldots,t^{m_r} z_r)$, where $m := (m_1,\ldots,m_r)$ is the unique solution to the linear system $B \cdot m = (1,\ldots,1)$. These are rational numbers because B has integer coefficients. Then one has $\psi_{(r)}(\Gamma(t,(z_1,\ldots,z_r))) = t\psi_{(r)}(z_1,\ldots,z_r)$ as wanted. It remains to prove that the solutions of this flow are transversal to all the spheres centred at 0 and they converge to 0 when the time goes to $-\infty$. This is a consequence of the following claim: $m_i > 0$ for every $i = 1,\ldots,r$. To prove this claim we first compute the m_i explicitly. We find:

$$m_1 = \frac{1 - a_1 + a_1 a_r - a_1 a_r a_{r-1} + \cdots + (-1)^{r-1} a_1 a_r a_{r-1} \ldots a_3}{1 + (-1)^{r-1} a_1 a_2 \ldots a_r},$$

$$m_2 = \frac{1 - a_2 + a_2 a_1 - a_2 a_1 a_r + \cdots + (-1)^{r-1} a_2 a_1 a_r \ldots a_4}{1 + (-1)^{r-1} a_1 a_2 \ldots a_r},$$

$$m_3 = \frac{1 - a_3 + a_3 a_2 - a_3 a_2 a_1 + \cdots + (-1)^{r-1} a_3 a_2 a_1 \ldots a_5}{1 + (-1)^{r-1} a_1 a_2 \ldots a_r},$$

and so on.

Assume r is odd, $r = 2h + 1$. It is clear that the denominator q_i of each $m_i = p_i/q_i$ is positive. We claim $p_i > 0$ for each i. This is a consequence of the following lemma:

4.6 Lemma. *Let $b_1, \ldots, b_{2h}$ be positive integers. Then*

$$1 - b_1 + b_1 b_2 - + \cdots + b_1 b_2 \ldots b_{2h} > 0.$$

Proof. We prove this by induction. If $h = 1$ we have $1 - b_1 + b_1 b_2 = 1 - b_1(1 - b_2)$. If $b_2 > 1$, then $b_1(1 - b_2) < 0$, so $1 - b_1 + b_1 b_2 > 1$. If $b_2 = 1$, then $1 - b_1 + b_1 b_2 = 1 > 0$. Now suppose (induction hypothesis) the claim holds when we have $2h > 2$ numbers, let us prove it for $2h + 2$. Set $\alpha = 1 - b_1 + b_1 b_2 - + \cdots + b_1 b_2 \ldots b_{2h+2}$. Then, $\alpha = 1 - b_1(1 - b_2(\beta))$, where $\beta = 1 - b_3 + b_3 b_4 - + \cdots + b_3 \ldots b_{2h+2}$. By the induction hypothesis one has $\beta > 0$. Hence $1 - b_2(\beta) \le 0$, so $\alpha \ge 1$. □

Now we take a cycle of even length $r = 2h$ as in Theorem 4.2, so $a_i \ge 1$ and at least two a_i, say a_{i_1} and a_{i_2} are strictly bigger than 1, with i_1 being odd and i_2 even in the cycle. In this case the denominator q_i of each m_i is negative; we claim p_i is also < 0, so that $m_i > 0$. We prove this by induction. If $h = 1$, one has $p_1 = 1 - a_1$, $p_2 = 1 - a_2$ and a_1, a_2 are both > 1 by hypothesis, so each p_i is negative. Similarly, for $r = 4$ one has $p_1 = 1 - a_1(1 - a_4(1 - a_3))$, and necessarily $a_1 > 1$ or $a_3 > 1$; in either case one has $p_1 < 0$, and similarly for p_2, p_3, p_4. Now assume (induction hypothesis) the claim holds for $2h > 2$, and we want to prove it for $r = 2h + 2$ numbers. We have: $p_1 = 1 - a_1(1 - a_r(\beta))$, where

$$\beta = 1 - a_{2h+1} + a_{2h+1} a_{2h} - + \cdots - a_{2h+1} \ldots a_3.$$

By hypothesis one a_{odd} is > 1 (notice that a_2 does not appear in the formula for p_1, so this could be the $a_{\mathrm{even}} > 1$ that we have by hypothesis). If this $a_{\mathrm{odd}} > 1$ is one of the a_i appearing in the expression for β, then by the induction hypothesis one has $\beta < 0$. Thus $1 - a_r(\beta) > 1$, hence $p_1 < 0$. If all the a_{odd} appearing in β are 1, then $\beta = 0$ and $a_1 > 1$ by hypothesis, so $p_1 < 0$ as claimed. The proof for the other p_i is similar. □

Proof of Lemma 4.3. We construct an action $\widehat{\Gamma} : \mathbb{R} \times \mathbb{C}^n \to \mathbb{C}^n$, $\widehat{\Gamma}(\lambda, (z_1, \ldots, z_n)) = (e^{i\lambda s_1} z_1, \ldots, e^{i\lambda s_n} z_n)$, $s_i \in \mathbb{Q}$, such that $\psi_\theta(\widehat{\Gamma}(\lambda, z)) = \psi_{\theta+\lambda}(z)$. This implies Lemma 4.3. On a 1-cycle, define $\widehat{\Gamma}(\lambda, (z_j)) = e^{\frac{i\lambda}{a_j - b_j}}$. To define it on an r-cycle, consider the $r \times r$-matrix A,

$$
\begin{pmatrix}
-1 & a_1 & 0 & \cdots & & \cdots & 0 \\
0 & -1 & a_2 & 0 & & \cdots & 0 \\
\vdots & \vdots & \vdots & \vdots & & & \vdots \\
\vdots & \vdots & \vdots & \vdots & & & \vdots \\
0 & 0 & 0 & 0 & -1 & & a_{r-1} \\
a_r & 0 & 0 & \cdots & 0 & & -1
\end{pmatrix}
$$

whose determinant is $(-1)^r (1 - a_1 \ldots a_r) \ne 0$. Let $s = (s_1, \ldots, s_r)$ be the unique solution to the linear system $A \cdot s = (1, \ldots, 1)$ and define $\widehat{\Gamma}$ as above, with these weights s_i, which are in $\mathbb{Q}$ because A is integral. Doing this for each cycle of σ we obtain a flow with the properties of Lemma 4.3. $\qquad\square$

4.7 Example. Consider the Pham-Brieskorn polynomial,

$$f(z_1, \ldots, z_{2n}) = z_1^{a_1} + \cdots + z_{2n}^{a_n} \ , \ a_i > 2 \text{ for all } i \, .$$

One has the Milnor fibration of f as given in [168]:

$$\frac{f}{\|f\|} : \mathbb{S}^{4n-1} - M \to \mathbb{S}^1 \, ,$$

where M is the link. Let F be one of the Hamiltonian vector fields obtained from the gradient vector field $(\frac{\partial f}{\partial z_1}, \ldots, \frac{\partial f}{\partial z_{2n}})$ by permuting by pairs the partial derivatives of f and changing the sign to one derivative in each pair. For example:

$$F(z_1, \ldots, z_{2n}) = (a_2 z_2^{a_2-1}, -a_1 z_1^{a_1-1}, \ldots, a_{2n} z_{2n}^{a_{2n}-1}, -a_{2n-1} z_{2n-1}^{a_{2n-1}-1}) \, .$$

Then F is holomorphic and its solutions are tangent to the fibres of f. Every such vector field F satisfies the hypotheses of Theorem 4.2, taking $X(z) = (z_1, \ldots, z_n)$, so it has an associated fibration "à la Milnor" given by Theorem 4.2. It would be interesting to understand how these fibrations are related to Milnor's original fibration.

VII.5 On the topology of the twisted Pham-Brieskorn singularities

We recall that a real singularity is a twisted Pham-Brieskorn singularity if it is defined in $\mathbb{C}^n$ by a polynomial map of the form

$$\psi(z) = \lambda_1 z_1^{a_1} \bar{z}_{\sigma_1} + \cdots + \lambda_n z_n^{a_n} \bar{z}_{\sigma_n} \, ,$$

where each $a_i \geq 2$, $i = 1, \ldots, n$, the λ_i are non-zero complex numbers and $\sigma = \{\sigma_1, \ldots, \sigma_n\}$ is a permutation of the set $\{1, \ldots, n\}$. We can re-label the variables $(z_1, \ldots, z_n)$ so that this singularity takes the form:

$$\widehat{\psi}(z) = \lambda_1 z_{\sigma_1}^{a_1} \bar{z}_1 + \lambda_2 z_{\sigma_2}^{a_2} \bar{z}_2 + \cdots + \lambda_n z_{\sigma_n}^{a_n} \bar{z}_n \, .$$

This can be regarded as being $0 \in \mathbb{C}^n$ union the polar variety in $\mathbb{C}^n - 0$ of the non-singular foliations $\mathcal{F}$ and $\mathcal{S}$, where $\mathcal{F}$ is the holomorphic 1-dimensional foliation defined by the vector field

$$F(z) = (\lambda_1 z_{\sigma_1}^{a_1}, \ldots, \lambda_n z_{\sigma_n}^{a_n}) \, ,$$

and $\mathcal{S}$ is the foliation given by the spheres around 0.

Given this vector field F, define as before

$$\phi_F(z) = \frac{\langle F(z), z \rangle}{|\langle F(z), z \rangle|} : \mathbb{S}^{2n-1} - M \longrightarrow \mathbb{S}^1 ,$$

where $M = V \cap \mathbb{S}^{2n-1}$ is the link, which can be regarded as the intersection of V with the unit sphere.

The theorem below is essentially a summary of results proved in Chapter VI (Sections 3, 4) and in the previous sections of this chapter.

5.1 Theorem.

(i) *The link M is a smooth, codimension 2, oriented submanifold of the sphere.*

(ii) *The map ϕ is the projection of a $\mathbb{C}^\infty$ (locally trivial) fibre bundle, which defines an open book decomposition of the sphere.*

(iii) *Each fibre is a parallelizable, open manifold of dimension $2n - 2$, that can be compactified by attaching its boundary M.*

(iv) *Each pair of antipodal fibres $\{E_\theta, E_{\theta+\pi}\}$ is naturally glued together along M forming the real analytic variety of points where the vector field $e^{i\theta} F(z)$ is tangent to the unit sphere (as a real vector field).*

(v) *In particular, the double of the fibre E_θ (compactified by attaching its boundary) is the link of the real hypersurface given by the real part of the function $\langle F(z), z \rangle$.*

The obvious next thing we want is to study the topology of the link of these singularities, as well as that of the corresponding fibrations. Alas this is not easy and the only known cases so far are: a) when the permutation σ is the identity; and b) when $n = 2$. Let us consider first case (a).

Following [214] we actually consider, more generally, F and X vector fields of the form $F = (z_1^{a_1}, \ldots, z_n^{a_n})$, $X = (z_1^{b_1}, \ldots, z_n^{b_n})$, such that the corresponding exponents satisfy $a_i > b_i \geq 1$ for each $i = 1, \ldots, n$. Let $\psi = \psi_{F,X}$ be defined as before,

$$\psi = \langle F(z), X(z) \rangle = \sum_{i=1}^{n} z_i^{a_i} \bar{z}_i^{b_i} .$$

By Theorem 4.2 we know that ψ satisfies the strong Milnor condition at 0, so one has an associated Milnor fibration. Let us set $c_i = a_i - b_i$ for each i. One has:

5.2 Theorem. *The singular variety $\widehat{V} = \psi^{-1}(0)$ is homeomorphic to the Brieskorn variety $V_{c_1, \ldots, c_n}$, and the corresponding Milnor fibrations are topologically equivalent. More precisely, there exists a homeomorphism $h : (\mathbb{C}^n, 0) \to (\mathbb{C}^n, 0)$ such that $\psi = f \circ h$, where f is the Pham-Brieskorn polynomial:*

$$f(z_1, \ldots, z_n) = z_1^{c_1} + \cdots + z_n^{c_n} .$$

The proof is straightforward. Let $E \subset \mathbb{C}^n$ be the divisor $\{ z_1 z_2 \ldots z_n = 0 \}$, and define $h : (\mathbb{C}^n - E) \to (\mathbb{C}^n - E)$ by: $h(z_1, \ldots, z_n) = (w_1, \ldots, w_n)$, where

$w_i = ||z_i||^{\frac{2b_i}{c_i}} z_i$. Then h is clearly a real analytic diffeomorphism that extends to a homeomorphism of $\mathbb{C}^n$ into itself. Furthermore, for each $i = 1, \ldots, n$ one has:

$$z_i^{a_i} \bar{z}_i^{b_i} = z_i^{c_i} ||z_i||^{2b_i} = (||z_i||^{\frac{2b_i}{c_i}} z_i)^{c_i} = w_i^{c_i} \,.$$

Hence the theorem.   □

So these singularities are real analytic singularities which are topologically equivalent to the usual complex analytic Pham-Brieskorn singularities, but they are not analytically equivalent because the corresponding maps have different algebraic multiplicity.

5.3 Remarks.

(i) As a special case of Theorem 5.2 we have that if some c_i is 1, then $\widehat{V}$ is actually homeomorphic to $\mathbb{C}^{n-1}$, because in this case the gradient of the polynomial f is never vanishing, so that the corresponding variety V is non-singular and it is a cone over the standard sphere. Furthermore, $\widehat{V}$ is also embedded in $\mathbb{C}^n$ as the cone over the standard $(2n-3)$-sphere, because the Milnor fibre is a $(2n-2)$-disc in this case.

(ii) This theorem also provides a new way for thinking of the Brieskorn manifolds and their Milnor fibres. For instance, Poincaré's homology sphere Σ is known to be the Brieskorn manifold $M_{(2,3,5)}$, i.e., the link of the singularity in $\mathbb{C}^3$ defined by $\{z_1^2 + z_2^3 + z_3^5 = 0\}$. By the previous theorem this manifold is homeomorphic (and hence, in this case, also diffeomorphic) to the set of points in $\mathbb{C}^3$ where the complex line field spanned by the vector field $F = (z_1^3, z_2^4, z_3^6)$ is tangent to the unit sphere $\mathbb{S}^5 \subset \mathbb{C}^3$. The corresponding Milnor fibre is the famous (open) manifold E_8 and by the previous results, its double is the set of points where the real line field spanned by F is tangent to $\mathbb{S}^5$.

Let us consider now the case when $n = 2$ and the permutation σ is not the identity. The singularities in question are, up to isomorphism of the form:

$$V = \{z_1^p \cdot \bar{z}_2 - z_2^q \cdot \bar{z}_1 = 0\} \,.$$

The simplest case is when $p = q = k > 1$. This was considered in [226] and one has:

5.4 Theorem. *The link L of the singularity*

$$V = \{z_2^k \cdot \bar{z}_1 - z_1^k \cdot \bar{z}_2 = 0\} \,,$$

consists of $k+3$ circles in the unit sphere $\mathbb{S}^3 \subset \mathbb{C}^2$, which are fibres of the Hopf fibration $\mathbb{S}^3 \mapsto \mathbb{S}^2$. The monodromy h of the corresponding fibre bundle is the periodic map $h(z_1, z_2) = (e^{-\frac{2\pi i}{k-1}} z_1, e^{-\frac{2\pi i}{k-1}} z_2)$, of period $k-1$, and the genus of the fibres E_θ is:

$$g(E_\theta) = \frac{(k-2)(k+1)}{2} \,.$$

Thus, the link of the hypersurface $\mathrm{Re}\left(z_2^k \cdot \bar{z}_1 - z_1^k \cdot \bar{z}_2\right) = 0$ *is a closed oriented surface in* $\mathbb{S}^3$ *of genus* k^2, *equal to the Poincaré-Hopf index of the vector field* $F = (z_2^k, -z_1^k)$.

The proof is straightforward and is given in detail in [226], so we do not include it here. The idea is to prove first that the link L is a union of Hopf fibres, which is done by giving an explicit parametrization of it. That the monodromy is as stated is immediate from the construction of the flow in Lemma 4.3. The interesting part is the computation of the genus of the fibres. Since L is a union of Hopf-fibres, the complement of the link L in $\mathbb{S}^3$ is a product $(\mathbb{S}^2 - \{\text{points}\}) \times \mathbb{S}^1$, so it also fibres over $\mathbb{S}^2$. Thus one has that $\mathbb{S}^3 - L$ is being fibred over $\mathbb{S}^2 - \{\text{points}\}$ in two different ways: one is via the Hopf fibration, the other via the fibration in Theorem 5.4. In both cases the fibres are transversal to the orbits of the Hopf flow. Using this we get a ramified covering projection from the fibres in Theorem 5.4 to the fibres of the Hopf fibration regarded as a map $\mathbb{S}^3 - L \to \mathbb{S}^2 - \{\text{points}\}$. Then Hurwitz's formula gives the genus of the fibres in Theorem 5.4. This result is anyhow a special case of the results in [202] that we describe in the following chapter, where we study the singularities

$$V = \{z_1^p \cdot \bar{z}_2 - z_2^q \cdot \bar{z}_1 = 0\}.$$

VII.6 Stability of the Milnor conditions under perturbations

A natural question to ask is whether the Milnor conditions are stable under perturbations by high order terms. More precisely, if a given real singularity, say defined by $f : \mathbb{R}^m \to \mathbb{R}^2$, satisfies the (strong) Milnor condition and f^* is obtained from f by adding to it terms of sufficiently high order, does f^* satisfy the (strong) Milnor condition, and are these singularities topologically equivalent? This was considered and answered affirmatively in [214], where precise orders for the perturbations allowed are also given for the quasi-homogeneous singularities, as for example the twisted Pham-Brieskorn real singularities. The key for this is to re-interpret these conditions in terms of finite-determinacy of map-germs, using [261, 117, 212, 213]. Here we only make a few comments about this and state some of the results in [214]. We refer to the literature for details.

Let $C(n, p)$ be the space of smooth map-germs from $(\mathbb{R}^n, 0)$ into $(\mathbb{R}^p, 0)$, and let $\mathcal{J}^k(n, p)$ be the set of k-jets $\{j^k\}$ of elements of $C(n, p)$. Let $\mathcal{R}$ be the group of germs of C^∞-diffeomorphisms $(\mathbb{R}^n, 0) \to (\mathbb{R}^n, 0)$; $\mathcal{R}$ acts on $C(n, p)$ by composition on the right. An element $f \in C(n, p)$ is k-$\mathcal{R}$-determined if the $\mathcal{R}$-orbit of f contains all germs whose k-jet at 0 coincides with the k-jet of f. Similarly, one has the group C^l-$\mathcal{R}$, of local diffeomorphisms of class C^l, $l > 0$, or homeomorphisms if $l = 0$, which acts on the corresponding space $C^l(n, p)$. We will consider rather the induced equivalence relation on $C(n, p)$ and the corresponding notion of

k-C^l-$\mathcal{R}$-determinacy. One says that f is C^l-$\mathcal{R}$-finitely determined if it is k-C^l-$\mathcal{R}$-determined for some k.

Let $\mathcal{I}_{\mathcal{R}}(f)$ be the ideal of $C(n)$ generated by the $p \times p$-minors of the Jacobian matrix of f and $N_{\mathcal{R}}(f) = det\{(df_x)(df_x)^t\} = $ sum of squares of $p \times p$-minors of df_x. We say that $N_{\mathcal{R}}(f)$ satisfies a Lojasiewicz condition of order r (> 0) if there exists a constant $c > 0$ such that $N_{\mathcal{R}}(f) \geq c|x|^r$.

The condition that f be C^l-$\mathcal{R}$-finitely determined for $0 \leq l < \infty$ is equivalent to the condition that $N_{\mathcal{R}}(f)$ satisfies a Lojasiewicz condition for some r. Moreover, this inequality provides precise estimates for the degree of C^l-$\mathcal{R}$-determinacy of f. This result was discovered by Kuo in [116] for the case p=1. The extension to higher dimensions was considered by many authors. (See [261] for a complete account of the subject, and [212] for precise estimates on the degree of C^l-determinacy of f.)

We recall that $f \in C(n,p)$ analytic satisfies the Milnor condition at 0 if its Jacobian matrix Df has rank p everywhere on a punctured neighborhood of $0 \in \mathbb{R}^n$.

The theorem below is implicit in [261] (see also [212], Proposition 2.4.d):

6.1 Theorem. *Let $f \in C(n,p)$ be analytic. Then f satisfies the Milnor condition at 0 if and only if f is C^l-$\mathcal{R}$-finitely determined for every $l \in [0, \infty)$.*

Proof. For analytic germs, $N_{\mathcal{R}}(f)$ satisfies a Lojasiewicz condition at zero if and only if the variety of the ideal $\mathcal{I}_{\mathcal{R}}(f)$ reduces to 0 ([261], Lemma 6.2), and this is clearly equivalent to the Milnor condition for f. Now the result follows, for instance, from [212], Proposition 2.4. □

As a consequence one has:

6.2 Corollary. *Let $f \in C(n,p)$ be analytic and assume it satisfies the Milnor condition at 0. Then every perturbation of f by terms of sufficiently high order also satisfies the Milnor condition at 0 and is topologically equivalent to f.*

When we consider a real analytic function from $\mathbb{C}^n$ into $\mathbb{C}$, then regarded as a function from $\mathbb{R}^{2n}$ into $\mathbb{R}^2$, its real and imaginary parts are $f_1 = \mathrm{Re}\ f = \frac{1}{2}(f + \overline{f})$ and $f_2 = \mathrm{Im}\ f = \frac{1}{2i}(f - \overline{f})$, respectively. So f_1 is weighted homogeneous if and only if f_2 is weighted homogeneous, and in this case they have the same weights and total degree. As mentioned above, for singularities as in Corollary 6.2 which are quasi-homogeneous, one can get rather precise estimates of the least order we need in the perturbations to guarantee that the Milnor condition is preserved (see [214]).

Regarding the strong Milnor condition, it was proved in [214] that if $f \in C(2n,2)$ is analytic and its component functions f_1 and f_2 satisfy conditions A and $B_{\mathbb{R}}$ at 0 (hence f satisfies the strong Milnor condition), then there exists a positive integer N such that for every map-germ f^* whose N^{th} jet at 0 coincides with that of f, $j^N f^*(0) = j^N f(0)$, the component functions f_1^* and f_2^* of the map-germ f also satisfy conditions A and $B_{\mathbb{R}}$ at 0.

VII.7   Remarks and open problems

We know from the previous discussion that the twisted Pham-Brieskorn singularities,

$$\psi(z) = \lambda_1 z_1^{a_1} \overline{z}_{\sigma_1} + \cdots + \lambda_n z_n^{a_n} \overline{z}_{\sigma_n} \, ,$$

are a class of real singularities which behave very much as complex singularities, in the sense that their links define open book decompositions of the spheres. It would be interesting to know more about the topology of these singularities.

As mentioned before, so far we only understand the cases when the perturbation σ is the identity (the corresponding singularity is topologically equivalent to a usual Pham-Brieskorn singularity) and the case $n = 2$ that we shall describe in the following chapter. A consequence of the results in [202] that we present in Chapter VIII is that for $n = 2$ the links are isotopic to the links of singularities of the form $f\bar{g}$, where both f, g are holomorphic functions $\mathbb{C}^2 \to \mathbb{C}$ with no common branch. However that can only happen in this dimension, because the twisted Pham-Brieskorn singularities have an isolated singularity, while the varieties $\{f = 0\}$ and $\{g = 0\}$, with f, g holomorphic, must intersect in a variety of dimension more than 0, if $n > 2$.

There are several open problems in this respect which I believe are interesting. Essentially everything we know that holds for complex singularities is likely to have a counterpart for these twisted singularities. For instance, one can stabilize them adding terms z_j^2 in new variables, and presumably the new link is the suspension over the previous one, one may look at the signature of the quadratic form of its Milnor fibre, etc. Here are some of the problems that I consider fundamental in this theory.

Problem 1: Study the topology of the link

   a) For general reasons, for $n > 2$ the link M^{2n-3} is necessarily connected; how connected is it? For instance, in the complex analytic case it is $(n - 3)$-connected. What about its homology? In particular, it would be interesting to know when the link is a homology (homotopy) sphere.

   b) Is one getting new manifolds in this way, which are not links of complex singularities? (either up homeomorphism or diffeomorphism) What about the algebraic knots $(\mathbb{S}^{2n-1}, M)$, where M is the link?

For instance, when $n = 1$ the link M is necessarily a finite union of circles embedded in the 3-sphere, so the topology of M by itself is not exciting; what is interesting is the corresponding algebraic knot $(\mathbb{S}^3, M)$. As we prove in the following chapter, the pairs we get in this way for the twisted Pham-Brieskorn singularities are isotopic to links coming from complex singularities, but the various components get different orientations and this causes the corresponding Milnor fibrations to be different.

In dimension $n = 2$ the link is a Seifert 3-manifold and one should be able to compute its Seifert invariants from the weights and the twisting σ (c.f. Chapter III). Then, using [186] one can decide whether or not these are links of complex singularities. But again, "almost" every Seifert manifold is homeomorphic (maybe reversing orientation) to a complex singularity link (see for instance [184]); however the Milnor fibrations one gets are sure not to be (in general) equivalent to the Milnor fibrations of complex singularities. On the other hand we know from the previous chapter that the 3-torus does arise from a real singularity of the type we are considering (though it does not have a Milnor fibration) and it is not a complex singularity link. And what about higher dimensions?

Of course a way for studying the topology of the link is to follow Milnor's program in [168], and study the link by looking at the Milnor fibre and the corresponding monodromy. This brings us to:

Problem 2: Study the topology of the fibres

The only cases I know so far are when the twisting σ is the identity, and the singularity is topologically equivalent to a complex one, so its fibre has the homotopy type of a bouquet of spheres of middle dimension, or when $n = 2$ and for general topological reasons the fibre must have the homotopy type of a bouquet of 1-spheres, i.e., circles. Are these special cases of a more general theorem?; has the fibre in general the homotopy of a cw-complex of middle dimension?

A hint may be given by a surprising result in VIII.3.2 below, which generalizes Theorem 5.4 above, stating that for $n = 1$ the double of the Milnor fibre of the singularity $z_1^p \bar{z}_2 + z_2^q \bar{z}_1 = 0$ is $p \cdot q$, the index of the vector field $(z_1^p, -z_2^q)$ that defines it.

On the other hand I believe it is sensible to think that in all these cases the fibres are homeomorphic to Stein manifolds and therefore have the homotopy type of CW-complexes of middle dimension (c.f. the next problem).

Problem 3: Understand the relation with the singularities $f\bar{g}$

It is obvious that every twisted Pham-Brieskorn singularity is defined by a linear combination of functions of the type $f\bar{g}$ with f, g holomorphic, and we know from Theorem I.6.4 that whenever $f\bar{g}$ has an isolated critical value, one has a Milnor fibration. Notice however that since f and g are holomorphic, their zero-sets are hypersurfaces in $\mathbb{C}^n$ and therefore they must intersect in an analytic space of complex codimension at most 2. This means that for $n > 2$ the hypersurface $V_{f\bar{g}} = f\bar{g}^{-1}(0)$ necessarily has non-isolated singularities. This emphasizes the remarkable regularity one has for the twisted Pham-Brieskorn singularities, where you always get isolated singularities. This is surely a special case of a more general theorem about linear combinations of functions of the type $f\bar{g}$.

Notice that for $n = 2$ the only possible cases are (up to scalars) $z_1^p \bar{z}_1 + z_2^q \bar{z}_2$ and $z_1^p \bar{z}_2 + z_2^q \bar{z}_1$. We already know that in the first case these singularities are topologically equivalent to the Pham-Brieskorn singularities $z_1^{p-1} + z_2^{q-1}$. In the next chapter we will see that the second type of singularities have links isotopic to the links defined by $\bar{z}_1 \bar{z}_2 (z_1^{p+1} + z_2^{q+1})$. Thus both cases yield links topologically equivalent to links of singularities $f\bar{g}$; but as remarked above, this cannot happen in higher dimensions.

The following problem is not fundamental for the study of these singularities, but if there is a positive answer to it, this would open a very interesting line of research, inspired in [151, 161, 162, 160]. I do not expect that answering Problem 4 should be too difficult, but I have not been able to do it.

Problem 4: Study the relation with complex geometry
More precisely, if M is the link of a twisted Pham-Brieskorn singularity and $N = M/\mathbb{S}^1$ is the quotient of M by its canonical $\mathbb{S}^1$-action, is N (canonically) a complex manifold? (maybe with some conditions on the weights and the twisting).

We recall from Chapter VI that if we consider the singularities

$$V = \{\, \lambda_1 z_1 \bar{z}_1 + \lambda_2 z_2 \bar{z}_2 + \cdots + \lambda_n z_n \bar{z}_n = 0 \,\}$$

with the λ_i satisfying certain conditions (that their convex hull in $\mathbb{C}$ contains the origin and they satisfy a weak hyperbolicity condition) then $V^* = V - 0$ is a real analytic manifold in $\mathbb{C}^n$ of codimension 2 and transversal to the holomorphic foliation $\mathcal{F}$ defined by the vector field $F = (\lambda_1 z_1, \ldots, \lambda_n z_n)$. Hence the transversal holomorphic structure of $\mathcal{F}$ endows V^* with a complex structure. The canonical $\mathbb{C}^*$-action on V^* is by holomorphic transformations and the quotient N is a complex manifold, diffeomorphic to the quotient $M/\mathbb{S}^1$, where M is the link and the $\mathbb{S}^1$-action is the restriction of the $\mathbb{C}^*$-action. These are the LVM-manifolds studied in VI.2.

For the twisted Pham-Brieskorn singularities one has a similar picture and the question is to decide whether (maybe with some extra conditions) the variety V^* is transversal to the corresponding holomorphic foliation.

Chapter VIII

Real Singularities and Open Book Decompositions of the 3-sphere

In the previous chapter we introduced the singularities of the form

$$\psi(z) = \lambda_1 z_1^{a_1} \overline{z}_{\sigma_1} + \cdots + \lambda_n z_n^{a_n} \overline{z}_{\sigma_n}$$

and proved that if all a_i are > 1, then these singularities behave in some respects as complex singularities, in the sense that they define open book decompositions of the sphere, and the corresponding projection map $(\mathbb{S}_\varepsilon^{2n-1} - M) \longrightarrow \mathbb{S}^1$, where M is the link, is given by $\psi/|\psi|$. We proved that if the permutation σ is the identity, then these singularities are topologically equivalent to usual Pham-Brieskorn singularities. This chapter is based on [202]. Here we study the other possible case when $n = 2$, i.e., singularities of the form,

$$z_1^p \overline{z}_2 + z_2^q \overline{z}_1 = 0 , \quad p, q > 1 .$$

We find the resolution graph of these singularities and from it, using [184, 200], we compute the genus of the pages of the corresponding Milnor fibration and its monodromy, thus giving the complete topological description of these fibrations. For this we find a "topological" resolution $\pi : X \to \mathbb{C}^2$ of the singularity $\{z_1^p \overline{z}_2 + z_2^q \overline{z}_1 = 0\}$ via the usual technique for studying complex plane curves, i.e., by performing appropriate blow-ups (Sections 4 and 5). The additional problem we face is that, since the singularities in question are only real analytic, we have to employ a "trick" to transform them by a homeomorphism, in some step of the resolution process, in order to get a "divisor" with normal crossings. In this way we obtain a topological resolution of f and then a decorated plumbing graph, which gives the isotopy class of its link and the topology of these fibrations (using [184, 200]).

Sections 1–3 are a brief exposition of results in [75, 184, 259, 200] about resolution of complex planes curves, the resolution graphs and the classification of horizontal open book fibrations, restricting the discussion to Seifert links, which is all we need (instead of the more general Waldhausen links).

The results of [202] that we present here prove that the link of the singularity $z_1^p \bar{z}_2 + z_2^q \bar{z}_1 = 0$ is orientation-preserving isotopic to the link of

$$\bar{z}_1 \bar{z}_2 \, (z_1^{p+1} + z_2^{q+1}) = 0 \,,$$

which is a singularity of the form $f\bar{g}$, with f, g holomorphic functions. Then [200] implies that the corresponding open book decompositions are topologically equivalent. In Section 7 we speak briefly about singularities of the type $f\bar{g}$ in general, and about the results of [199] in this respect.

VIII.1   On the resolution of embedded complex plane curves

We recall that in IV.5 we spoke about resolution of singularities, and particularly of curves and surfaces. Here we review more carefully the case of resolution of plane curves, i.e., complex curves in $\mathbb{C}^2$. This material is classic and there are several excellent expositions of the topic, as for instance [249], [44, III.8.4,5] and Chapter V in [75], so we only sketch the arguments, both for completeness and to introduce notation and arguments that we use later in this chapter.

Let $X = \mathbb{C}^2$ and consider its germ at 0. The set of lines through 0 is $\mathbb{CP}^1$. The blowup of X at 0 is the space $\widetilde{X}$ given by

$$\widetilde{X} = \{(z, \ell) \in \mathbb{C}^2 \times \mathbb{CP}^1 \mid z \in \ell\} \,,$$

with projection $\pi : \widetilde{X} \to X$. If $z \neq 0$, then there is exactly one $\ell = \ell_z \in \mathbb{C}^2$ passing through z and 0, so $\pi^{-1}(z)$ consists of the single point (z, ℓ_z), while one has $\pi^{-1}(0) = E \cong \mathbb{CP}^1$. Notice that if we let $z = (z_1, z_2)$ denote the coordinates in $\mathbb{C}^2$ and $[w_1 : w_2]$ the homogeneous coordinates of points in $\mathbb{CP}^1$, then $\widetilde{X}$ is the subset of $(z, \ell) \in \mathbb{C}^2 \times \mathbb{CP}^1$ defined by the equation:

$$z_1 w_2 = z_2 w_1 \,,$$

which has no critical points since w_1, w_2 cannot be both zero. Hence $\widetilde{X}$ is a smooth submanifold of $\mathbb{C}^2 \times \mathbb{CP}^1$.

It is customary to cover $\widetilde{X}$ with two coordinate charts, corresponding to the usual charts for $\mathbb{CP}^1$. These are:

$$U_1 = \{((z_1, z_2), [w_1, w_2]) \in \widetilde{X} \mid w_1 \neq 0 \,; \text{ with coordinates } v_1 = z_1 \,; u_1 = w_2/w_1 \};$$

$$U_2 = \{((z_1, z_2), [w_1, w_2]) \in \widetilde{X} \mid w_2 \neq 0 \,; \text{ with coordinates } v_2 = z_2 \,; u_2 = w_1/w_2 \}.$$

Both charts U_1 and U_2 are biholomorphic to $\mathbb{C}^2$. This means that $\widetilde{X}$ is obtained by taking two copies of $\mathbb{C}^2$, U_1 and U_2, and gluing them away from the axis by the transition (or attaching) function for these two charts:

$$u_2 = u_1^{-1} \; ; \text{ and } \; v_2 = u_1 v_1 \, .$$

Notice that the equation $u_2 = u_1^{-1}$ is the transition function for the usual decomposition of $\mathbb{S}^2 \cong \mathbb{CP}^1$ in two coordinate charts.

This process of blowing up is of course local, so that we may now pick a preferred point $(z_1, \ell_1) \in \mathbb{CP}^1 \subset \widetilde{X}$ and blow it up to obtain a new complex surface $\widetilde{X}_2$ with a projection $\pi_2 : \widetilde{X}_2 \to X_1 = X$, and so on.

We now want to use this process of blowing ups to describe the embedded resolution of singularities of curves in $\mathbb{C}^2$. By this we mean a non-singular complex surface $\widetilde{X}_s$ obtained by a finite number of blow-ups as above, together with the projection map $\pi_s : \widetilde{X}_s \to X_1 = X$, such that the proper (or strict) transform $\widetilde{C} \subset \widetilde{X}_s$ of the germ at 0 of a curve $C \subset \mathbb{C}^2$ is non-singular; in a neighborhood of the exceptional divisor $E = \pi_s^{-1}(0)$ it is a disjoint union of r discs, where r is the number of branches of C at 0. Moreover, we want E to be a "normal crossings divisor", as explained in IV.5, and that each component of $\widetilde{C}$ meets E transversally in a smooth point of E.

1.1 Example. Consider the map $f : \mathbb{C}^2 \to \mathbb{C}$ defined by $(z_1, z_2) \to z_1 \cdot z_2$ and let $C = f^{-1}(0)$, so that C consists of the two coordinate axis $\{z_1 = 0\}$ and $\{z_2 = 0\}$. The intersection $C \cap \mathbb{S}^3$ is the *Hopf link* in $\mathbb{S}^3$. This is the standard *double point*, which is resolved by a single blow-up, since this transformation separates the lines through the origin. To see this analytically, let U_1, U_2 be the above coordinate charts of $\widetilde{X}$, the blow-up of $\mathbb{C}^2$ at 0, and let π be the projection map. In the chart U_1 corresponding to $u_1 \neq 0$, the projection π is given by $\pi(v_1, u_1) = (v_1, u_1 \cdot v_1)$, so that $\pi^{-1}(C) \cap U_1$ is given by the equations $u_1^2 v_1 = 0$, and $u_1 \neq 0$ by hypothesis, so one has $v_1 = 0$. Similarly, on U_2 one has $\pi(v_2, u_2) = (u_2 \cdot v_2, v_2)$ and $\pi^{-1}(C) \cap U_2$ is given by $u_2 v_2^2 = 0$ with $u_2 \neq 0$, so $v_2 = 0$. Thus the proper transform of C consists of the curve $v_1 = 0$ in $U_1 \subset \widetilde{X}$, which is a copy of $\mathbb{C}$, union $v_2 = 0$ in $U_2 \subset \widetilde{X}$, which is another copy of $\mathbb{C}$. Both of these copies of $\mathbb{C}$ in $\widetilde{X}$ are disjoint and they meet the divisor $E = \pi^{-1}(0)$ transversally at each point.

It is clear that if we had, instead of the double crossing at 0 given by the two coordinate axis of $\mathbb{C}^2$, r lines passing in $\mathbb{C}^2$ through the origin, then we could resolve this singularity with a single blow-up; we would get r disjoint copies of $\mathbb{C}$ embedded in $\widetilde{X}$ and meeting the divisor E transversally.

1.2 Example. Consider now the singularity in $\mathbb{C}^2$ given by:

$$\{ (z_1, z_2) \,|\, z_1^2 + z_2^3 = 0 \} \, ,$$

whose link is the trefoil knot. It is an exercise to show that in this case the singularity is again resolved with a single blow-up. However this resolution is not entirely satisfactory because the proper transform $\widetilde{C}_1$ is tangent to the exceptional curve

E_1. By performing a second blow-up one gets a resolution where the exceptional divisor E_2 consists of two copies of $\mathbb{CP}^1$ meeting transversally in a double point P, and the proper transform $\widetilde{C}_2$, which is of course smooth, meets E precisely at P; i.e., the total transform consists of three non-singular curves meeting at one point. With a further blow-up we remove this triple crossing. We get a normal crossings divisor E_3 with three components, and the proper transform intersects E_3 transversally at a smooth point of E_3.

1.3 Example. Consider, more generally, the singularity in $\mathbb{C}^2$ given by:

$$\{\, (z_1, z_2) \,|\, z_1^p + z_2^q = 0 \,\} \quad ; 2 \leq p \leq q \,.$$

We consider the blow-up $\widetilde{X}$ as before. Notice that if $p = q$, then C consists of p lines through the origin and therefore a blow-up resolves them as in the previous Example 1.1. So we may assume $p < q$. The equation for $\pi^{-1}(C) \cap U_1$ is:

$$v_1^p + (u_1 v_1)^q = v_1^p (1 + u_1^q v_1^{q-p}) = 0 \,;$$

and the equation for $\pi^{-1}(C) \cap U_2$ is:

$$(u_2 v_2)^p + v_2^q = v_2^p (u_2^p + v_2^{q-p}) = 0 \,.$$

(Notice that this also applies for the case $p = q$ and confirms that the resolution consists of p disjoint complex lines that meet the divisor transversally.) We set $C^1 = \widetilde{C}$, $C_1^1 = C^1 \cap U_1$ and $C_2^1 = C^1 \cap U_2$. Since $p < q$, C_1^1 does not meet the line $\{v_1 = 0\}$ and therefore it is non-singular.

Let us see what happens at the second coordinate chart U_2. There are three cases to consider according as $q - p$ is $<$, $=$, or $> p$. When $q - p = p$, the curve C_2^1 consists of p lines which meet at $0 \in U_2$ and this can be resolved by a blow-up as in Example 1.1.

In the other two cases the curve C_2^1 is in general singular, or (as in the above case $p = 2$, $q = 3$, it is smooth but tangent to the divisor), but in all cases one can prove that the singularities have been "simplified" in a certain way that we will not make precise here, and we refer to the literature for this. The point is that one arrives at the following theorem, already stated in IV.5:

1.4 Theorem. (*Embedded good resolution*) *Let $C \subset \mathbb{C}^2$ be a germ of a complex analytic curve with a singularity at 0. Then there exists a non-singular complex surface $\widetilde{X}$ and a proper morphism $\pi : \widetilde{X} \to \mathbb{C}^2$ which is a composition of a finite number of blow-ups, such that:*

(i) *The exceptional divisor E has non-singular irreducible components, all copies of $\mathbb{CP}^1$, which meet normally; thus each singularity of E, if there is more than one irreducible component, is an ordinary double point.*

(ii) *The proper transform $\widetilde{C}$ of C in $\widetilde{X}$ (i.e., the closure of $\pi^{-1}(C-0)$) consists of a finite number of disjoint complex lines that meet E transversally at smooth points of E.*

There are several methods for describing the (embedded) resolution of a singularity, one of them is by their *resolution graphs*. These are similar to the graphs used in Chapter IV to describe surface singularities. Given the germ at 0 of a complex plane curve $C \subset \mathbb{C}^2$, let $\pi : \widetilde{X} \to \mathbb{C}^2$ be a good embedded resolution, obtained by a sequence of blow-ups; let $E = \pi^{-1}(0)$ be the exceptional divisor, $E = E_1 \cup \cdots \cup E_r$ its decomposition in irreducible components, and let $\widetilde{C}$ be the proper transform of C, with $\widetilde{C} = \widetilde{C}_1 \cup \cdots \cup \widetilde{C}_s$ its connected components. So each E_i is a copy of $\mathbb{CP}^1$ embedded in $\widetilde{X}$ with a certain self-intersection number $-w_i = E_i^2$, determined by the stage in which E_i appeared in the sequence of resolutions that define π; this weight, an integer, measures the Euler number of the normal bundle of E_i in $\widetilde{X}$. Each $\widetilde{C}_i$ is a complex line in $\widetilde{X}$ that meets E transversally at a smooth point in one of the components E_i. Given this information, we form *the resolution graph* of $\widetilde{X}$ as follows: to each E_i we associate a vertex v_i with a weight, which is $-w_i = E_i^2$; two vertices are joined by a line if the corresponding curves intersect. All vertices represent now curves of genus 0. By performing plumbing according to this graph, as in Chapter IV, we recover the topology of $\widetilde{X}$ and the way the divisor E is embedded in $\widetilde{X}$. Blowing down E to a point we obtain $\mathbb{C}^2$ and a projection $\pi : \widetilde{X} \to$ which is the resolution map.

It remains to describe the curve $\widetilde{C} = \widetilde{C}_1 \cup \cdots \cup \widetilde{C}_s$: for each $\widetilde{C}_j$ we draw an arrow based at the vertex v_i that represents the curve E_i where $\widetilde{C}_j$ meets E. When we blow down E, the lines $\widetilde{C}_1, \ldots, \widetilde{C}_s$ project into the s branches of C. Since the projection π is a biholomorphism away from E, one has that for every sufficiently small sphere $\mathbb{S}_\varepsilon^3$ around $0 \in \mathbb{C}^2$, the pair $(\mathbb{S}_\varepsilon^3, L)$, where $L = C \cap \mathbb{S}_\varepsilon^3$ is the link of C, is diffeomorphic to the pair $(\partial \widetilde{X}, \partial \widetilde{X} \cap \widetilde{C})$, where $\partial \widetilde{X} = \pi^{-1}(\mathbb{S}_\varepsilon^3)$ is the boundary of a regular neighborhood of the exceptional divisor E. Therefore, the topology of the link L in $\mathbb{S}_\varepsilon^3$ is entirely determined by the resolution graph.

Thus, the resolution graph consists of a plumbing graph, where all genera are 0, which describes the topology of the space $\widetilde{X}$, and a number of arrows that represent the components of the link. Thus a resolution graph is called in [75, p. 134] a *decorated plumbing graph* (a tree in this case).

The point now is how to construct the resolution graph of a plane curve. In general, this can be done using the *Puiseux parameterization* of (irreducible) plane curves: it is well known that each germ of an irreducible complex analytic curve C (or more generally, each branch of such curve) is homeomorphic to $(\mathbb{C}, 0)$. Furthermore, if C is defined by $f(x, y) = 0$ then, making a change of coordinates if necessary, one can solve this equation for y in terms of x, obtaining a solution of the form:

$$ y = a_1 \, x^{m_1/n_1} + c_2 \, x^{m_2/n_2} + \cdots, $$

called a *Puiseux series* for C, or a *Puiseux parameterization* $\mathbb{C} \mapsto C$, where the a_i are non-zero complex numbers, each pair (m_i, n_i) are relatively prime positive integers and $m_1/n_1 < m_2/n_2 < \cdots$. The existence of these parameterizations was first shown by Newton, using what is now called the *Newton polygon*, but it was Puiseux who investigated these parameterizations with more detail. The (m_i, n_i)

are the *Puiseux pairs* of the irreducible curve C. The *Puiseux pairs* determine the topology of C: we know already from Chapter I that C intersects transversally every small sphere around 0 and C is locally embedded in $\mathbb{C}^2$ as the cone over its intersection $K = C \cap \mathbb{S}^3_\varepsilon$, which is a knot. The *Puiseux pairs* tell us exactly what type of knot it is: the first *Puiseux pair* gives a torus knot K_1 of type (m_1, n_1), the second gives a *cable knot* of the previous one, i.e., a knot K_2 embedded in the boundary of a tubular neighborhood of K_1, and the pair (m_2, n_2) describes this embedding, and so on. One has that all but finitely many terms of the power series can be removed without changing the topology of the knot. Thus we conclude that the knots one gets in this way are all *iterated torus knots*, described by the *Puiseux pairs* (see for instance [44, 8.3] or the Appendix to Chapter I in [75] where the construction of these knots out of the *Puiseux pairs* is beautifully explained.)

The *Puiseux pairs* determine a resolution graph of the singularity (see, for instance [44, 8.4, Th. 15]). In the following sections we give examples of this.

VIII.2   The resolution and Seifert graphs

This section is a brief summary of material in [75, 184], which are our basic references here.

We recall (see III.1.10) that a compact oriented 3-dimensional manifold M is a *Seifert manifold* if it admits an effective $\mathbb{S}^1$-action. The orbits of this action give a fibreing $\pi : M \to B = M/\mathbb{S}^1$, where B is a compact, connected 2-manifold with or without boundary (according as M has boundary or not). We also recall that a link in M means a disjoint, finite union of circles embedded in M. We assume further that the base B is orientable.

A link L in M is *reducible* if M can be expressed as a non-trivial disjoint sum, i.e., M can be expressed as a connected sum $M = M_1 \# M_2$ and this decomposition separates L. Otherwise we say that L is *irreducible*.

2.1 Definition. An irreducible link L in a manifold M is a *Seifert link* if it is a union of Seifert fibres of some Seifert fibration of M (c.f. [75, Chapter II]).

In the sequel, we avoid considering Seifert links whose complement in M is a solid torus or a product (torus $\times [0, 1]$). These degenerate cases do not appear among the links of singularities considered in this work, except in the special cases discussed in the remark at the end of of Section 5.

2.2 The Seifert graph. Given a Seifert link L, the uniqueness theorem of Waldhausen ([259] or [109, Theorem VI. 18]) implies that there exists a unique Seifert fibration of M, up to isotopy, for which L is a union of Seifert fibres, and the isotopy class of L is characterized by its *Seifert graph* $G(M, L)$. This graph $G(M, L)$ has a single vertex, to which one associates a weight and attaches arrows and stalks, each with an associated pair of numbers (α, β). This is done as follows.

For each component of L (respectively for each exceptional Seifert fibre which is not a component of L), one attaches to the vertex an arrow (respectively a stalk), whose extremity is weighted by the pair (α, β) of its corresponding normalized Seifert invariants $(0 \leq \beta < \alpha)$. These numbers are defined by the equation $\alpha a + \beta b = 0$ in $H_1(N, \mathbb{Z})$, where N is a small tubular neighborhood of the component of the link (or of the corresponding exceptional fibre), saturated with Seifert fibres, b is an oriented Seifert fibre on ∂N and a is an oriented curve on ∂N such that the intersection $a \cdot b$ is $+1$ on ∂N, oriented as the boundary of N.

The vertex of the graph is endowed with a weight, which is *the rational Euler number $e_0 = e(M \to B)$* of the Seifert fibration, introduced in (III.1.10). To define e_0 one defines first the usual *Euler number e* of the Seifert fibration. We recall that if E is an oriented $\mathbb{S}^1$ bundle over an oriented 2-dimensional, compact, connected, manifold B, then its usual Euler class is the primary (*i.e., non-automatically zero*) obstruction for constructing a section of E (c.f. IV.1). This class lives in $H^2(B; \mathbb{Z})$ and it becomes a number when we evaluate it on the orientation class of B. If B has non-empty boundary, then $H^2(B; \mathbb{Z}) \cong 0$, so the bundle is trivial. However, if we fix a choice of a trivialization of E over ∂B, *i.e.*, a section of $\tau : \partial B \to E$, then one has an *Euler class of E relative to τ, $e(E; \tau) \in H^2(B, \partial B; \mathbb{Z}) \cong \mathbb{Z}$*; evaluating $e(E; \tau)$ on the orientation cycle of the pair $(B, \partial B)$ we obtain an integer, which is by definition, the *Euler number of E relative to τ*. Now, given an oriented Seifert fibration $\pi : M \to B$ on a 3-manifold M, let us remove from B small, pairwise disjoint, open discs around the points corresponding to the special fibres, and denote by B_0 what is left. Let E be $\pi^{-1}(B_0)$, which is M minus a union of open solid tori. This is an $\mathbb{S}^1$ bundle over B_0. On each boundary torus T_i, one can choose a unique (up to isotopy) oriented curve a which intersects each Seifert fibre in exactly one point and satisfies that $m = \alpha[a] + \beta[b]$, where m is a meridian of T_i, (α, β) are the corresponding reduced Seifert invariants, and $[b]$ is the homology class represented by one Seifert fibre. This curve a determines a section of $E|_{T_i}$. Doing this for each boundary torus we obtain a section of E over ∂B_0. The *Euler number $e = e(M)$* of the Seifert fibration $\pi : M \to B$ is defined to be the *Euler number of E relative to* the given trivialization over ∂B_0.

Then the *rational Euler number* of the Seifert fibration, which is the weight of the vertex in the Seifert graph, is defined by:

$$e_0 = e - \sum_{i=1}^{d} \frac{\beta_i}{\alpha_i}.$$

The number e_0 has important geometric properties and it has been used by several authors. For instance, in this work it appeared already in III.2.ii, where it was denoted by $e(M \to B)$, following the notation of [186]. This number was introduced in [187, 204] for Seifert manifolds in general and in [204, 186, 184] it is shown that a closed Seifert manifold is the link of a surface singularity iff e_0 is negative.

The Seifert graphs provide a more compact way of describing the resolution graphs of Seifert links; all the data of a resolution graph is encoded in the Seifert graph, and the way for passing from one to the other is standard, and very well explained in [184, Th. 5.6].

For example, Figure 12 represents the Seifert graph of the torus knot $(2,3)$ obtained from the complex singularity $z_1^2 + z_2^3 = 0$.

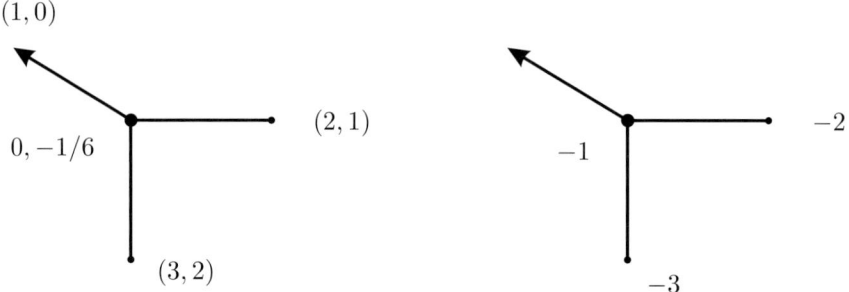

Figure 12: Seifert graph   Resolution graph.

These constructions generalize naturally to *Waldhausen links*, of which the Seifert links are a special case; we refer to [259, 184, 200] for this.

VIII.3   Seifert links and horizontal fibrations

Let us present now some concepts and results of [200] that we need in the sequel.

3.1 Definition. Let L be a link in a 3-dimensional compact oriented manifold M. An *open book fibration* of L is a C^∞ locally trivial fibration $\Phi : M - L \longrightarrow \mathbb{S}^1$ which equips M with an open book decomposition with binding L. (See I.5 for the definition and basic properties of open books.)

In other words, for each component K of the link L there exists an open tubular neighborhood $N(K)$ of K in $M - (L - K)$ and a homeomorphism $\tau : \mathbb{S}^1 \times \mathbb{D}^2 \longrightarrow N(K)$ such that for all $(t, z) \in \mathbb{S}^1 \times \mathbb{D}^2$ one has:

$$(\Phi \circ \tau)(t, z) = \frac{z}{|z|} .$$

In this case the fibres of Φ are called *the pages*; each page is a 2-dimensional open surface whose closure in M is a compact surface with boundary L. The link L is called the *binding* of the open book. In the sequel, all the open book fibrations considered have connected pages.

3.2 Definition. If (M, L) is a Seifert link and if one has an open book fibration Φ of L, then such a fibration is said to be *horizontal* if the fibres of Φ are transverse to the Seifert fibres, which are circles.

According to [259], this transversality is automatically realized for all Seifert links up to isotopy, except in the degenerated cases avoided at the beginning of this section.

Now, let $\Phi : M - L \longrightarrow \mathbb{S}^1$ be an open book fibration of the link $L \in M$ and let K be a component of L. Let us choose an orientation $\overrightarrow{K}$ of K. Let D be a meridian disk of $N(K)$ oriented so that one has intersection $D \cdot \overrightarrow{K} = +1$ in M, and let us equip its boundary m with the induced orientation. One denotes by $\epsilon(\overrightarrow{K}) \in \{-1, +1\}$ the degree of the restriction of Φ to the oriented meridian $\overrightarrow{m}$. Note that if $-\overrightarrow{K}$ denotes K equipped with the opposite orientation, then $\epsilon(-\overrightarrow{K}) = -\epsilon(\overrightarrow{K})$.

If L is a Seifert link and if Φ is horizontal, then there exist two natural orientations of the link L. The first one, denoted by $\overrightarrow{L}_{\text{flow}}$ is obtained as follows: as the Seifert fibres of $M - L$ are transversal to the fibres of Φ, one can orient each Seifert fibre b by a flow which lifts, via Φ, the unit tangent vector field of $\mathbb{S}^1 = \{z \in \mathbb{S}; |z| = 1\}$ compatible with the complex orientation. As the base of the Seifert fibration is orientable, this orientation $\overrightarrow{b}_{\text{flow}}$ is the same for each Seifert fibre b of $M - L$ and extends to an orientation $\overrightarrow{L}_{\text{flow}}$ of L as a union of Seifert fibres. In other words, we may think of the Seifert fibres as being the orbits of an $\mathbb{S}^1$-action on M, and therefore the usual orientation on $\mathbb{S}^1 \subset \mathbb{C}$ induces an orientation on the Seifert fibres.

On the other hand, we may equip the fibres F of Φ with their natural orientation, defined by requiring $F \cdot \overrightarrow{b}_{\text{flow}} = +1$. Then the second natural orientation of L, denoted by $\overrightarrow{L}_{\text{bound}}$, is the orientation of L as boundary of F.

It follows from the definitions that $\epsilon(\overrightarrow{K}_{\text{bound}}) = +1$ for each component K of L, while $\epsilon(\overrightarrow{K}_{\text{flow}})$ can be ± 1. In [200], only the open book fibrations such that $\overrightarrow{L}_{\text{bound}} = \overrightarrow{L}_{\text{flow}}$, i.e., $\epsilon(\overrightarrow{K}_{\text{flow}}) = +1$, are studied.

3.3 Definition. Two horizontal fibrations $\Phi : M - L \longrightarrow \mathbb{S}^1$ and $\Phi' : M' - L' \longrightarrow \mathbb{S}^1$ are *topologically equivalent* if there exist orientation preserving homeomorphisms $H : (M, L) \longrightarrow (M', L')$ and $\rho : \mathbb{S}^1 \longrightarrow \mathbb{S}^1$ such that:

(i) $\rho \circ \Phi = \Phi' \circ H|_{(\mathbb{S}^3 - L)}$;

(ii) For each component K of L, $\epsilon(\overrightarrow{K}_{\text{flow}}) = \epsilon(\overrightarrow{H(K)}_{\text{flow}})$.

Notice that these conditions imply that $H(\overrightarrow{L}_{\text{flow}}) = \overrightarrow{L'}_{\text{flow}}$, i.e., that for each component K of L, the orientation that the flow obtained via Φ induces on K corresponds to the orientation on $H(K)$ given by the flow obtained via Φ'. We remark that Lemma 4.5 of [200] extends easily to the situation we consider here and provides a classification of horizontal fibrations of Seifert links where the $\epsilon(\overrightarrow{K}_{\text{flow}})$ are not necessarily $+1$. In fact, [200] deals with Waldhausen links in general, while we consider only Seifert links. Thus condition (i) can be replaced by:

(i′) The fibres of Φ and Φ' are diffeomorphic and their monodromies are conjugated in the mapping-class group of the fibre.

3.4. The Nielsen Graph. Given a Seifert link (M, L) and a horizontal fibration $\Phi : M - L \to \mathbb{S}^1$ with connected fibre $F := \Phi^{-1}(t)$, considered as an oriented surface in M with boundary L, the *monodromy* of Φ is a diffeomorphism $h : F \to F$ defined by the first return map on F of the Seifert fibres, oriented by the flow. This is a periodic diffeomorphism and one can use the work of Nielsen [191] to study it. The *Nielsen graph* $\mathcal{G}(h)$ of h is a complete invariant defined in [200] which classifies the monodromy up to conjugation. We recall that open book decompositions are completely characterized, up to equivalence, by the topology of the fibres and the monodromy maps (see Chapter I). So the Nielsen graph $\mathcal{G}(h)$ describes fully the corresponding open book decomposition.

Let us recall briefly the construction of the graph $\mathcal{G}(h)$. Given an oriented surface S and an orientation-preserving periodic diffeomorphism $\tau : S \to S$ with order N, denote by $\mathcal{O}$ the space of orbits of τ. The projection $p : F \to \mathcal{O}$ is an n-sheeted branched cyclic cover, ramified over a finite number of points $P_1, \ldots, P_{d'}$, called the exceptional orbits. Let $D_1, \ldots, D_{d'}$ be disjoint open discs in $\mathcal{O}$ such that $P_i \in D_i$ for all $i = 1, \ldots, d'$; let us set $C_i = \partial D_i$ for all $i = 1, \ldots, d'$, and $\breve{\mathcal{O}} = \mathcal{O} - \coprod_{i=1}^{d'} D_i$. Denote also by C_i, $i = d' + 1, \ldots, d' + d$, the boundary components of $\mathcal{O}$. To each exceptional orbit P_i, $i = 1, \ldots, d'$, and to each boundary component C_i, $i = d'+1, \ldots, d'+d$, of $\mathcal{O}$ one associates a triple $(m_i, \lambda_i, \sigma_i)$ defined as follows. Let us endow $\mathcal{O}$ and $\breve{\mathcal{O}}$ with the orientations induced, via p, from that on F, and we equip each C_i, $i = 1, \ldots, d'+d$, with the orientation opposite to that of the boundary of $\breve{\mathcal{O}}$. The integer m_i is the number of connected components of $p^{-1}(C_i)$; then define λ_i by $\lambda_i m_i = N$, and σ_i is the integer modulo λ_i defined by $\rho([C_i]) = m_i \sigma_i$, where $\rho : H_1(\breve{\mathcal{O}}, \mathbb{Z}) \longrightarrow \mathbb{Z}/N\mathbb{Z}$ is the homomorphism associated to the N-sheeted cyclic cover $p_{|p^{-1}(\breve{\mathcal{O}})}$. Then the Nielsen graph $\mathcal{G}(\tau)$ consists of a single vertex, weighted by N and by the genus of the quotient surface $\mathcal{O}$, to which d' *stalks* and d *boundary-stalks* are attached, representing respectively the exceptional orbits and the boundary components of $\mathcal{O}$. The extremity of each stalk or boundary-stalk is equipped with the corresponding triple $(m_i, \lambda_i, \sigma_i)$.

Figure 13 below tells us how to pass from the resolution (or Seifert) graph $G(\mathbb{S}^3, L)$ to the Nielsen graph of the corresponding monodromy. This is implied by the following result of [202] which is an extension of ([200], Lemme 4.4), and its proof is exactly the same as that in [200]. The theorem asserts that the Nielsen graph is determined by the Seifert invariants of (M, L) and by the orientation of the components of L regarded as boundaries of the fibres of Φ.

3.5 Theorem. *Let* $\Phi : M - L \to \mathbb{S}^1$ *be a horizontal fibration of a Seifert link* (M, L), *with connected fibre* F, *and let* $h : F \to F$ *be the periodic representative of the monodromy of* Φ *obtained by the first return on* F *of the Seifert fibres. Assume the resolution graph is as in Figure* 13 *above. Then:*

(i) *The order* N *of the monodromy* h *is:*

$$N = -\frac{1}{e_0} \sum_{i=d'+1}^{d'+d} \frac{\epsilon_i}{\alpha_i} \, .$$

(ii) *The points in the orbits space* $\mathcal{O} = F/h$ *that correspond to the exceptional orbits, are the intersection points of the fibre* F *with the exceptional fibres of*

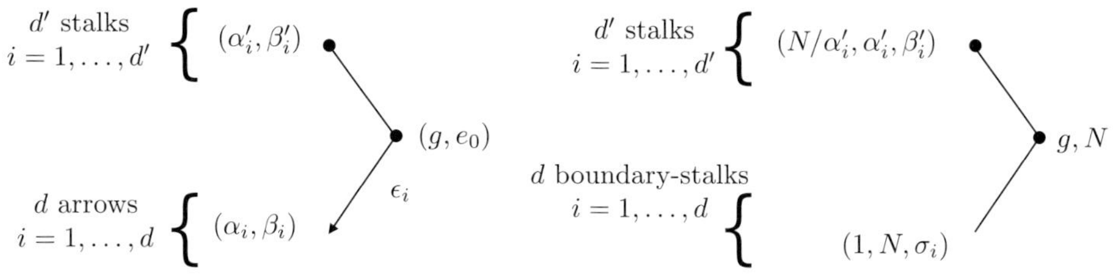

$$d' \text{ stalks} \atop i = 1, \dots, d'} \left\{ (\alpha'_i, \beta'_i) \quad \bullet$$

$$d' \text{ stalks} \atop i = 1, \dots, d'} \left\{ (N/\alpha'_i, \alpha'_i, \beta'_i) \quad \bullet$$

$$(g, e_0) \quad \bullet \qquad \qquad \qquad \qquad \qquad \qquad \bullet \; g, N$$

$$\epsilon_i$$

$$d \text{ arrows} \atop i = 1, \dots, d} \left\{ (\alpha_i, \beta_i) \quad \searrow$$

$$d \text{ boundary-stalks} \atop i = 1, \dots, d} \left\{ \quad (1, N, \sigma_i) \right.$$

Figure 13: Seifert graph $G(\mathbb{S}^3, L)$ Nielsen graph $\mathcal{G}(h)$.

the Seifert fibration; and the triple $(m_i, \lambda_i, \sigma_i)$ attached to each exceptional orbit of is given by the corresponding triple $(N/\alpha_i, \alpha_i, \beta_i)$.

(iii) *The boundary component of $\mathcal{O}$ corresponding to the component K_i of L is equipped with the triple $(1, N, \sigma_i)$, where σ_i, modulo N, is given by the equality:*

$$\alpha_i \sigma_i - N\beta_i + \epsilon(\overrightarrow{K_{i\text{flow}}}) = 0,$$

where (α_i, β_i) are the corresponding reduced Seifert invariants.

In particular the projection $\pi : F \to F/h$ provides a description of the fibre F as an n-sheeted cyclic cover over the surface with genus g, branched over d' points with branching indices $\alpha_i, i = 1, \dots, d'$ and N for the d remaining points. Thus we know the topology of the fibre F: it has d boundary components and its genus is obtained from the Hurwitz formula. One gets:

3.6 Corollary. *With the notation of Theorem 3.5, the genus of the fibres of Φ is:*

$$g_F = \frac{1}{2}\left((N-1)(d-2) + Nd' - \sum_{i=1}^{d'} \frac{N}{\alpha'_i} \right).$$

So we have that the Nielsen graph $\mathcal{G}(h)$ of the horizontal fibration describes fully the open book decomposition, giving us the monodromy map of the fibration and the genus of the fibres. And $\mathcal{G}(h)$ is determined by the Seifert invariants of (M, L) together with the orientation of the components of L regarded as boundaries of the fibres of Φ.

VIII.4 An example

As a motivation for the following section, we study here the topology of the singularity

$$f(z_1, z_2) = z_2^3 \cdot \overline{z}_1 + z_1^5 \cdot \overline{z}_2.$$

Following the method used for studying the topology of complex plane curves (see for instance [75, 134]), let us try to describe the topology of the link L by performing a composition $\pi : X \longrightarrow \mathbb{C}^2$ of a finite number of blow-ups of points, starting with the blow-up of the origin in $\mathbb{C}^2$. We will see that we need to consider also a homeomorphism θ in such a way that the total transform $(f \circ \pi)^{-1}(0)$ has normal crossings (c.f. IV.5). One then identifies the 3-sphere $\mathbb{S}^3_\epsilon$ with the boundary of a small semi-algebraic tubular neighborhood W of the exceptional divisor $E = \pi^{-1}(0)$ and L with the intersection of the strict transform of f with the boundary ∂W of W.

Let $\pi_1 : X_1 \to \mathbb{C}^2$ be the blow-up of $0_{\mathbb{C}^2}$ and set $\Psi_1 = f \circ \pi_1$. In the first coordinate chart of X_1, i.e., over $\mathbb{C}^2 - \{z_2 = 0\}$, the blow-up is given by: $z_1 \mapsto z_1 z_2$ and $z_2 \mapsto z_2$. The exceptional divisor E_1 has equation $z_2 = 0$ and one has:

$$\Psi_1(z_1, z_2) = (f \circ \pi_1)(z_1, z_2) = f(z_1 z_2, z_2) = z_2^3 \bar{z}_1 \bar{z}_2 + z_1^5 z_2^5 \bar{z}_2 = z_2^3 \bar{z}_2 (\bar{z}_1 + z_1^5 z_2^2).$$

In the total transform $\Psi_1(z_1, z_2) = 0$, the factor $z_2^3 \bar{z}_2$ corresponds to the equation of $E_1 := \{z_2 = 0\}$, while $(\bar{z}_1 + z_1^5 z_2^2) = 0$ is the equation of a smooth branch S_1 of the strict transform of f, namely the complex curve with equation $z_1 = 0$.

In the second chart of X_1, i.e., over $\mathbb{C}^2 - \{z_1 = 0\}$, the blow-up is given by $z_1 \mapsto z_1$ and $z_2 \mapsto z_1 z_2$; E_1 has equation $z_1 = 0$ and

$$(f \circ \pi_1)(z_1, z_2) = z_1^3 z_2^3 \bar{z}_1 + z_1^5 \bar{z}_1 \bar{z}_2 = z_1^3 \bar{z}_1 (z_2^3 + z_1^2 \bar{z}_2).$$

We notice that $z_2^3 + z_1^2 \bar{z}_2 = 0$ is the equation of a singular real surface S, so we have to resolve this singularity.

One branch of S has equation $z_2 = 0$. Unfortunately, the presence of both z_2 and $\bar{z}_2$ in the equation of S does not allow one to factorize either z_2 or $\bar{z}_2$. Therefore it is useless trying to separate the branch $z_2 = 0$ from the other branches of S by performing additional blow-ups. Indeed, let us try an additional blow-up $\pi' : X' \to X$. In the second chart we have,

$$(\Psi_1 \circ \pi')(z_1, z_2) = \Psi_1(z_1, z_1 z_2) = z_1^5 \bar{z}_1 (z_2^3 z_1 + \bar{z}_1 \bar{z}_2),$$

and the factor $(z_2^3 z_1 + \bar{z}_1 \bar{z}_2)$ still involves z_2 and $\bar{z}_2$ with the same exponents as before, so this singularity cannot be resolved by blow-ups.

Thus we start again with the term $z_2^3 + z_1^2 \bar{z}_2$, defining S, and we use a trick: we compose π_1 with an orientation-preserving homeomorphism $\theta : X_1 \to X_1$ in order to replace S by a complex plane curve. So we start with the term $(z_2^3 + z_1^2 \bar{z}_2)$ and we write it as $\bar{z}_2 \left(\left(\frac{z_2}{|z_2|^{\frac{1}{2}}} \right)^4 + z_1^2 \right)$; now define the map $\theta : X_1 \to X_1$ by:

$$\theta(z_1, z_2) = \left(z_1, \frac{z_2}{|z_2|^{\frac{1}{2}}} \right)$$

in the second chart. This is well defined away from the two complex lines transverse to E_1 with equations $z_2 = 0$ and $z_2 = \infty$ and it extends in the obvious way to a homeomorphism from X_1 to X_1. This homeomorphism coincides with the identity map on the two lines and it is a diffeomorphism on their complement. In the second chart, the inverse map is $\theta^{-1}(z_1, z_2) = (z_1, z_2|z_2|)$, and the image of S by θ has equation:

$$\overline{z}_2|z_2|(z_2{}^4 + z_1{}^2) = 0\,,$$

or equivalently $\overline{z}_2(z_2{}^4 + z_1{}^2) = 0$. The term $z_2{}^4 + z_1{}^2$ defines a complex analytic plane curve, which can be resolved by a finite sequence of blow-ups of points in the usual way. Let $\pi_2 : X_2 \longrightarrow X_1$ be the blow-up of the point $(z_1, z_2) = (0, 0)$ of X_1. In the second chart, the exceptional divisor $E_2 = \pi_2^{-1}(0, 0)$ has equation $z_1 = 0$ and the strict transform of $\theta(S)$ by π_2 has equation:

$$\overline{z}_2\,|z_2|\,(z_1^2 z_2^4 + 1) = 0\,.$$

The factor $\overline{z}_2$ corresponds to a smooth branch S_2 of the strict transform of f by $f \circ \pi_1 \circ \theta^{-1} \circ \pi_2$. The term $z_1^2 z_2^4 + 1 = 0$ does not intersect the exceptional divisor, so it has no contribution for the topology of L. In the first chart E_2 has equation $z_2 = 0$ and the strict transform of $\theta(S)$ by π_2 has equation

$$z_2{}^2 + z_1{}^2 = 0\,,$$

which corresponds to the equation of two transverse smooth complex curves S_3 and S_4, which are separated by performing the blow-up $\pi_3 : X_3 \to X_2$ of their common point.

Therefore, if we let $\pi = \pi_3 \circ \pi_2 \circ \theta \circ \pi_1$, then the total transform $\overline{(f \circ \pi)^{-1}(0)}$ has normal crossings and the strict transform $\pi^{-1}(f^{-1}(0) - \{0\})$ consists of the four smooth curves $S_i, i = 1, \ldots, 4$. The configuration of the divisor $(f \circ \pi)^{-1}(0)$ is represented in Figure 14, each irreducible compact component E_j being weighted by its self intersection in X.

Let us now identify $\pi^{-1}(\mathbb{S}_\epsilon^3)$ with a small tubular neighborhood W of $\pi^{-1}(0)$ in X obtained by a plumbing process. The link $L = f^{-1}(0) \cap \mathbb{S}_\epsilon^3$ is, up to isotopy, the intersection of $S_1 \cup S_2 \cup S_3 \cup S_4$ with the boundary of W. Therefore L has four components $K_i = S_i \cap \partial W, i = 1, \ldots, 4$, and its isotopy class is encoded in the dual plumbing graph Γ of the divisor $(f \circ \pi)^{-1}(0)$, also represented in Figure 14. As this graph has a single *rupture vertex* (*i.e.*, a vertex with more than three incident edges or arrows), then the link L is a Seifert link. By using the plumbing calculus of [184], one computes from Γ the Seifert graph $G(\mathbb{S}^3, L)$, also represented in Figure 14.

Before computing the degrees $\epsilon(\overrightarrow{K_{i\,\mathrm{flow}}})$, let us introduce some definitions and make some remarks.

Let C be an irreducible component of the total transform $(f \circ \pi)^{-1}(0)$. As in [134, 1.3.2], define a *curvette* of C as a smooth complex curve in X intersecting transversally C at a smooth point of $(f \circ \pi)^{-1}(0)$. One defines the multiplicity $m(C)$ of f along C as the degree of the restriction of f to a curvette of C.

Remember that in a neighborhood of S_1 in X, $f \circ \pi$ has the following local expression:

$$(f \circ \pi)(z_1, z_2) = (f \circ \pi_1)(z_1, z_2) = z_2^3 \overline{z}_2(\overline{z}_1 + z_1^5 z_2^2)\,,$$

where $z_2 = 0$ is the equation of E_1 and where $\bar{z}_1 + z_1^5 z_2^2 = 0$ is that of S_1. Therefore $m(E_1)$ is the degree of the map $z_2 \mapsto z_2^3 \bar{z}_2$, that is $m(E_1) = 3 - 1 = +2$, and $m(S_1)$ is the degree of $z_1 \mapsto \bar{z}_1$, so $m(S_1) = -1$. Similarly, $m(S_2) = -1$, and $m(S_3) = m(S_4) = +1$ as S_3 and S_4 appear through holomorphic factors in the local expression of $f \circ \pi$.

Now, one remarks that, as in the usual resolution of complex plane curves, the multiplicity of a compact irreducible component of the divisor created by the blow-up of a point P is the sum of the multiplicities of the components of the total transform passing through P. Therefore $m(E_2) = m(E_1) + m(S_2) + m(S_3) + m(S_4) = 3$, and $m(E_3) = m(E_1) + m(E_2) + m(S_3) + m(S_4) = 7$.

At this step we can compute already the order of the periodic monodromy h of $\frac{f}{|f|}$. Indeed, by definition, the periodic monodromy is the self-diffeomorphism on a fibre of $\frac{f \circ \pi}{|f \circ \pi|}$ given by the first return map of the flow determined by the Seifert fibration. Therefore its degree is equal, up to sign, to the degree of $\frac{f \circ \pi}{|f \circ \pi|}$ restricted to a regular Seifert fibre of $(\mathbb{S}_3, L)$ disjoint from L. But one of the main ideas of the plumbing calculus is that a regular Seifert fibre of $(\mathbb{S}_3, L)$ is, up to isotopy, the intersection with ∂W of a curvette γ of the rupture component of the exceptional divisor (*i.e.*, that which corresponds to the rupture vertex of the dual resolution graph). Then the order N of h is, up to sign, the degree of the restriction of f to a curvette γ of E_3, *i.e.*, $N = m(E_3) = 7$.

Let us now compute the degree $\epsilon(\overrightarrow{K_{1\,\mathrm{flow}}})$. Remember again that in a neighborhood of S_1,

$$(f \circ \pi)(z_1, z_2) = z_2^3 \bar{z}_2 (\bar{z}_1 + z_1^5 z_2^2).$$

In this local chart, $W = \{(z_1, z_2) \; ; \; |z_2| \leq \eta\}$, where $\eta \ll 1$. Thus,

$$K_1 = \mathbb{S}_1 \cap \partial W = \{(z_1, z_2) \; ; \; z_1 = 0, \; |z_2| = \eta\}.$$

Let $\overrightarrow{K_{1\mathbb{C}}}$ be the knot K_1 oriented as the boundary of the complex curve $S_1 \cap W$. Let us compute first $\epsilon(\overrightarrow{K_{1\mathbb{C}}})$. By definition, it is the degree of the restriction of $(f \circ \pi)/|f \circ \pi|$ to a small meridian of K_1, which is nothing but the degree of the restriction of $f \circ \pi$ to a curvette of S_1. Therefore $\epsilon(\overrightarrow{K_{1\mathbb{C}}}) = m(S_1) = -1$.

Let us now compare the orientations $\overrightarrow{K_{1\mathbb{C}}}$ and $\overrightarrow{K_{1\,\mathrm{flow}}}$. Let γ be a curvette of E_3 and let us orient the Seifert fibre $b = \partial W \cap \gamma$ as the boundary of the complex curve $W \cap \gamma$. As $m(E_3) = 7$ is positive, $\overrightarrow{b}_{\mathbb{C}} = \overrightarrow{b}_{\mathrm{flow}}$. Moreover, by plumbing calculus, one knows that the orientation of $\overrightarrow{K_{1\mathbb{C}}}$ as a Seifert fibre is compatible with that of $\overrightarrow{b}_{\mathbb{C}}$. Therefore, $\overrightarrow{K_{1\mathbb{C}}} = \overrightarrow{K_{1\,\mathrm{flow}}}$, and then, $\epsilon(\overrightarrow{K_{1\,\mathrm{flow}}}) = m(S_1) = -1$. Similarly, $\epsilon(\overrightarrow{K_{2\,\mathrm{flow}}}) = m(S_2) = -1$; $\epsilon(\overrightarrow{K_{3\,\mathrm{flow}}}) = m(S_3) = +1$, and $\epsilon(\overrightarrow{K_{4\,\mathrm{flow}}}) = m(S_4) = +1$.

Using Theorem 3.5 one computes the Nielsen graph $\mathcal{G}(h)$ of the monodromy of $\frac{f}{|f|}$ from the Seifert graph $G(\mathbb{S}^3, L)$ and from the degrees $\epsilon(\overrightarrow{K_{i\,\mathrm{flow}}})$, $i = 1, \ldots, 4$. This is represented in Figure 14. In particular, one can recover from Theorem 3.5.i that the order of the monodromy is 7.

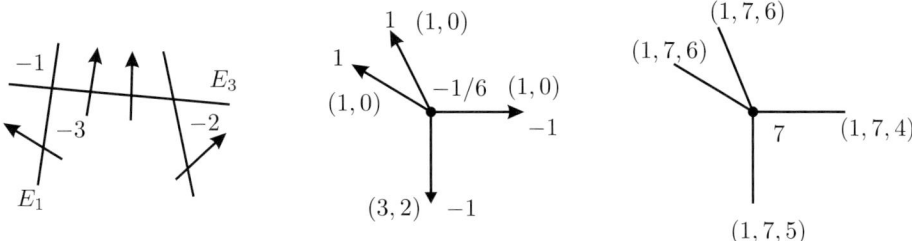

Figure 14: The divisor, the Seifert graph and the Nielsen graph.

VIII.5   Resolution and topology of the singularities $z_1^p \bar{z}_2 + z_2^q \bar{z}_1 = 0$

We now consider the general case of the function $f : (\mathbb{C}^2, 0) \to (\mathbb{C}, 0)$ defined by $f(z_1, z_2) = z_1^p \bar{z}_2 + z_2^q \bar{z}_1$, with p and q integers, $p \geq q \geq 2$. The arguments of the previous section generalize to the following result:

5.1 Theorem. *Given f as above one has:*

(i) *The link $L = f^{-1}(0) \cap \mathbb{S}^3$ is a Seifert link with $k + 2$ components, where $k = \gcd(p+1, q+1)$. Two of these components are the Hopf link $(\{z_1 z_2 = 0\}) \cap \mathbb{S}^3$, while the others are a torus link of type $(p+1, q+1)$.*

(ii) *The degree $\epsilon(\vec{K}_{\mathrm{flow}})$ equals -1 for the two components of the Hopf link, and $+1$ for each of the k remaining components.*

(iii) *The monodromy of the fibration $\frac{f}{|f|} : \mathbb{S}^3 - L \longrightarrow \mathbb{S}^1$ has a periodic representative h whose order is $N = kp'q' - p' - q' = \frac{1}{k}(pq - 1)$, where $p' = \frac{p+1}{k}$ and $q' = \frac{q+1}{k}$.*

(iv) *Each fibre $F_\theta = (\frac{f}{|f|})^{-1}(e^{i\theta})$ has genus $\frac{1}{2}k(N-1) = \frac{1}{2}(pq - 1 - k)$.*

(v) *The plumbing graph of $(\mathbb{S}^3, L)$, the Seifert graph $G(\mathbb{S}^3, L)$ and the Nielsen graph of h are represented in Figure 15, where $p'\sigma_1 - N\beta_1 - 1 = 0$ and $q'\sigma_2 - N\beta_2 - 1 = 0$.*

We remark that we stated the theorem above considering the unit sphere $\mathbb{S}^3 \subset \mathbb{C}^2$ for simplicity, but one can replace this by any sphere centred at 0, of arbitrary positive radius, by [227, Proposition 2.1] (see the previous chapter, where we prove that these singularities are quasi-homogeneous). Chapter VII also gives us an explicit representative of the monodromy of this fibration: this is given by the map $(z_1, z_2) \mapsto (e^{-\frac{2\pi i(p+1)}{pq-1}} z_1, e^{-\frac{2\pi i(q+1)}{pq-1}} z_2)$.

Proof. Let $\pi_1 : X_1 \to \mathbb{C}^2$ be the blow-up of $0_{\mathbb{C}^2}$. In the first chart, $\mathbb{C}^2 - \{z_2 = 0\}$, one has:

$$(f \circ \pi_1)(z_1, z_2) = f(z_1 z_2, z_2) = z_2^q \bar{z}_2 (\bar{z}_1 + z_1^p z_2^{p-q}),$$

k components

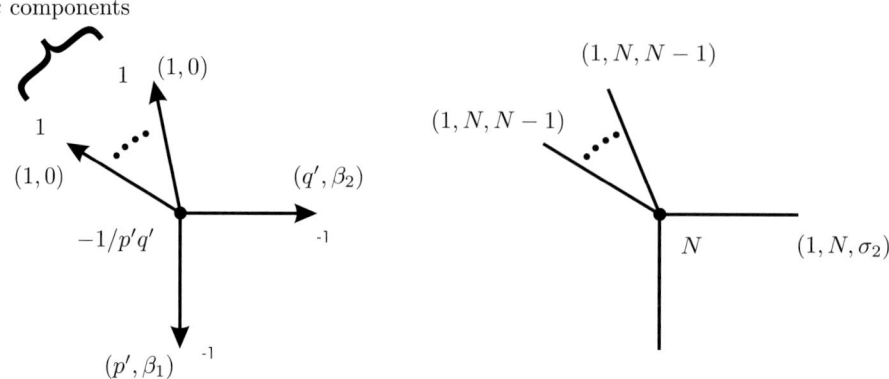

Figure 15: The Seifert graph and the Nielsen graphs.

and $\bar z_1 + z_1^p z_2^{p-q} = 0$ is the equation of a smooth branch S_1 of the strict transform of f, namely the complex curve with equation $z_1 = 0$. Its multiplicity is the degree of the map $z_1 \mapsto \bar z_1$, then $m(S_1) = -1$. In the second chart one has,

$$(f \circ \pi_1)(z_1, z_2) = z_1^q \bar z_1 \left(z_2^q + z_1^{p-q}\bar z_2\right),$$

and, just as in Section 4, it is useless to keep making additional blow-ups to resolve the singularity:

$$S = \left\{ z_2^q + z_1^{p-q}\bar z_2 = 0 \right\},$$

due to the presence of both z_2 and $\bar z_2$ in the equation. Thus we use the same trick we did before. Let us write

$$z_1^q \bar z_1 \left(z_2^q + z_1^{p-q}\bar z_2\right) = z_1^q \bar z_1 \bar z_2 \left(\left(\frac{z_2}{|z_2|^{\frac{2}{q+1}}}\right)^{q+1} + z_1^{p-q}\right).$$

We now compose π_1 with the homeomorphism $\theta : X_1 \to X_1$ defined in the second chart by

$$\theta(z_1, z_2) = \left(z_1, \frac{z_2}{|z_2|^{\frac{2}{q+1}}}\right)$$

out of the two complex lines $z_2 = 0$ and $z_2 = \infty$; this extends as the identity map on these two lines. In the second chart, the inverse map is $\theta^{-1}(z_1, z_2) = (z_1, z_2|z_2|^{\frac{2}{q-1}})$, and $\theta(S)$ has equation:

$$\bar z_2 |z_2|^{\frac{2}{q-1}} \left(z_2^{q+1} + z_1^{p-q}\right) = 0,$$

or equivalently $\bar z_2 (z_2^{q+1} + z_1^{p-q}) = 0$. This is the equation of a complex plane curve, so it can be resolved by a finite sequence $\pi' : X \longrightarrow X_1$ of blow-ups of points in the classical way. As in Section 4, after one blow-up the term $\bar z_2$ gives rise to a smooth branch S_2 transverse to the new component of the exceptional divisor which has multiplicity $m(S_2) = -1$. After a finite sequence of additional blow-ups one obtains $k := \gcd(p+1, q+1)$ other branches $S_3, \ldots, S_{k+2}$, all transverse to the same component C of the exceptional divisor. To identify the link in $\mathbb{S}^3$ defined by these k components of L we observe that in the chart $\{z_1 \neq 0\}$ of the blow-up X_1, the equation $z_2^{q+1} + z_1^{p-q} = 0$ is the equation of the strict transform of the holomorphic curve $z_2^{q+1} + z_1^{p+1} = 0$, whose link is well known to be a torus link of type $k(\frac{p+1}{k}, \frac{q+1}{k})$, by [38] (see also [168]). Therefore, identifying $\pi^{-1}(\mathbb{S}_\varepsilon^3)$ with a tubular neighborhood W of the divisor $\pi^{-1}(0)$, the link L has $k+2$ components $K_i = S_i \cap \partial W$, two of them, K_1 and K_2 consisting of the Hopf link, and the remaining components $K_3, \ldots, K_{k+2}$ form a torus link of type $k(\frac{p+1}{k}, \frac{q+1}{k})$.

This determines the resolution (plumbing) graph of the singularity. Again, using classical computations of Seifert invariants and the plumbing calculus of [184], one obtains from this the Seifert graph $G(\mathbb{S}^3, L)$: there are two exceptional Seifert fibres, which are K_1 and K_2, with Seifert invariants respectively $\alpha_1 = (p+1)/k = p'$ and $\alpha_2 = (q+1)/k = q'$. As the ambient space is the 3-sphere, then the rational Euler class of the Seifert fibration is $e_0 = -\frac{1}{p'q'}$ and the two classes $\beta_1 \bmod p'$ and $\beta_2 \bmod q'$ are related by $p'\beta_2 + q'\beta_1 = 1$ (see [110] for more details).

As in Section 4, one obtains the weights $\epsilon(\overrightarrow{K_{i\,\text{flow}}}), i = 1, \ldots, k+2$ by computing the multiplicities $m(S_i), i = 1, \ldots, k+2$ of the branches of the strict transform of f by π. These are $\epsilon(\overrightarrow{K_{1\,\text{flow}}}) = \epsilon(\overrightarrow{K_{2\,\text{flow}}}) = -1$, and $\epsilon(\overrightarrow{K_{i\,\text{flow}}}) = +1$, $i = 3, \ldots, k+2$.

As in Section 4, we can already compute the order N of the periodic monodromy of $f/|f|$ by computing the multiplicity of the rupture component of the exceptional divisor. In the complex case $z_1 z_2 (z_1^{k+1} + z_2^{k+1})$ this multiplicity equals $kp'q' + p' + q'$, the term $kp'q'$ coming from the k branches of $z_1^{k+1} + z_2^{k+1}$, and the terms p' and q' from the two branches z_1 and z_2. In our real case, the computations of the multiplicities following the sequence of blow-ups are the same as in the complex case, except that the multiplicities of the branches corresponding to the Hopf link (i.e., the branches z_1 and z_2) are counted negatively. Therefore, $N = kp'q' - p' - q'$. Using Section 3, one obtains the Nielsen graph and the genus of the fibre. In particular, one can recover the order N of the monodromy from the formula (i) from Theorem 3.5. We notice that in this case, the corresponding Seifert decomposition of the 3-sphere has only two exceptional fibres (the Hopf link), and both of them are components of L. Hence, in the Nielsen graph of the monodromy, all the stalks are boundary-stalks. $\square$

In order to state the following result we recall the constructions of the previous chapter and notice that the function $f(z_1, z_2) = z_1^p \bar{z}_2 + z_2^q \bar{z}_1$ in Theorem 5.1 can be regarded as the Hermitian product of the vector fields $\xi = (z_2^q, z_1^p)$ and $\iota = (z_1, z_2)$.

5.2 Corollary. *With the hypothesis and notation of Theorem 5.1, each pair of antipodal fibres F_θ and $F_{\theta + \pi}$ is naturally glued together along the link L forming a smooth real analytic surface S_θ in $\mathbb{S}^3$. The genus of S_θ is $p \cdot q$, the Poincaré-Hopf* *index at 0 of the vector field ξ, and S_θ is diffeomorphic to the set of points where the real line field spanned by ξ is tangent to $\mathbb{S}^3$.*

Proof. The only new statement here is the claim that the genus of S_θ is $p \cdot q$. This follows from Theorem 5.1 since each fibre F_θ has genus $\frac{1}{2}(pq - 1 - k)$ and it has $k + 2$ boundary components. When we glue two such fibres along their boundary to get S_θ, we create $k + 1$ handles. Hence the genus of S_θ is:

$$g(S_\theta) = 2g(F_\theta) + (k + 1) = p \cdot q \,,$$

as stated. Finally, that this number is the Poincaré-Hopf index of the vector field ξ follows from the fact that this vector field is holomorphic, so its index at 0 equals the dimension of the vector space $\mathcal{O}_{\mathbb{C}^2,0}/(z_2^q, z_1^p)$. $\qquad\square$

5.3 Remark. There are two special cases which are not envisaged in the previous discussion, these are when $p = q = 1$ or $p > q = 1$. These are studied in detail in [202]. When $p = q = 1$ one has (see §4 in the previous Chapter VII) that if $|\lambda_1| = |\lambda_2|$, then f is not a submersion in a punctured neighborhood of $0 \in \mathbb{C}^2$ and the corresponding link $L = f^{-1}(0) \cap \mathbb{S}_\varepsilon^3$ is not fibred. In fact, this link is not even a 1-dimensional manifold (see [202]). If $|\lambda_1| \neq |\lambda_2|$ the situation is more interesting. In this case, as well as when $p > q = 1$, the results of [214] (see VII. 3, 4) imply that f satisfies the Milnor condition at 0, but the methods used there do not say anything about the strong Milnor condition and one cannot decide with those techniques whether or not the map $\frac{f}{|f|}$ defines an open book decomposition of S_ε^3. In [202] it is proved that they do; this is done by making a careful geometric study of these singularities in a neighborhood of the link. It is proved that in both cases the link L is the Hopf link $\{z_1 z_2 = 0\}$ and the map $\frac{f}{|f|}$ is, on the complement of L, the projection of a fibre bundle; the fibres are annuli and the monodromy is the identity map.

VIII.6   On singularities of the form $f\bar{g}$

We know from the previous sections that the links of the singularities

$$\overline{V}_{p,q} = \{z_1^p \bar{z}_2 + z_2^q \bar{z}_1 = 0\} \,, \; p, q \geq 2 \,,$$

are Seifert links in $\mathbb{S}^3$, being orbits of the $\mathbb{S}^1$-action on $\mathbb{C}^2$ given by:

$$(e^{it}, (z_1, z_2)) \longmapsto (e^{-is_1 t} z_1, e^{-is_2 t} z_2) \,,$$

where $s_1 = \frac{1+p}{pq-1}$ and $s_2 = \frac{1+q}{pq-1}$. The Hopf link $\mathbb{S}^3 \cap (\{z_1 z_2 = 0\})$ gives two components of the link

$$L = \overline{V}_{p,q} \cap \mathbb{S}^3 \,,$$

the others being a torus link (or knot) of type $(p + 1, q + 1)$. In fact these are the links of type 5 in the classification of Seifert links in [75, 7.3, p. 63]. These links can also be obtained via complex singularities: they are isotopic to the links in $\mathbb{S}^3$ determined by

$$z_1 z_2 (z_1^{p+1} + z_2^{q+1}) = 0\,.$$

So it is natural to ask whether the corresponding Milnor fibrations, and the open book decompositions of $\mathbb{S}^3$, are topologically equivalent. Theorem 5.1 shows that this is not the case. In fact one has that the open book fibration provided by $f(z_1, z_2) = z_1^p \bar{z}_2 + z_2^q \bar{z}_1$ induces the "negative" orientation around the two components of the Hopf link. As a consequence of this fact we obtained (using [200]) that if we let $k = \gcd(p+1, q+1)$ be the greatest common divisor of these numbers, and if we set $p' = \frac{p+1}{k}$ and $q' = \frac{q+1}{k}$, then the genus of the fibres of $\frac{f}{|f|}$ is

$$\frac{1}{2} k(kp'q' - p' - q' - 1) = \frac{1}{2}(pq - 1 - k)\,,$$

while the same method tells us that the genus of the fibres in the holomorphic case is

$$\frac{1}{2} k(kp'q' + p' + q' + 1)\,.$$

Similarly, the monodromy for the above open book decompositions has period $kp'q' - p' - q' = \frac{1}{k}(pq - 1)$, whereas in the holomorphic case it has period $kp'q' + p' + q'$.

One actually has that the link of $f(z_1, z_2) = z_1^p \bar{z}_2 + z_2^q \bar{z}_1$ is orientation-preserving isotopic to the link of the singularity:

$$(z_1, z_2) \longmapsto \bar{z}_1 \bar{z}_2 \, (z_1^{p+1} + z_2^{q+1})\,.$$

This fact, together with A'Campo's example (c.f. VII.1 above)

$$(z_1, z_2) \longmapsto z_1 z_2 (\bar{z}_1 + \bar{z}_2)\,,$$

motivated the study in [199] of real singularities in $\mathbb{R}^4$ of the form $f\bar{g}$, with f and g holomorphic functions $(\mathbb{C}^2, 0) \to (\mathbb{C}, 0)$. It is clear that if the function $f\bar{g}$ has an isolated critical point at 0, so that it satisfies the Milnor condition for having an associated Milnor fibration (see Chapter VII), then the link of this function is $L = L_f \cup L_g$, where L_f and L_g are the links of f and g. However, due to complex conjugation, as in the previous case, the components of L_g get the opposite orientation to the one induced by the Milnor fibration of g, thus the corresponding fibrations differ in general. Moreover, that these singularities satisfy the Milnor condition at 0 does not necessarily imply (a priori) that they satisfy the strong Milnor condition and therefore define open book decompositions (c.f. Chapter VII).

In fact this type of singularities already appeared in L. Rudolph's article [215], and his work suggested that given

$$f, g : (\mathbb{C}^2, 0) \to (\mathbb{C}, 0)\,,$$

holomorphic with no common branch, one has that the real analytic map,

$$f\bar{g} : (\mathbb{R}^4, 0) \to (\mathbb{R}^2, 0) \,,$$

has an isolated critical point at 0 iff the link $L = L_f \cup L_g$ is fibred. This was recently proved by A. Pichon in [199]. She further proved that if this condition holds, then the underlying Milnor fibration is an open book decomposition with binding $L_f \cup L_g$, and the projection map is $f\bar{g}/|f\bar{g}|$ in a tubular neighborhood of this link. She also found, using the resolution of the singularity of the holomorphic map-germ $f\,g$, a very simple combinatorial argument that allows one to know whether or not the link $L_f \cup L_g$ is fibred. Using these results she was able to show that a large family of fibrations of Waldhausen links in $\mathbb{S}^3$ arise as the Milnor fibrations of real analytic germs $f\bar{g}$, other than those coming from complex singularities. These results have been recently extended and improved in [201]. In order to explain this we recall (c.f. [75]) that a *multilink* in $\mathbb{S}^3$ is the data of an oriented link $L = K_1 \cup \cdots \cup K_l$ together with a multiplicity $n_i \in \mathbf{Z}$ associated with each component K_i. We denote such a multilink by

$$L = n_1 K_1 \cup \cdots \cup n_l K_l \,,$$

and we fix the convention that $n_i K_i = (-n_i)(-K_i)$, where $-K_i$ means K_i with the opposite orientation.

Now, given a holomorphic map $f : (\mathbb{C}^2, 0) \to (\mathbb{C}, 0)$, we may decompose it into irreducible analytic factors,

$$f = \prod_{i=1}^{l} f_i^{n_i} \,,$$

where the n_i are all positive integers; in this case the multilink associated with f is

$$L_f = \bigcup_{i=1}^{l} n_i L_{f_i} \,,$$

where $L_{f_i} = f_i^{-1}(0) \cap \mathbb{S}_e$ is the link of f_i. One has a similar decomposition for the holomorphic map $g = \prod_{j=1}^{m} g_j^{n_j}$ and the multilink associated to $\bar{g}$ is $L_{\bar{g}} = \bigcup_{i=1}^{m} -n_j L_{g_j}$. The link $L_{f\bar{g}}$ of $f\bar{g}$ is $L_f \cup -L_g$.

6.1 Definition. A multilink $L = n_1 K_1 \cup \cdots \cup n_l K_l$ in $\mathbb{S}^3$ is *fibred* if there exists a map $\Phi : \mathbb{S}^3 \setminus L \longrightarrow \mathbf{S}^1$ which satisfies the following two conditions:

(i) The map Φ is a C^∞ locally trivial fibration.

(ii) For each $i = 1, \ldots, l$, there exist a regular neighborhood $N(K_i)$ of K_i in $M \setminus (L \setminus K_i)$, an orientation-preserving diffeomorphism $\tau : \mathbf{S}^1 \times \mathbf{D}^2 \to N(K_i)$ such that $\tau(\mathbf{S}^1 \times \{0\}) = K_i$ and an integer $k_i \in \mathbf{Z}$ such that for all $(t, z) \in \mathbf{S}^1 \times (\mathbf{D}^2 \setminus \{0\})$ one has:

$$(\Phi \circ \tau)(t, z) = \left(\frac{z}{|z|}\right)^{n_i} t^{k_i}.$$

In this case we say that Φ is a *fibration of the multilink L*.

Notice that this is much stronger than just asking the link to be fibred in the sense of Chapter I, which is the first condition in the definition above. Condition (ii) demands that near the binding, the projection map ϕ takes into account the multiplicity of each component of the multilink. In fact in [201] there is an example showing a link which is fibred in the sense of Chapter I (it is the link D_4) to which different sets of weights are attached, making it a multilink in various ways. It is shown that with some weights it fibres and with others it does not.

One has the following result [201, Theorem 2]:

6.2 Theorem. *Let $f : (\mathbb{C}^2, 0) \to (\mathbf{C}, 0)$ and $g : (\mathbb{C}^2, 0) \to (\mathbf{C}, 0)$ be two holomorphic germs without common branches. Then the following two conditions are equivalent:*

(i) *the real analytic germ $f\bar{g} : (\mathbb{C}^2, 0) \to (\mathbf{R}^2, 0)$ has an isolated critical value at $0 \in \mathbb{C}$;*

(ii) *the multilink $L_{f\bar{g}}$ is fibred.*

Moreover, if these conditions hold, then the Milnor fibration of $f\bar{g}$,

$$\frac{f\bar{g}}{|f\bar{g}|} : \mathbb{S}^3_e \setminus (L_f \cup L_g) \to \mathbb{S}^1,$$

is a fibration of the multilink $L_f - L_g$.

In fact the theorem in [201] is stronger because it gives a third condition which is equivalent to the two conditions above and is surprisingly easy to verify, making it easy to know whether or not a given multilink is fibred. This improves the results of [199] in two ways: first allowing the multiplicities to be $\neq \pm 1$, i.e., asking only for $0 \in \mathbb{C}$ to be an isolated critical value instead of demanding $0 \in \mathbb{C}^2$ to be an isolated critical point; and second giving that the projection map of the fibration is $f\bar{g}/|f\bar{g}|$ everywhere and not only on a neighborhood of the link. This last improvement is possible thanks to Theorem I.6.4 (also proved in [201]). We remark that Theorem 6.2 allows us to use the machinery of [200] to study the topology of these fibrations, just as in the previous examples in Sections 4 and 5 above.

Bibliography

[1] R. Abraham and J. Robin. *Transversal mappings and flows*. W.A. Benjamin, Inc., 1967.

[2] N. A'Campo. Le nombre de Lefschetz d'une monodromie. *Indag. Math*, 35:113–118, 1973.

[3] M. Aguilar, J. Seade, and A. Verjovsky. Indices of vector fields and topological invariants of real analytic singularities. *Crelle's, J. Reine u. Ange. Math.*, 504:159–176, 1998.

[4] S. Akbulut and H.C. King. The topology of real algebraic sets with isolated singularities. *Ann. of Math.*, 113:425–446, 1981.

[5] P. Aluffi. Chern classes for singular hypersurfaces. *Trans. Amer. Math. Soc.*, 351:3989–4026, 1999.

[6] A. Andreotti and Th. Frankel. The Lefschetz theorem on hyperplane sections. *Ann. of Math.*, 69:713–717, 1959.

[7] A. Andreotti and H. Grauert. Théorème de finitude pour la cohomologie des espaces complexes. *Bull. Soc. Math. France*, 90:193–259, 1962.

[8] V.I. Arnold. On the distribution of ovals of real plane algebraic curves, involutions of four-dimensional manifolds and the arithmetic of integral quadratic forms. *Funct. Anal. and Appl.*, 5, No. 3:169–176, 1971.

[9] V.I. Arnold. Critical points of smooth functions and their normal forms. *Russ. Math. Surv.*, 30, No.5:1–75, 1975.

[10] V.I. Arnold. A branched covering of $\mathbb{CP}^2 \to S^4$, hyperbolicity and projective topology. *Siberian Math. J.*, 29:717–726, 1988.

[11] V.I. Arnold. Topological content of the Maxwell theorem on multipole representation of spherical functions. *Topological Methods in Nonlinear Analysis*, 7:205–217, 1996.

[12] V.I. Arnold. Relatives of the quotient of the complex projective plane by complex conjugation. *Proc. Steklov Inst. Math.*, 224, no. 1:46–56, 1999.

[13] V.I. Arnold and A.B. Givental. Symplectic geometry. *in "Dynamical Systems IV", Encycl. Math. Sci. 4; Springer Verlag*, pages 1–136, 1990.

[14] M. Artin. On isolated rational singularities of surfaces. *Amer. J. Math.*, 88:129–136, 1966.

[15] L. Astey, S. Gitler, E. Micha, and G. Pastor. On the homotopy type of complete intersections. *Topology*, 44:249–260, 2005.

[16] M.F. Atiyah. Riemann surfaces and spin structure. *Ann. Scient. Ec. Norm. Sup.*, 4:47–62, 1971.

[17] M.F. Atiyah and J. Berndt. Projective planes, Severi varieties and spheres. *Surv. Differ. Geom.*, 8:1–27, 2003. In Surveys in differential geometry. "Lectures on geometry and topology held in honor of Calabi, Lawson, Siu, and Uhlenbeck at Harvard University", ed. S.-T. Yau.

[18] M.F. Atiyah and F. Hirzebruch. Riemann-Roch theorems for differentiable manifolds. *Bull. Amer. Math. Soc.*, 65:276–281, 1959.

[19] M.F. Atiyah and E. Rees. Vector bundles on projective 3-space. *Invent. Math.*, 35:131–153, 1976.

[20] M.F. Atiyah and I.M. Singer. The index of elliptic operators. I. *Ann. of Math.*, 87:484–530, 1968.

[21] M.F. Atiyah and I.M. Singer. The index of elliptic operators. III. *Ann. of Math.*, 87:546–604, 1968.

[22] M.F. Atiyah and E. Witten. M-theory dynamics on a manifold of g_2 holonomy. *Adv. Theor. Math. Phys.*, 6, No.1:1–106, 2002.

[23] L. Auslander, L. Green, and F. Hahn. Flows on some 3-dimensional homogeneous spaces. *Princeton Univ. Press*, 53, 1963. Chapter III in Annals of Math. Study.

[24] W. Barth, C. Peters, and A. Van de Ven. *Compact complex surfaces.* Springer Verlag, 1984.

[25] S. Bechtluft-Sachs. The computation of η-invariants on manifolds with free circle action. *J. Funct. Anal.*, 174:251–263, 2000.

[26] K. Bekka. Regular quasi-homogeneous stratifications. *Travaux en cours, Hermann*, 55:1–14, 1997. In "Stratification, singularities and diferential equations II Stratifications and Topology of Singular Space", ed. D. Trotman and L.C. Wilson.

[27] K. Bekka and S. Koike. The Kuo condition, an inequality of Thom's type and (c)-regularity. *Topology*, 37:45–62, 1998.

[28] L. Bers. Uniformization, moduli, and Kleinian groups. *Bull. London Math. Soc.*, 4:257–300, 1972.

[29] Ch. Bonatti and X. Gómez-Mont. The index of a holomorphic vector field on a singular variety. *Astérisque*, 222:9–35, 1994.

[30] A. Borel. *Automorphic forms on $SL_2(\mathbb{R})$.* Camb. Univ. Press, 1997.

[31] F. Bosio. Variétés complexes compactes: une généralisation de la construction de Meersseman et López de Medrano-Verjovsky. *Ann. Inst. Fourier*, 51:1259–1297, 2001.

[32] F. Bosio and L. Meersseman. Real quadrics in $\mathbb{C}^n$, complex manifolds and convex polytopes. To appear in J. Diff. Geom.

[33] R. Bott. Nondegenerate critical manifolds. *Ann. of Maths.*, 69:248–261, 1954.

[34] J.P. Brasselet. From Chern classes to Milnor classes – a history of characteristic classes for singular varieties. *Adv. Stud. Pure Math.*, 29:31–52, 2000. in "Singularities - Sapporo 1998".

[35] J.-P. Brasselet, D. Lehmann, J. Seade, and T. Suwa. Milnor classes of local complete intersections. *Trans. Amer. Math. Soc.*, 354:1351–1371, 2002.

[36] J.-P. Brasselet, D.B. Massey, A.J. Parameswaran, and J. Seade. Euler obstruction and defects of functions on singular varieties. *J. London Math. Soc.*, 70, No. 2:59–76, 2004.

[37] J.-P. Brasselet and M.H. Schwartz. Sur les classes de Chern d'un ensemble analytique complex. *Astérisque S.M.F.*, 83:93–147, 1981.

[38] K. Brauner. Zur Geometrie der Funktionen zweier komplexen Veränderlichen III, IV. *Abh. Math. Sem. Hambuerh*, 6:8–24, 1928.

[39] E. Brieskorn. Beispiele zur Differentialtopologie von Singularitäten. *Invent. Math*, 2:1–14, 1966.

[40] E. Brieskorn. Examples of singular normal complex spaces which are topological manifolds. *Proc. Nat. Acad. Sci. U.S.A*, 55:1395–1397, 1966.

[41] E. Brieskorn. Die Monodromie der isolierten Singularitäten von Hyperflächen. *Manuscripta Math.*, 2:103–161, 1970.

[42] E. Brieskorn. Sur les groupes de tresses [d'après V.I. Arnold]. *Springer Lecture Notes in Math.*, 317:21–44, 1973. Séminaire Bourbaki, 24ème année (1971/1972), Exp. No. 401.

[43] E. Brieskorn. Milnor lattices and Dynkin diagrams. *Proc. Sympos. Pure Math. A. M. S.*, 40:153–165, 1983. In "Singularities", Part 1 (Arcata, Calif., 1981).

[44] E. Brieskorn and H. Knörrer. *Plane algebraic curves*. Birkhäuser Verlag, Basel, 1986.

[45] E. Brieskorn and K. Saito. Artin-Gruppen und Coxeter-Gruppen. *Invent. Math.*, 17:245–271, 1972.

[46] E. Brieskorn and A. van de Ven. Some complex structures on products of homotopy spheres. *Topology*, 7:389–393, 1968.

[47] R.-O. Buchweitz and G.-M. Greuel. The Milnor number and deformations of complex curve singularities. *Invent. Math.*, 58:241–281, 1980.

[48] D. Burghelea and A. Verona. Local homological properties of analytic sets. *Manuscripta Math.*, 7:55–66, 1972.

[49] E. Calabi and B. Eckmann. A class of compact, complex manifolds which are not algebraic. *Ann. of Math. 58*, pages 494–500, 1953.

[50] C. Camacho, N. Kuiper, and J. Palis. The topology of holomorphic flows with singularity. *Inst. Hautes Études Sci. Publ. Math.*, 48:5–38, 1978.

[51] E. Cartan. Sur la géométrie pseudo-conforme des hypersurfaces de deux variables complexes. *Œuvres II, Paris*, pages 1231–1304, 1953.

[52] H. Cartan. Quotient d'un espace analytique par un groupe d'automorphisms. *Princeton Univ. Press*, pages 90–102, 1957. In "Algebraic geometry and topology" (Lefschetz symposium volume), edit. by R.H. Fox, D.C. Spencer and A.W. Tucker.

[53] P.T. Church and K. Lamotke. Non-trivial polynomial isolated singularities. *Indag. Math*, 37:149–154, 1975.

[54] P. Conner and F. Raymond. Holomorphic Seifert fiberings. *Springer Verlag, Lecture Notes in Maths.*, 124-204:160–161, 1974.

[55] H. S. M. Coxeter. *Generators and relations for discrete groups*. Springer Verlag, 1980.

[56] H.S.M. Coxeter. *Regular complex polytopes*. Cambridge University Press, 1991.

[57] I. Dolgachev. Quotient conical singularities on complex surfaces. *Funct. Anal. Appl.*, 8:160–161, 1974.

[58] I. Dolgachev. Automorphic forms and quasihomogeneous singularities. *Funct. Anal. Appl.*, 9:149–151, 1975.

[59] I. Dolgachev. On the link space of a Gorenstein quasihomogeneous surface singularity. *Math. Ann.*, 265:529–540, 1983.

[60] A. Douady. Variétés à bords anguleux et voisinages tubulaires. *Séminaire Henri Cartan 14*, 1961-62.

[61] H.B. Duan and E. Rees. Functions whose critical set consists of two connected manifolds. *Bol. Soc. Mat. Mexicana*, 37:139–149, 1992. in "Papers in honor of José Adem".

[62] A. Durfee. Fibered knots and algebraic singularities. *Topology*, 13:47–59, 1974.

[63] A. Durfee. Knot invariants of singularities. *Proc. Sympos. Pure Math., A. M. S.*, 29:441–448, 1975. In Algebraic geometry.

[64] A. Durfee. The signature of smoothings of complex surface singularities. *Math. Ann.*, 232:85–98, 1978.

[65] A. Durfee. Fifteen characterizations of rational double points and simple critical points. *Enseign. Math.*, 25:131–163, 1979.

[66] A. Durfee. Neighborhoods of algebraic sets. *Trans. Amer. Math. Soc*, 276:517–530, 1983.

[67] A. Durfee and H.B. Lawson Jr. Fibered knots and foliations of highly connected manifolds. *Invent. Math*, 17:203–215, 1972.

[68] W. Ebeling. *The monodromy groups of isolated singularities of complete intersections.* Springer Verlag, 1987.

[69] W. Ebeling and S.M. Gusein-Zade. On the index of a vector field at an isolated singularity. *Fields Inst. Commun.*, 24:141–152, 1999. In "The Arnoldfest", edited by E. Bierstone et al.

[70] W. Ebeling and S.M. Gusein-Zade. On the index of a holomorphic 1-form on an isolated complete intersection singularity. *Doklady Math.*, 64:221–224, 2001.

[71] W. Ebeling and S.M. Gusein-Zade. Indices of 1-forms on an isolated complete intersection singularity. *Mosc. Math. J.*, 3:439–455, 742–743, 2003.

[72] W. Ebeling, S.M. Gusein-Zade and J. Seade, Homological index for 1-forms and a Milnor number for isolated singularities. *Int. J. Math.* 15 (2004), 895–905.

[73] F. Ehlers, W.D. Neumann, and J. Scherk. Links of surface singularities and CR space forms. *Comment. Math. Helv.*, 62:240–264, 1987.

[74] Ch. Ehresmann. Sur les spaces fibrés différentiables. *Compt. Rend. Acad. Sci. Paris*, 224:1611–1612, 1947.

[75] D. Eisenbud and W. Neumann. *Three-dimensional link theory and invariants of plane curve singularities.* Princeton University Press, 1985.

[76] H. Esnault, J. Seade, and E. Viehweg. Characteristic divisors on complex manifolds. *J. Reine Angew. Math.*, 424:17–39, 1992.

[77] J. Fernández, I. Luengo, A. Melle, and A. Nemethi. On rational cuspidal projective plane curves. November, 2004. Preprint, arXiv:math.AG/0410611.

[78] R. Fintushel and R.J. Stern. Instanton homology of Seifert fibred homology three spheres. *Proc. London Math. Soc.*, 61:109–137, 1990.

[79] A. Floer. An instanton-invariant for 3-manifolds. *Comm. Math. Phys.*, 118:215–240, 1988.

[80] R. Fox. On Fenchel's conjecture about f-groups. *Mat. Tidsskrift B*, pages 61–65, 1952.

[81] M. Freedman and R. Kirby. A geometric proof of Rochlin's theorem. *Proc. Sympos. Pure Math. A.M.S.*, XXXII:85–97, 1978. In "Algebraic and geometric topology".

[82] R. Friedman. *Algebraic surfaces and holomorphic vector bundles*. Springer Verlag, 1998.

[83] S. Fukuhara, Y. Matsumoto, and K. Sakamoto. Casson's invariant of Seifert homology 3-spheres. *Math. Ann.*, 287:275–285, 1990.

[84] M. Furuta and B. Steer. Seifert fibred homology 3-spheres and the Yang-Mills equations on Riemann surfaces with marked points. *Adv. Math. 96*, pages 38–102, 1992.

[85] A.M. Gabrielov. Polar curves and intersection matrices of singularities. *Invent. Math.*, 54:15–22, 1979.

[86] T. Gaffney. The integral closure of modules and Whitney equisingularity. *Inv. Math.*, 102:301–322, 1992.

[87] X. Gómez-Mont. An algebraic formula for the index of a vector field on a hypersurface with an isolated singularity. *J. Alg. Geometry*, 7:731–752, 1998.

[88] X. Gómez-Mont, J.A. Seade, and A. Verjovsky. On the topology of a holomorphic vector field in a neighborhood of an isolated singularity. *Funct. Anal. Appl.*, 27:97–103, 1993.

[89] M. Goresky and R. MacPherson. *Stratified Morse Theory*. Springer Verlag, 1987.

[90] V. Goryunov. Functions on space curves. *J. London Math. Soc.*, 61:807–822, 2000.

[91] H. Grauert. Über Modifikationen und exzeptionelle analytische Mengen. *Math. Ann.*, 146:331–368, 1962.

[92] G.-M. Greuel and E. Looijenga. The dimension of smoothing components. *Duke Math. J.*, 52:263–272, 1985.

[93] G.-M. Greuel and J. Steenbrink, On the topology of smoothable singularities, Proc. Symp. Pure Math. A.M.S. 40, Part 1, (1983) 535–545.

[94] A. Haefliger. Deformations of transversely holomorphic flows on spheres and deformations of Hopf manifolds. *Compositio Math.*, 55:241–251, 1985.

[95] H. Hamm. Lokale topologische Eigenschaften komplexer Räume. *Math. Ann.*, 191:235–252, 1971.

[96] H. Hamm. Exotische Sphären als Umgebungsränder in speziellen komplexen Räumen. *Math. Ann.*, 197:44–56, 1972.

[97] R. Hartshorne. *Residues and duality*. Springer Verlag, 1966.

[98] E. Heintze, R. Palais, C.-L. Terng, and G. Thorbergsson. *Hyperpolar actions on symmetric spaces*, pages 214–245. Edited by S.-T. Yau, International Press, Cambridge, 1995.

[99] H. Hironaka. Normal cones in analytic Whitney stratifications. *Publ. Math. Inst. Hautes Études Sci.*, 36:127–138, 1969.

[100] H. Hironaka. *Stratification and Flatness*, pages 199–265. Sijthoff & Noordhoff Int. Publishers, 1977.

[101] F. Hirzebruch. *Topological Methods in algebraic geometry*. Springer Verlag, 1956.

[102] F. Hirzebruch. The topology of normal singularities of an algebraic surface (after D. Mumford). *Soc. Math. France*, 8 Exp. No. 250:129–137, 1965. Séminaire Bourbaki.

[103] F. Hirzebruch. Singularities and exotic spheres. 314, 1966/67. Sém. Bourbaki 19ème année.

[104] F. Hirzebruch. Hilbert modular surfaces. *L'Enseignement Math.*, 19:183–281, 1973.

[105] F. Hirzebruch, W.D. Neumann, and S.S. Koh. *Differentiable manifolds and quadratic forms*. Marcel Dekker, Inc., New York, 197.

[106] W.Y. Hsiang and B.H. Lawson. Minimal submanifolds of low cohomogeneity. *J. Differential Geometry*, 5:1–38, 1971.

[107] D. Husemoller. *Fibre bundles*. Springer Verlag, 1994.

[108] T. Izawa and T. Suwa. Multiplicity of functions on singular varieties. *Internat. J. Math.*, 14:541–558, 2003.

[109] W. Jaco. *Lectures on three-manifold topology*. A.M.S., 1980.

[110] M.J. Jankins and W.D. Neumann. *Lectures on Seifert manifolds*. Brandeis Univ. Lecture Notes vol. 2, 1983.

[111] A. Jaquemard. Fibrations de Milnor pour des applications réelles. *Boll. Un. Mat. Ital.*, 3-B:591–600, 1989.

[112] M. Kervaire. Courbure intégrale généralisée et homotopie. *Math. Ann.*, 131:219–252, 1956.

[113] M. Kervaire and J. Milnor. Groups of homotopy spheres I. *Ann. of Math.*, 77:504–537, 1963.

[114] F. Klein. *Lectures on the icosahedron and the solution of equations of the fifth degree*. Dover, 1956.

[115] P. Kronheimer and T.S. Mrowka. Embedded surfaces and the structure of Donaldson's polynomial invariants. *J. Differential Geom.*, 41:573–734, 1995.

[116] N. Kuiper. The quotient space of $\mathbb{CP}(2)$ by complex conjugation is the 4-sphere. *Math. Ann.*, 208:175–177, 1974.

[117] T.C. Kuo. On C^0-sufficiency of jets of potential function. *Topology*, 8:167–171, 1969.

[118] K. Lamotke. *Regular solids and isolated singularities*. Friedr. Vieweg & Sohn, Braunschweig, 1986.

[119] F. Larrión and J. Seade. Complex surface singularities from the combinatorial point of view. *Topology and Appl.*, 66:251–265, 1995.

[120] H.B. Laufer. *Normal two-dimensional singularities*. Princeton Univ. Press, 1971.

[121] H.B. Laufer. Taut two-dimensional singularities. *Math. Ann.*, 205:131–164, 1973.

[122] H.B. Laufer. Minimally elliptic singularities. *Am. J. Math.*, 99:1257–1295, 1977.

[123] H.B. Laufer. On μ for surface singularities. *Proc. Symp. Pure Math. AMS*, XXX, Part 1:45–49, 1977. In "Several complex variables".

[124] H.B. Lawson and M.-L. Michelsohn. *Spin geometry*. Princeton University Press, 1989.

[125] D.T. Lê. Deux noeuds algébriques FM-équivalents sont égaux. *C. R. Acad. Sci. Paris Sér. A*, 272:214–216, 1971.

[126] D.T. Lê. Singularités isolées d'hypersurfaces. *Acta Scientiarum Vietnamicarum*, VII:24–33, 1971.

[127] D.T. Lê. Sur les noeuds algébriques. *Compositio Math.*, 25:281–321, 1972.

[128] D.T. Lê. Noeuds algébriques. *Ann. Inst. Fourier (Grenoble)*, 23:117–126, 1973. In Colloque International sur l'Analyse et la Topologie Différentielle, CNRS, Strasbourg.

[129] D.T. Lê. Some remarks on relative monodromy. In P. Holm, editor, *Real and complex singularities*, Proc. Nordic Summer School, pages 397–403. Sijthoff and Noordhoff, 1977.

[130] D.T. Lê. Le concept de singularité isolée de fonction analytique. *Adv. Stud. Pure Math.*, 8:215–227, 1986.

[131] D.T. Lê. *Polyèdres évanescents et effondrements*, pages 293–329. Academic Press, 1988.

[132] D.T. Lê. Complex analytic functions with isolated singularities. *J. Algebraic Geometry*, 1:83–100, 1992.

[133] D.T. Lê and D. Cheniot. Remarques sur les deux exposés précédents. *Astérisque* 7, 8, pages 253–252, 1973. in Singularités à Cargèse.

[134] D.T. Lê, F. Michel, and C. Weber. Courbes polaires et topologie des courbes planes. *Ann. Sci. École Norm. Sup.*, 24:141–169, 1991.

[135] D.T. Lê, J. Seade, and A. Verjovsky. Quadrics, orthogonal actions and involutions in complex projective spaces. *Enseign. Math., II. Sér.*, 49:173–203, 2003.

[136] D.T. Lê and B. Teissier. Cycles evanescents et conditions de Whitney. *in Proc. Symp. Pure Math*, 40 (Part 2):65–103, 1983.

[137] J. Levine. Polynomial invariants of knots of codimension two. *Ann. of Math.*, 84:537–554, 1966.

[138] A. Libgober. Theta characteristics on singular curves, spin structures and Rochlin theorem. *Ann. Sci. École Norm. Sup.*, 21:623–635, 1988.

[139] A. Libgober. Homotopy groups of the complements to singular hypersurfaces II. *Ann. of Math.*, 139:117–144, 1994.

[140] A. Libgober and S.S.-T. Yau. An obstruction for smoothing of Gorenstein surface singularities. *Comment. Math. Helv.*, 65:413–433, 1990.

[141] A. Libgober and S.S.-T. Yau. Erratum: "An obstruction to smoothing of Gorenstein surface singularities". *Comment. Math. Helv.*, 67:670, 1992.

[142] Y. Lim. The equivalence of Seiberg-Witten and Casson invariants for homology 3-spheres. *Math. Res. Lett.*, 6:631–643, 1999.

[143] J.J. Loeb and M. Nicolau. On the complex geometry of a class of non-Kählerian manifolds. *Isr. J. Math.*, 110:371–379, 1999.

[144] S. Lojasiewicz. Triangulation of semi-analytic sets. *Annali Sc. Norm. Sup. de Pisa*, 18:449–474, 1964.

[145] E. Looijenga. A note on polynomial isolated singularities. *Indag. Math.*, 33:418–421, 1971.

[146] E. Looijenga. *Isolated Singular Points on Complete Intersections*. Cambridge Univ. Press, Cambridge, London, New York, New Rochelle, Melbourne, Sydney, 1984.

[147] E. Looijenga. The smoothing components of a triangle singularity. II. *Math. Ann.*, 269:357–387, 1984.

[148] E. Looijenga. Riemann-Roch and smoothings of singularities. *Topology*, 25:293–302, 1986.

[149] S. López de Medrano. The space of Siegel leaves of a holomorphic vector field. *Springer-Verlag Lecture Notes in Math.*, 1345:233–245, 1988. In "Holomorphic dynamics" (Mexico, 1986), ed. X. Gomez-Mont et al.

[150] S. López de Medrano. Topology of the intersection of quadrics in $\mathbb{R}^n$. *Springer Verlag Lecture Notes in Math.*, 1370:280–292, 1989. In "Algebraic topology" (Arcata, CA, 1986), ed. Carlsson et al.

[151] S. López de Medrano and A. Verjovsky. A new family of complex, compact, non-symplectic manifolds. *Bol. Soc. Brasil. Mat.*, 28:253–269, 1997.

[152] R. MacPherson. Chern classes for singular varieties. *Ann. of Math*, 100:423–432, 1974.

[153] A. Marin. $\mathbb{C}P^2/\sigma$ *ou Kuiper et Massey au pays des coniques*, pages 141–152. Progress in Mathematics, Volume 62, Birkhäuser, 1986.

[154] D. Massey. *Introduction to perverse sheaves and vanishing cycles.* In Singularity theory (Trieste, 1991), edited by D. T. Lê and K. Saito and B. Teissier, World Sci. Publishing, p. 487–508, 1995.

[155] D. Massey. *An introduction to the Lê cycles of a hypersurface singularity.* In Singularity theory (Trieste, 1991), edited by D. T. Lê and K. Saito and B. Teissier, World Sci. Publishing, p. 468–486, 1995.

[156] D. Massey. *Lê cycles and hypersurface singularities*, volume 1615. Springer-Verlag, Lecture Notes, 1995.

[157] D. Massey. *Milnor fibres of non-isolated hypersurface singularities.* In Singularity theory (Trieste, 1991), edited by D.T. Lê and K. Saito and B. Teissier, World Sci. Publishing, p. 458–467, 1995.

[158] D. Massey. Hypercohomology of Milnor fibers. *Topology*, 35:969–1003, 1996.

[159] W. Massey. The quotient space of the complex projective plane under conjugation is a 4 sphere. *Geom. Dedicata*, 2:371–374, 1973.

[160] L. Meersseman. A new geometric construction of compact complex manifolds in any dimension. *Math. Ann.*, 317:79–115, 2000.

[161] L. Meersseman and A. Verjovsky. Structures algébriques exotiques de $\mathbb{R}^4$ et conjectures de Poincaré. *C. R. Acad. Sci. Paris Sér. I Math.*, 332:63–66, 2001.

[162] L. Meersseman and A. Verjovsky. Holomorphic principal bundles over toric varieties. J. Reine Angew. Math., 2004.

[163] F. Michel and A. Pichon. On the boundary of the Milnor fibre of nonisolated singularities. *Int. Math. Res. Not.*, 43:2305–2311, 2003.

[164] J. Milnor. Construction of universal bundles II. *Annals of Math.*, 63:430–436, 1956.

[165] J. Milnor. On manifolds homeomorphic to the 7-sphere. *Ann. of Math.*, 64:399–405, 1956.

[166] J. Milnor. *Topology from the Differentiable Viewpoint.* Univ. Press of Virginia, Charlottesville, 1965.

[167] J. Milnor. On isolated singularities of hypersurfaces. 1966. Unpublished.

[168] J. Milnor. *Singular Points of Complex Hypersurfaces.* Annals of Maths. Study 61, Princeton University Press, Princeton, 1968.

[169] J. Milnor. *Morse Theory.* Ann. of Math. Study 51, Princeton Univ. Press, 5th printing, 1973.

[170] J. Milnor. *Characteristic classes.* Princeton University Press, 1974.

[171] J. Milnor. *On the 3-dimensional Brieskorn manifolds $M(p,q,r)$.* In Knots, links and 3-manifolds, ed. L. Neuwirth, Ann. of Maths. Studies 84, Princeton Univ. Press, 1975.

[172] J. Milnor. Curvatures of left invariant metrics on Lie groups. *Advances in Math.*, 21:293–329, 1976.

[173] D. Mond and D. Van Straten. Milnor number equals Tjurina number for functions on space curves. *J. London Math. Soc.*, 63:177–187, 2001.

[174] P.S. Mostert. On a compact lie group acting on a manifold. *Ann. of Maths.*, 65:447–455, 1957.

[175] J. Mostovoy. Algebraic cycles and antiholomorphic involutions on projective spaces. *Bol. Soc. Mat. Mex.*, 6:151–170, 2000.

[176] D. Mumford. The topology of normal singularities of an algebraic surface and a criterion for simplicity. *Publ. Math. I.H.E.S.*, 9, 1961.

[177] D. Mumford. Theta characteristics of an algebraic curve. *Ann. Sci. École Norm. Sup.*, 4:181–192, 1971.

[178] D. Mumford, J. Fogarty, and F. Kirwan. *Geometric invariant theory. Third edition.* Springer Verlag, Berlin, 1994.

[179] R. Narasimhan. *Introduction to the theory of analytic spaces.* Springer Verlag, 1966.

[180] A. Némethi. Casson invariant of cyclic coverings via eta-invariant and Dedekind sums. *Topology Appl.*, 102:181–193, 2000.

[181] A. Némethi and L.I. Nicolaescu. Seiberg-Witten invariants and surface singularities. *Geom. Topol.*, 6:269–328, 2002.

[182] A. Némethi and L.I. Nicolaescu. Seiberg-Witten invariants and surface singularities. II. Singularities with good $\mathbb{C}^*$-action. *J. London Math. Soc.*, 69:593–607, 2004.

[183] W.D. Neumann. Brieskorn complete intersections and automorphic forms. *Invent. Math.*, 42:285–293, 1977.

[184] W.D. Neumann. A calculus for plumbing applied to the topology of complex surface singularities and degenerating complex curves. *Trans. Amer. Math. Soc.*, 268:299–344, 1981.

[185] W.D. Neumann. Abelian covers of quasihomogeneous surface singularities. *Proc. Sympos. Pure Math., A.M.S.*, 40, Part 2:233–243, 1983. In "Singularities" (Arcata, Calif., 1981).

[186] W.D. Neumann. Geometry of quasihomogeneous surface singularities. *Proc. Sympos. Pure Math., A.M.S.*, 40, Part 2:245–258, 1983. In "Singularities" (Arcata, Calif., 1981).

[187] W.D. Neumann and F. Raymond. Seifert manifolds, plumbing, μ-invariant and orientation reversing maps. *Lecture Notes in Math., Springer Verlag*, 664:163–196, 1978. In "Algebraic and geometric topology".

[188] W.D. Neumann and J. Wahl. Casson invariant of links of singularities. *Comment. Math. Helv.*, 65:58–78, 1990.

[189] L. Neuwirth. On Stallings fibrations. *Proc. Amer. Math. Soc.*, 14:380–381, 1963.

[190] L.I. Nicolaescu. Eta invariants of Dirac operators on circle bundles over Riemann surfaces and virtual dimensions of finite energy Seiberg-Witten moduli spaces. *Israel J. Math.*, 114:61–123, 1999.

[191] J. Nielsen. Surface transformation classes of algebraically finite type. *Mat.-Fys. Medd. Danske Vid. Selsk.*, 21, 1944. Collected Papers 2, Birkhäuser, 1986.

[192] S. Ochanine. Signature modulo 16, invariants de Kervaire généralisés et nombres caractéristiques dans la K-théorie réelle. *Mém. Soc. Math. France (N.S.)*, 5, 1980/81.

[193] P. Orlik. *Seifert manifolds*, volume 291. Springer Verlag Lecture Notes, 1972.

[194] P. Orlik and Ph. Wagreich. Isolated singularities of algebraic surfaces with $\mathbb{C}^*$ action. *Ann. of Math.*, 93:205–228, 1971.

[195] A. Parusiński. A generalization of the Milnor number. *Math. Ann.*, 281:247–254, 1988.

[196] A. Parusiński. Characteristic classes of hypersurfaces and characteristic cycles. *J. Algebraic Geom.*, 10:63–79, 2001.

[197] L. Paunescu. The topology of the real part of a holomorphic function. *Math. Nachr.*, 174:265–272, 1995.

[198] F. Pham. Formules de Picard-Lefschetz généralisées et ramification des intégrales. *Bull. Soc. Math. France*, 93:333–367, 1965.

[199] A. Pichon, Real analytic germs $f\bar{g}$ and open-book decompositions of the 3-sphere, *Int. J. Math.* 16 (2005), 1–12.

[200] A. Pichon. Fibrations sur le cercle et surfaces complexes. *Ann. Inst. Fourier (Grenoble)*, 51:337–374, 2001.

[201] A. Pichon and J. Seade, Fibered multilinks and real singularities $f\bar{g}$, preprint math. AG/0505312.

[202] A. Pichon and J.A. Seade. Real singularities and open-book decompositions of the 3-sphere. *Ann. Fac. Sci. Toulouse*, XII:245–265, 2003.

[203] H. Pinkham. Deformations of algebraic varieties with g_m action. *Astérisque*, 20, 1974.

[204] H. Pinkham. Normal surface singularities with $\mathbb{C}^*$-action. *Math. Ann.*, 227:183–193, 1977.

[205] A. Ranicki. *High-dimensional knot theory*. Springer Verlag Monographs in Mathematics, 1998.

[206] F. Raymond. Classification of the actions of the circle on 3-manifolds. *Trans. A.M.S.*, 131:51–78, 1968.

[207] F. Raymond and A. Vasquez. 3-manifolds whose universal coverings are Lie groups. *Topology Appl.*, 12:161–179, 1981.

[208] E. Rees. On a question of Milnor concerning singularities of maps. *Proc. Edinburgh Math. Soc*, 43:149–153, 2000.

[209] V.A. Rochlin. Proof of Gudkov's hypothesis. *Funct. Anal. Appl.*, 6:136–138, 1972.

[210] D. Rolfsen. *Knots and links*. Publish or Perish inc., 1976.

[211] C. Rourke and D.P. Sullivan. On the Kervaire obstruction. *Ann. of Math.*, 94:397–413, 1971.

[212] M.A.S. Ruas. On the degree of C^ℓ-determinacy. *Math. Scand.*, 59:59–70, 1986.

[213] M.A.S. Ruas and M.J. Saia. C^ℓ-determinacy of weighted homogeneous germs. *Hokkaido Math. Journal*, XXVI, no.1:89–99, 1997.

[214] M.A.S. Ruas, J. Seade, and A. Verjovsky. On real singularities with a Milnor fibration. *In Trends in Singularities*, pages 191–213, 2002. edited by A. Libgober and M. Tibăr.

[215] L. Rudolph. Isolated critical points of mappings from R^4 to R^2 and a natural splitting of the Milnor number of a classical fibered link. I. basic theory; examples. *Comment. Math. Helv.*, 62:630–645, 1987.

[216] K. Saito. Quasihomogene isolierte Singularitäten von Hyperflächen. *Invent. Math.*, 14:123–142, 1971.

[217] R.N.A. Santos, Topological Triviality of Families of Real Isolated Singularities and their Milnor Fibrations, *Mathematica Scandinavica* 96 (2005), 96–106.

[218] R.N.A. Santos and M.A. Ruas, Real Milnor Fibration and C-Regularity, *Manus. Math.* 117 (2005), 207–218.

[219] N. Saveliev. Floer homology of Brieskorn homology spheres. *J. Differential Geom.*, 53:15–87, 1999.

[220] J. Scherk. CR structures on the link of an isolated singular point. *CMS Conf. Proc., Amer. Math. Soc.*, 6:397–403, 1986. Proceedings of the 1984 Vancouver conference in algebraic geometry.

[221] J. Schürmann. *Topology of singular spaces and constructible sheaves*, volume 63. Birkhäuser Verlag, Basel, 2003.

[222] M.H. Schwartz, Classes caractéristiques définies par une stratification d'une variété analytique complexe, *C. R. Acad. Sci. Paris* 260, 1965, 3262–3264, 3535–3537.

[223] J. Seade. On the η-function of the dirac operator on $\gamma \backslash s^3$. *An. Inst. Mat. Univ. Nac. Autónoma de México*, 21:129–147, 1981.

[224] J. Seade. Singular point of complex surfaces and homotopy. *Topology*, 21:1–8, 1982.

[225] J. Seade. A cobordism invariant for surface singularities. *Proc. Symp. Pure Maths.*, 40 part 2:479–484, 1983.

[226] J. Seade. Fibred links and a construction of real singularities via complex geometry. *Boletim Soc. Bras. Mat.*, 27:199–215, 1996.

[227] J. Seade. Open book decompositions associated to holomorphic vector fields. *Bol. Soc. Mat. Mexicana*, 3:323–335, 1997.

[228] J. Seade. On the topology of hypersurface singularities. *Lecture Notes in Pure and Applied Mathematics, Marcell Decker Inc. vol. 232*, 2003. In"Real and Complex Singularities", ed. by D. Mond and M. Saia,.

[229] J. Seade and B. Steer. The elements of π_s^3 represented by invariant framings of quotients of $SL_2(\mathbb{R})$ by certain discrete subgroups. *Adv. in Math.*, 46:221–229, 1982.

[230] J. Seade and B. Steer. A note on the eta function for quotients of $PSL_2(r)$ by co-compact fuchsian groups. *Topology*, 26:79–91, 1987.

[231] J. Seade and B. Steer. Complex singularities and the framed cobordism class of compact quotients of 3-dimensional Lie groups by discrete subgroups. *Comment. Math. Helv.*, 65:349–374, 1990.

[232] J. Seade and T. Suwa. A residue formula for the index of a holomorphic flow. *Math. Ann.*, 304:345–360, 1994.

[233] J. Seade and T. Suwa. An adjunction formula for local complete intersections. *Internat. J. Math.*, 9:759–768, 1998.

[234] J. Seade, M. Tibǎr, and A. Verjovsky. Milnor numbers and Euler obstruction. To appear in Bull. Braz. Math. Soc.

[235] H. Shimizu. On discontinuous groups operating on the product of upper half-planes. *Ann. of Maths.*, 77:33–71, 1963.

[236] D. Siersma. A bouquet theorem for the Milnor fibre. *J. Algebraic Geom.*, 4:51–66, 1995.

[237] D. Siersma. The vanishing topology of non isolated singularities. *NATO Sci. Ser. II Math. Phys. Chem.*, 21:447–472, 2001. In "New developments in singularity theory" (Cambridge, 2000).

[238] S. Smale. Generalized Poincaré's conjecture in dimensions greater than four. *Ann. of Maths.*, 74:391–406, 1961.

[239] E. Spanier. *Algebraic Topology*. Springer Verlag, First edition 1966.

[240] J. Stallings. *On fibering certain 3-manifolds*. In "Topology of 3-manifolds and related topics", p. 95–100, Prentice-Hall, Englewood Cliffs, N.J., 1962.

[241] J. Steenbrink. Mixed hodge structures associated with isolated singularities. *Proc. Sympos. Pure Math., A.M.S.*, 40:513–536, 1983. In Singularities, Part 2 (Arcata, Calif., 1981).

[242] N. Steenrod. *The topology of fibre bundles*. Princeton University Press, reprint of the 1957 edition, 1999.

[243] J. Stevens. *Deformations of singularities*. Springer Verlag, Berlin, 2003.

[244] R.E. Stong. *Notes on cobordism theory*. Princeton University Press, Princeton, N.J., 1968.

[245] E. Straume. *Compact connected Lie transformation groups on spheres with low cohomogeneity. I (II)*, volume 119 (125). Mem. Amer. Math. Soc., 1996 (1997).

[246] D.P. Sullivan. On the intersection ring of compact three manifolds. *Topology*, 14:275–277, 1975.

[247] I. Tamura. Spinnable structures on differentiable manifold. *Proc. Japan Acad.*, 48:293–296.

[248] B. Teissier. Introduction to equisingularity problems. *AMS Proc. Symp. Pure Maths*, 9:593–632, 1975.

[249] B. Teissier. *Introduction to curve singularities*. In Singularity theory (Trieste, 1991), edited by D.T. Lê and K. Saito and B. Teissier, World Sci. Publishing, p. 866–893, 1995.

[250] R. Thom. Quelques propriétés globales des variétés différentiables. *Comm. Math. Helv.*, 28:17–86, 1954.

[251] R. Thom. Généralisation de la théorie de Morse et variétés feuilletées. *Ann. Inst. Fourier*, 14:175–190, 1964.

[252] M. Tibăr. Bouquet decomposition of the Milnor fibre. *Topology*, 35:227–241, 1996.

[253] M. Tibăr. Vanishing cycles of pencils of hypersurfaces. *Topology*, 43:619–633, 2004.

[254] V.A. Vassiliev. A geometric realization of the homologies of classical lie groups and complexes. *St. Petersburg Math. J. (formerly Leningrad Math. J.)*, 3:809–815, 1992.

[255] A. Verona. *Stratified Mappings: Structure and Triangulability*. Springer Verlag Lecture Notes in Maths. 1102, 1984.

[256] P. Wagreich. The structure of quasihomogeneous singularities. *In Proc. Symp. Pure Math*, 40 (Part 2):593–611, 1983.

[257] Ph. Wagreich. Algebras of automorphic forms with few generators. *Trans. A.M.S.*, 262:367–389, 1980.

[258] J. Wahl. Smoothings of normal surface singularities. *Topology*, 20:219–246, 1981.

[259] F. Waldhausen. Eine Klasse von 3-dimensionalen Mannifaltigkeiten. *Invent. Math.*, 3:308–333 [4 (1967), 97–117], 1967.

[260] C.T.C. Wall. Stability, pencils and polytopes. *Bull. London Math. Soc.*, 12:401–421, 1980.

[261] C.T.C. Wall. Finite determinacy of smooth map-germs. *Bull. London. Math. Soc.*, 13:481–539, 1981.

[262] C.T.C. Wall. Duality of real projective plane curves: Klein's equation. *Topology*, 35:355–362, 1996.

[263] H. Whitney. Local properties of analytic varieties. *Symposium in honor of M. Morse, Princeton Univ. Press, edited by S. Cairns*, 1965.

[264] E. Winkelnkemper. Manifolds as open books. *Bull. A.M.S.*, 79:45–51, 1973.

[265] S.S.T. Yau. Hypersurface weighted dual graphs of normal singularities of surfaces. *Amer. J. Math.*, 101:761–812, 1979.

[266] S. Yokura. On characteristic classes of complete intersections. *Contemp. Math. A.M.S.*, 241:349–369, 1999. In "Algebraic geometry: Hirzebruch 70" (Warsaw, 1998).

[267] O. Zariski. *Algebraic surfaces*. Springer Verlag, 2nd edition, 1972.

[268] J.J. Loeb and M. Nicolau. Holomorphic flows and complex structures on products of odd-dimensional spheres. *Math. Ann.*, 306:781–817, 1996.

[269] S. Gusein-Zade, I. Luengo, A. Melle-Hérnandez, Zeta functions of germs of meromorphic functions, and the Newton diagram, *Funct. Anal. Appl.* 32 (1998), no. 2, 93–99.

[270] S. Gusein-Zade, I. Luengo, A. Melle-Hérnandez, On the topology of germs of meromorphic functions and its applications, *St. Petersburg Math. J.* 11 (2000), no. 5, 775–780.

Index

Progress in Mathematics

Your Specialized Publisher in Mathematics

Birkhäuser

Ferran Sunyer i Balaguer Prize Winners since 1998

Ferran Sunyer i Balaguer (1912–1967) was a self-taught Catalan mathematician who, in spite of a serious physical disability, was very active in research in classical mathematical analysis, an area in which he acquired international recognition. His heirs created the Fundació Ferran Sunyer i Balaguer inside the Institut d'Estudis Catalans to honor the memory of Ferran Sunyer i Balaguer and to promote mathematical research.
Each year, the Fundació Ferran Sunyer i Balaguer and the Institut d'Estudis Catalans award an international research prize for a mathematical monograph of expository nature. The prize-winning monographs are published in this series.

Prize 2005, PM 241: Seade, J.
On the Topology of Isolated Singularities in Analytic Spaces (2005)
ISBN 3-7643-7322-9

The aim of this book is to give an overview of selected topics on the topology of singularities, with emphasis on its relations to other branches of geometry and topology. The first chapters are mostly devoted to complex singularities and a myriad of results spread in a vast literature, including recent research. The second part of the book studies real analytic singularities which arise from the topological and geometric study of holomorphic vector fields and foliations. In the low dimensional case these turn out to be related to fibred links in the 3-sphere defined by meromorphic functions.

Prize 2005, PM 240: Ambrosetti, A. /
Malchiodi, A.
Perturbation Methods and Semilinear Elliptic Problems in R^n (2005)
ISBN 3-7643-7321-0

This monograph addresses perturbation methods in critical point theory. It particularly emphasizes applications such as semilinear elliptic problems on R^n, bifurcation from the essential spectrum, the prescribed scalar curvature problem, nonlinear Schroedinger equations, and singularly perturbed elliptic problems in domains.

Prize 2004, PM 233: David, G.
Singular Sets of Minimizers for the Mumford–Shah Functional (2005)
ISBN 3-7643-7182-X

The Mumford–Shah functional was introduced in the 1980s as a tool for automatic image segmentation, but its study gave rise to many interesting questions of analysis and geometric

measure theory. The main object under scrutiny is a free boundary K where the minimizer may have jumps. The book presents an extensive description of the known regularity properties of the singular sets K, and the techniques to get them. It is largely self-contained, and should be accessible to graduate students in analysis. The core of the book is composed of regularity results that were proved in the last ten years and which are presented in a more detailed and unified way.

Prize 2003, PM 223: Andreu-Vallio, F. /
Caselles, V. / Mazòn, J.M.
Parabolic Quasilinear Equations Minimizing Linear Growth Functionals (2004)
ISBN 3-7643-6619-2

This book contains a detailed mathematical analysis of the variational approach to image restoration based on the minimization of total variation submitted to the constraints given by the image acquisition model. This model had a strong influence on the development of variational methods for image denoising and restoration, and pioneered the use of the BV model in image processing. After a full analysis of the model, the minimizing total variation flow under different boundary conditions is studied and its main qualitative properties are exhibited.

Prize 2002, PM 212: Lubotzky, A. /
Segal, D.
Subgroup Growth (2003)
ISBN 3-7643-6989-2

Subgroup growth studies the distribution of subgroups of finite index in a group as a function of the index. In the last two decades this topic has developed into one of the most active areas of research in infinite group theory; this book is a systematic and comprehensive account of

the substantial theory which has emerged. As well as determining the range of possible "growth types", for finitely generated groups in general and for groups in particular classes such as linear groups, a main focus of the book is on the tight connection between the subgroup growth of a group and its algebraic structure.

Prize 2002, PM 209: Unterberger, A.
Automorphic Pseudodifferential Analysis and Higher-Level Weyl Calculi (2002)
ISBN 3-7643-6909-4

The subject of this book is the study of automorphic distributions, by which is meant distributions on R^2 invariant under the linear action of SL$(2, Z)$, and of the operators associated with such distributions under the Weyl rule of symbolic calculus. Researchers and postgraduates interested in pseudodifferential analyis, the theory of non-holomorphic modular forms, and symbolic calculi will benefit from the clear exposition and new results and insights.

Prize 2001, PM 200: Golubitsky, M. /
Stewart, I.
The Symmetry Perspective. From Equilibrium to Chaos in Phase Space and Physical Space (2002) ISBN 3-7643-6609-5

Prize 2000, PM 222: Ortega, J.-P. /
Ratiu, T.S.
Momentum Maps and Hamiltonian Reduction (2004)
ISBN 0-8176-4307-9

Prize 1999, PM 192: Dehornoy, P.
Braids and Self-Distributivity (2000)
ISBN 3-7643-6343-6

Prize 1998, PM 179: Morales Ruiz, J.J.
Differential Galois Theory and Non-Integrability of Hamiltonian Systems (1999)
ISBN 3-7643-6078-X

Progress in Mathematics

*Your Specialized
Publisher in
Mathematics*
Birkhäuser

Progress in Mathematics

*Your Specialized
Publisher in
Mathematics*
Birkhäuser